REF
QH
311
.C644 Contemporary
1986    classics in the
        life sciences

| DATE | | | |
|---|---|---|---|
| | | | |
| | | | |
| | | | |
| | | | |
| | | | |
| | | | |
| | | | |
| | | | |
| | | | |
| | | | |
| | | | |
| | | | |

© THE BAKER & TAYLOR CO.

# Contemporary Classics in the Life Sciences

# Volume 2: The Molecules of Life

# Contemporary Classics in Science
EUGENE GARFIELD, *Editor-in-Chief*

---

This volume is one of a series published by ISI Press®. The series is designed to bring together analyses of papers that have been designated Citation Classics because they are influential and widely quoted.

---

**Books published in this series:**

*Contemporary Classics in the Life Sciences*
    Volume 1: *Cell Biology*
    Volume 2: *The Molecules of Life*
    edited by JAMES T. BARRETT

**Books to be published in this series:**

*Contemporary Classics in Clinical Medicine*

*Contemporary Classics in Plant, Animal, and Environmental Sciences*

*Contemporary Classics in Physical, Chemical, and Earth Sciences*

*Contemporary Classics in Engineering and Applied Science*

*Contemporary Classics in the Social and Behavioral Sciences*

Contemporary Classics
in Science

# Contemporary Classics in the Life Sciences

## Volume 2: The Molecules of Life

Edited by
James T. Barrett

With a Foreword by
Joshua Lederberg

Preface by
Eugene Garfield

**iSi PRESS**®

Philadelphia

Published by

# ISI PRESS®   A Subsidiary of the Institute for Scientific Information®
3501 Market Street, Philadelphia, Pennsylvania 19104 U.S.A.

© 1986 ISI Press

Library of Congress Cataloging in Publication Data

Contemporary classics in the life sciences.

   (Contemporary classics in science)
   Selected articles originally appearing in *Current Contents*.
      Includes bibliographical references and indexes.
      Contents: v. 1. Cell biology—v. 2. The molecules of life.
      1. Life sciences—Addresses, essays, lectures.
   2. Biology—Addresses, essays, lectures.  I. Barrett, James T., 1927–   II. Series.
   QH311.C644    1986     574     85-24840

ISBN 0-89495-053-3 (v. 1)
ISBN 0-89495-075-4 (v. 2)

All rights reserved. No part of this publication may be reproduced, stored in a retrieval system, or transmitted, in any form or by any means, including electronic, mechanical, photographic, or magnetic, without prior written permission of the publisher.

Printed in the United States of America
92  91  90  89  88  87  86    7  6  5  4  3  2  1

# Contents

| | |
|---|---|
| Foreword | vii |
| Preface | xi |
| Introduction | xvii |
| 1. Carbohydrates | 1 |
| 2. Lipids and Related Compounds | 13 |
| 3. Nucleic Acids | 58 |
| 4. Proteins and Amino Acids | 83 |
| 5. Enzymes | 138 |
| 6. Physical Analysis and Instrumentation | 196 |
| 7. Chemical Analysis and Preparative Methods | 225 |
| 8. Statistics | 249 |
| Index of Authors | 269 |
| Index of Subjects | 273 |
| Index of Institutions | 279 |

# Foreword

The interest shown by authors and readers in these vignettes is an evident response to a gap in contemporary scientific communication: that of personal historical reflection. However debatable the criterion of citations as a measure of intellectual value, they do speak to the intertextual continuity of scientific and methodological effort. Doubtless some uncited, even unpublished, productions are equally worthy of such reflection; their collection awaits new invention of how to unearth such cryptic gems. Meanwhile citation statistics, coupled with the good judgment of authors and editors, have helped in the selection of a panorama of discovery reminiscences unparalleled in compilations of the contemporary history of science.

ISI's instructions to authors account for the emphasis placed on certain themes: the inspiration of the work, the obstacles to its publication and acceptance, the most recent work by the author or a disciple elaborating on the same work. The multifarious glimpses of how so many scientists have perceived acts of discovery in day-to-day science, or more precisely put, how they tell the story today, are raw material for abundant reflection on the actual mechanism of science. It is especially rich on such sociological and psychological issues as academic organization, risk taking, the gatekeeping functions of journal publication, resistance to innovation, priority and credit. Between the lines, and sometimes in them, are many of the ambivalences and stresses adumbrated for the scientific career.[1,2] Some exhibit the tension between imagination and criticism, between the creation and destruction of worlds, that characterizes the most trenchant of intellectual and artistic advances.

The pieces are too brief to give more than hints of the internal technical history of the science itself, but these are also abundant. Especially interesting are the authors' own reflections on the fruition and elaboration of their earlier work, and their citations to current sources on the same subjects.

As useful as these vignettes are, especially as starting points for further inquiry, the collection has certain intrinsic limitations as serious history of science. The constraints include those that must be associated with any form of biography, especially autobiography.[3,4,5] The most meticulous of self-chronicles are tainted with the conflicting interests of authors likely to be over-involved with their biographical subject. The commentaries are casual productions, limited in space and documentation, and subjected to a bare minimum of editorial scrutiny. They are fascinating documents in themselves; they are an opening, not a final word. With the best of intentions, personal recollections may be fraught with conflation of timing and of motive. These are notoriously unreliable with respect to attributions of internal mental processes.[6,7] The commentaries should not bear an unwonted burden of being regarded as finished critical studies; they are valuable enough as vernacular, intimate statements of transparent face validity.

Still lacking in existing organs of scientific communication is any relaxed channel for the extended discussion of historical controversy. Anguish about misplaced priority of discovery[8] is fiercely felt, but mutedly published. Many authors of scientific articles may cite relevant prior work in the most patchy way—they do not always regard themselves as intellectual historians—and there is no place to repair the deficit without turning it into a federal case. There is no usable medium today to continue discussions of "what may really have happened" beyond "Contemporary Classics." Lacking such a critical forum, our authors may feel liberated to express their feelings more authentically than they have under the gauntlet of customary peer criticism; and we should be grateful that they have had that opportunity. Conversely, each article stands outside the discipline, characteristic of science, of public critical discourse. Whether an extended forum can be built without stifling easy self-expression is a challenge to architects of our communication system in science.

Meanwhile, there is much to savour and more to ponder in the menu before us.

*Joshua Lederberg*
*Rockefeller University*

## REFERENCES

1. **Merton R K.** *Sociological ambivalence.* New York: Free Press, 1976.
2. **Eiduson B T & Beckman L, eds.** *Science as a career choice.* New York: Russell Sage Foundation, 1973.
3. **Runyan W M.** *Life histories and psychobiography.* Oxford: Oxford University Press, 1982.
4. **Zuckerman H A & Lederberg J.** From schizomycetes to bacterial sexuality: a case study of discontinuity in science. (Unpublished ms.), 1978.
5. **Woolgar S W.** Writing an intellectual history of scientific development. The use of discovery accounts. *Soc. Stud. Sci.* 6:395-422, 1976.

6. **Zimmerman D R.** *Rh: the intimate history of a disease and its conquest.* New York: Macmillan, 1973.
7. **Nisbett R E & Wilson T D.** Telling more than we know: verbal reports on mental processes. *Psychol. Rev.* 84:231-59, 1977.
8. **Merton R K.** Priorities in scientific discovery. *Am. Soc. Rev.* 22:635-59, 1957 (reprinted in: Merton R K. *The sociology of science.* Chicago: University of Chicago Press, 1973. p. 286-324).

# Preface

For almost 20 years I've been writing about Citation Classics, my term for highly cited papers and books that are classics in their fields. In 1977 we began publishing in *Current Contents*® (*CC*®) the feature "This Week's Citation Classic"—an invited 500-word commentary by the author of a Citation Classic. Over 2,100 autobiographical commentaries have appeared so far. In requesting these commentaries, we asked authors of Citation Classics to describe their research, its genesis, and circumstances that affected its progress and publication. We encouraged them to include the type of personal details that are rarely found in formal scientific publication, such as obstacles encountered and byways taken. We also asked that they mention the contributions of co-authors, any awards or honors they received for their research, and any new terminology arising from their work. Finally, we asked them to speculate on the reasons for their paper or book having been cited so often.

With Volume One and Two we inaugurate the series "Contemporary Classics in Science." These two volumes contain the commentaries published in *CC/Life Sciences* from the beginning of 1977 to the end of 1984. The first volume, *Cell Biology,* includes Citation Classics from research on the cellular level, while the second volume, *The Molecules of Life,* includes those from research on the molecular level. Volume Two also contains commentaries on physical analysis and instrumentation, chemical analysis and preparative methods, and statistics—an artifact of publication from the years 1977 and 1978, when the same commentary, whatever the subject, appeared in each edition of *CC*. Volumes containing the commentaries published in other editions of *CC* from 1979 on are planned. The next volume will collect those published in *CC/Clinical Practice.* Subsequent volumes will cover the physical sciences and applied sciences, as well as plant, animal, and environmental sciences.

Although I have previously described how we choose a paper or book as a Citation Classic,[1,2,3] it is useful to review these procedures in order to make the purpose of this new monographic series clear. Not every scientific publication that deserves the designation "classic" is included in this series, which is limited to *Citation* Classics. There are other types of classics, including those that are rarely or only occasionally cited. More about these later. Our primary criterion for inviting a researcher to contribute a commentary is the number of citations that a particular work has accumulated.

We begin the selection process in gross terms by singling out the 300,000 papers and books most cited in our *Science Citation Index®* (*SCI®*) and *Social Sciences Citation Index®* (*SSCI®*) files (which presently span the years 1955 to 1985 and 1966 to 1985, respectively). We do not, however, rely solely on the highest number of citations in the entire population of publications, but make our selections by fields, in some of which, such as ecology, engineering, or mathematics, 100 citations or even fewer may qualify a work as a Citation Classic.

Searching for Citation Classics, therefore, is like fishing with nets. We seine the waters of the scientific literature in search of the biggest fish in a school. The big fish are the relatively few papers and books with the highest number of citations *in their field*. Now it's plain that the biggest fish in one school will be dwarfed by even the smallest fish in another school. So, too, the number of citations necessary to make a work a Citation Classic in radio astronomy, with its much smaller population of researchers and papers, is smaller than the number giving a work status as a Citation Classic in biochemistry. Realizing that one discipline is more populated than another, we have used different nets when searching different waters. In the Sea of Biochemistry we expect to encounter many giant fish, since in these waters the population of published papers is so great. The net we select has a wide mesh that captures only the big fish and allows the little ones to swim through. But the net we drop into the Bay of Radio Astronomy has a much smaller mesh and is designed to catch the largest of the relatively small fish found there. The mesh of the nets corresponds to the different thresholds of citation frequency we set depending on the fields in which we are searching for Citation Classics.

Along with the initial search for Citation Classics in terms of absolute and relative frequency of citation, we extend the search by creating a separate file for each journal to identify the most-cited papers published in that journal. If one assumes that a journal uniquely defines a field or specialty, then the list of most-cited papers for that journal will include many of the classics for that field. We have found that many classics were published in the first volumes of a specialty journal associated with the emergence of the then new field. But we have also found that the classic paper for a new field was sometimes published in a multidisciplinary journal such as *Nature* or *Science*. Thus, the most-cited paper published in a specialty journal may have been cited only 50 times, whereas the primordial paper for the same field may have appeared in *Nature* and received 100 citations. In cases such as these, we ask both the author of the paper published in the multidisciplinary journal and the author of the paper published in the specialty journal for commentaries. Not surprisingly, we often find that both papers were written by the same author(s).

We are further refining our selection process by relying increasingly on research-front data derived from co-citation analysis.[4] An analysis of most-cited, or core, papers in a research front provides a more sensitive classification of subjects than does citation analysis by journal.

Furthermore, and as a supplement to analytical methods, we ask for nominations from *CC* readers of works which they believe may qualify as Citation Classics.

\* \* \* \* \*

We emphasize that this collection represents only a sample from the larger group we have identified as Citation Classics. About half of the authors invited to write commentaries actually do so. It follows, then, that the omission of a paper in this volume in no way signifies that it is not a Citation Classic. In the recent series of essays in *CC* devoted to the 1,000 most-cited papers in the *SCI* from 1961 to 1982, I identified papers for which we have received and published a *Citation Classics*® commentary.[5] Parts nine and ten of this ten-part series will appear in *CC* within the next few months. In Appendix A of Volume One I have provided a sampling of 100 most-cited works for which we have not yet received a commentary. I hope that some of the authors of these Citation Classics will put pen to paper to reminisce and interpret the citation impact of their research. Still, a remarkable number of Nobel laureates and other scientists I have described as *of Nobel class* have contributed commentaries appearing in both volumes. These are balanced by many from hundreds of other scientists who have received rather little formal recognition for their contributions.

\* \* \* \* \*

Finally, a word about those classics I have reported as rarely or only occasionally cited. I have taken to describing these papers and books as uncited classics, and there are many reasons for a classic work having few or no citations. I have already described how a relatively low citation count can qualify a work as a classic when weighed against its companions in other fields, and how we use a flexible, composite, and, we hope, intelligent algorithm to compensate for this problem. It is most difficult, however, to compensate for the lifetime citation counts of papers and books published as long as 50 years ago or even more recently than that. We know that most older works receive fewer citations since typically most citations are received in the first decade after publication. Moreover, the exponential growth of the literature has changed the significance of any fixed threshold. In 1955 we processed 80,000 papers in about 600 journals in the *SCI,* whereas in 1985 we indexed about 600,000 from 3,000 journals in the natural and physical sciences. These numbers reflect the rapid expansion of modern science, which in turn affects the size of bibliographies and the potential for citation impact.

Furthermore, a recognized classic work may fail to be a Citation Classic because it has suffered what Robert K. Merton terms "obliteration by incorporation": "the obliteration of the source of ideas, methods, or findings, by their incorporation in currently accepted knowledge."[6,7] Thus, some works are no longer cited because their substance has been absorbed in the literature.[8] Just as neologisms and eponyms become part of scientific language, obliterated works become the common knowledge within a field, and explicit citation to them is viewed as unnecessary or pedantic. Quite often, a classic work is cited for a few years, excites rapid advances in its subject, and is then superseded by reviews or other papers containing new information. For a variety of reasons including little understood citation behavior, the primordial work is not mentioned; it nevertheless remains classic. On such matters I urge the reader to examine Joshua Lederberg's pithy remarks on obliteration in connection with the discovery that DNA is involved in the genetic transformation in bacteria.[9] The commentary published by Maclyn McCarty[10] on that milestone paper recently appeared and will be included in a future volume in this series for the life sciences.

* * * * *

I have always believed that these commentaries contribute to future historiography by preserving important biographical and behind-the-scenes information, otherwise generally unavailable. Working scientists reading this book will learn about unfamiliar aspects of otherwise familiar and classic research. These commentaries provide grist for the mill of historians and sociologists of science. They also help sensitize students and the public to the diverse nature and methods of science. The publication of these commentaries in this collected form adds to their value. Perhaps, too, the appearance of this book will stimulate other scientists to write *Citation Classics* commentaries. Since this is a continuing monographic series, their contributions will always be welcome.

**Eugene Garfield**
*Institute for Scientific Information*

## REFERENCES

1. **Garfield E.** *Citation Classics*—four years of the human side of science. *Essays of an information scientist.* Philadelphia: ISI Press, 1982. Vol. 5. p. 123–34.
2. ———. The 100 most-cited papers ever and how we select *Citation Classics. Ibid.,* 1985. Vol. 7. p. 175–81.
3. ———. *Contemporary Classics in the Life Sciences:* an autobiographical feast. *Current Contents* (44):3–8, 4 November 1985.
4. ———. ABCs of cluster mapping. Parts 1 & 2. Most active fields in the life and physical sciences in 1978. *Essays of an information scientist.* Philadelphia: ISI Press, 1981. Vol. 4. p. 634–49.
5. ———. The articles most-cited in the *SCI*, 1961–1982. Pts. 1–5. *Ibid.,* 1985. Vol. 7. p. 175–81; 218–27; 270–6; 306–12; 325–35. Pts. 6–8. *Current Contents* (14):3–10; (20):3–12; (33):3–11. Pts. 9–10 (in press).

6. **Merton R K.** *Social theory and social structure.* New York: Free Press, 1968. pp. 27–9, 35–8.
7. ———. Foreword. (Garfield E) *Citation indexing—its theory and application in science, technology, and humanities.* New York: John Wiley & Sons, 1979. p. vii–xi.
8. **Garfield E.** The "obliteration phenomenon" in science—and the advantage of being obliterated. *Essays of an information scientist.* Philadelphia: ISI Press, 1977. Vol. 2. p. 396–8.
9. **Lederberg J.** Foreword. (Garfield E) *Ibid.,* 1977. Vol. 1. p. xiv.
10. **McCarty M.** *Citation Classic.* Commentary on Avery O T, MacLeod C M & McCarty M. Studies on the chemical nature of the substance inducing transformation of pneumococcal types. Induction of transformation by a desoxyribonucleic acid fraction isolated from pneumococcus Type III. *J. Exp. Med.* 79:139–58, 1944. *Current Contents/Life Sciences* 28(50):26, 16 December 1985.

# Introduction

Biographies and autobiographies of scientists have proliferated in recent years. Few of these, however, supply detailed information on a specific experiment or publication. Exceptional, therefore, is the paper published by the Nobel Prize-winning British physiologist Sir Alan L. Hodgkin entitled "Chance and Design in Electrophysiology: An Informal Account of Certain Experiments on Nerve Carried Out between 1934 and 1952." Hodgkin recalls the books, papers, people, and environments that determined his choice of research. He attempts to estimate the portion of his results obtained by planning and the portion obtained by accident. At the outset he explains why such an account is needed:

> I believe that the record of published papers conveys an impression of directness and planning which does not at all coincide with the actual sequence of events. The stated object of a piece of research often agrees more closely with the reason for continuing or finishing the work than it does with the idea which led to the original experiments. In writing papers, authors are encouraged to be logical, and, even if they wished to admit that some experiment was done for a perfectly dotty reason, they would not be encouraged to "clutter-up" the literature with irrelevant personal reminiscences. But over a long period I have developed a feeling of guilt about suppressing the part which chance and good fortune played in what now seems to be a rather logical development.[1]

The biologist Sir Peter B. Medawar, another British Nobel Prize winner, puts it rather more directly by asking if the scientific paper is not a fraud. His question is not meant to cast suspicion on the facts that are published in a scientific paper; rather, he asks if the universally accepted format of the scientific article, generally presented in terms of a fictional inductive method, does not systematically misrepresent the thought processes that led to scientific discoveries.[2] Medawar believes it does.

The sociologist Robert K. Merton notes that

> Typically, the scientific paper or monograph presents an immaculate appearance which reproduces little or nothing of the intuitive leaps, false starts, mistakes, loose ends, and happy accidents that actually cluttered up the inquiry. The public record of science therefore fails to provide many of the source materials needed to reconstruct the actual course of scientific developments.... This practice of glossing over the actual course of inquiry results largely from the mores of scientific publication which call for a passive idiom and format of reporting which imply that ideas develop without benefit of human brain and that investigations are conducted without benefit of human hand.[3]

He then goes on to observe that Bacon, Leibniz, and Mach all took note of the difference between logical casuistry, based on the Euclidean and Cartesian ideals, and the often nonrational, nearly always circuitous course of discovery. Since the scientific article was invented 300 years ago there has evolved a format embracing these ideals and leading to typically immaculate, linear, and flawed accounts of discovery. There is room, therefore, for accounts, however brief, of the actual paths of inquiry.

The feature "This Week's Citation Classic" published in each weekly issue of *Current Contents®* (*CC®*) since 1977 has provided scientists with a forum in which they are permitted to describe their discoveries as they recall their having happened. In this way, *Citation Classics* commentaries continue to fill a lacuna in the history of science.

## REFERENCES

1. **Hodgkin A L.** Chance and design in electrophysiology: an informal account of certain experiments on nerve carried out between 1934 and 1952. *The pursuit of nature: informal essays on the history of physiology.* Cambridge: Cambridge University Press, 1977. p. 1.
2. **Medawar P B.** Is the scientific paper a fraud? (Edge D, ed.) *Experiment.* London: BBC, 1964. p. 7-12.
3. **Merton R K.** On the history and systematics of sociological theory. *Social theory and social structure.* New York: Free Press, 1968. p. 4-6.

# Chapter 1

# Carbohydrates

The commentaries on these few Citation Classics concerned with methods in carbohydrate chemistry record circumstances that will evoke nostalgic thoughts about the conditions and reagents used in many early post-World War II laboratories. Many will remember the need to crystallize and recrystallize their own reagents. Anthrone (see p. 6) was usually received as a yellow to dark-orange crystalline product that was simply unusable without recrystallization. When Rondle states that his paper "came into being with the arrival of a 'proper' spectrophotometer," he really means that he was able to discard his cherished cigar box colorimeter. Brown apparently built his own Geiger counter since these were not commercially available in 1946. Trevelyan's drain pipe chromatography chamber would perhaps compare favorably with the assortment of cylinders and aquaria used for the same purpose by his contemporary American scientists.

Although the equipment and reagents were primitive in those days (the last Classic in this group was published in 1963), the scientists were obviously first rate. Nelson's respectful visit with Somogyi and the latter's contribution to Nelson's manuscript published solely under Nelson's name is a credit to both of them. Reissig's year without salary for the purpose of working in the laboratory of "el Dire" Leloir (a later Nobel Laureate) is unheard of today. Rondle, Gordon, and others who generously compliment the acumen of their coauthors also exemplify a character trait that is vanishing behind a screen of multiauthorship in which one coauthor may not even be the passing acquaintance of another.

Many of the papers described by these "This Week's Citation Classic" (TWCC) commentaries were genuinely outstanding. The modified Somogyi method for the quantitation of glucose authored by Nelson was cited at a rate of about 250 times a year for a total of 4,485 between 1961 and 1982. Warren's important contribution of the thiobarbituric acid assay for sialic acid was quoted at a rate of nearly 190 times per year (total 2,656), and sugar chromatography as developed by Trevelyan, Procter, and Harrison had an annual citation rate of 160 (total 2,248). Carbohydrate methods were obviously needed and when perfected found great popularity.

The progression from simple colorimetry or titrimetry to spectrophotometry and gas–liquid chromatography as these new procedures and instrumentation were developed can be noted in these TWCCs. These narratives also chart the course of methods for the quantitative measurement of simpler carbohydrates, such as the hexoses and pentoses, to the more complicated sugar amines, acetylated amino sugars, sialic acids, etc.

These TWCCs present a compact summary about scientists and their science in a unique environment and are thus equally important for the human qualities that the authors express as for their science.

Number 3  January 17, 1977

# Citation Classics

Nelson N. A photometric adaptation of the Somogyi method for the determination of glucose. *J. Biol. Chem.* 153: 375-80, 1944.

...The author reports a modification of the Somogyi method "for glucose determination in biological material." Somogyi's "copper reagents" are adapted for "colorimetric use" by omitting "iodide and iodate in their preparation." The author developed "a new arsenomolybdate reagent" which, when used with "Somogyi's micro reagent," gives "satisfactory stability and reproducibility of color."

Professor Norton Nelson
New York University Medical Center

"It is a thoroughly pleasant but surprising experience to have one's paper, especially a very early one, referred to as a "Citation Classic." Many things have happened since that was published and many papers preceded and followed it, but apparently none has served as broad a practical end.

"I have always regarded the effort reported in that paper as a bit of emergency "engineering" to solve a major, and another secondary problem. In 1938, '39 and '40, I was working with a very bright and prolific colleague, Arthur Mirsky, at the May Institute for Medical Research in Cincinnati; and another talented investigator, Samuel Rappaport, at the Cincinnati Children's Hospital Research Foundation; and, with a brother, Waldo E. Nelson, also at the Cincinnati Children's Hospital (and also obviously talented). Work with these three collaborators, in one way or another, involved measurements of reducing carbohydrates, especially glucose, in a variety of ways. Thus, there were many irons in the fire; glycogenase, the movement of glucose and glycogen storage, kidney threshold studies, the role of phosphate in carbohydrate metabolism, and the control of diabetic children.

"Somewhere along the line, perhaps about 1940, all refrigerators in all available institutes were jammed with specimens awaiting analysis with which our staff, even with weekend work, were unable to keep abreast. Obviously something needed to be done and that was to improve the efficiency of glucose analysis, without losing the advantages of the Somogyi procedure.

"I was ready to tackle the problem; I needed only the time. It was clear that almost certainly the approach would have to be photometric, but that existing color reagents were unsuitable in a number of regards. However, the range of possibilities was not all that large. So, over the next few weekends, through a series of cross sectional trials, I systematically varied components, and composition and methods of preparation of candidate color reagents as to their suitability for the quantitative estimation of the reduced copper oxide. It turned out that I could get the properties I needed from an arsenomolybdate and was able to devise two quite reproducible ways of preparing the reagent. These produced stable color densities directly proportional to the cuprous oxide, which were reproducible and with an absorption peak at a predictable wave length. This procedure was readily transferrable to the basic reduction technique of Somogyi, wherein it replaced the reliable but tedious titration method; in this way it greatly accelerated the analysis. Within a few weeks our refrigerators, though not precisely empty, were clearly ready to receive more specimens....

"There was a secondary problem which had to do with an annoying error which we detected in bloods deproteinized by the Somogyi procedure. So, back to the weekend chores. Actually the problem was fairly simple once I had clearly defined the difficulty. The solution was to use a barium rather than sodium hydroxide to precipitate the zinc, with the advantage that a surplus could be used, assuring removal of zinc while the excess barium was precipitated, leaving no nasty postfiltration residues to foul up the reduction process. I did feel it important enough to speak directly to Dr. Somogyi about this and went to St. Louis to meet with him. It was a delightful visit: Dr. Somogyi had the predictable Hungarian charm. He agreed that he was troubled with the same problem and that he had developed a remedy which in principle was identical to mine. His procedure, though similar, was more convenient than mine, and with his permission I included it in this paper."[1]

1. Nelson N. Personal communication, November 12, 1976.

3

Number 32

# Citation Classics

August 8, 1977

Rondle, Charles J M & Morgan, Walter T J. The determination of glucosamine and galactosamine. *Biochemical Journal* 61:586-9, 1955.

This paper describes an updating of the earlier method of Elson and Morgan for the determination of hexosamines by treatment with acetyl acetone followed by addition of Ehrlich's reagent. Emphasis is placed on the importance of using purified reagents and rigidly controlled conditions. Details are given of the accuracy likely to be obtained in any test or series of tests. Evidence is presented on the stability of the coloured substances produced. [The *SCI*® indicates that this paper was cited 604 times in the period 1961-1975.]

---

Dr. Charles J.M. Rondle
London School of Hygiene
and Tropical Medicine
Keppel Street
London WC1E 7HT
England

February 28, 1977

"I am astonished but delighted to learn that this paper has been cited so frequently. Perhaps some readers would like to know how it came to be written. As a raw Cambridge graduate I was lucky to join Walter Morgan's team as a Ph.D. student. At that time (1951) we were all working on the isolation and chemical composition of 'human' blood group substances obtained from one source or another--from hog mucin to human ovarian cysts. It had been apparent for many years that hexosamines were important constituents of blood group substances. At the Lister Institute of Preventive Medicine [London, England] they were detected and measured in acid hydrolysates of material by methods based on that introduced by Elson and Morgan in 1933. My first acquaintance with a 'modification' was in the use of a colourimeter made, in true Lister fashion, from a bell transformer, a discarded cigar box and a flying spot galvanometer which had to be used in a dark room. I would add for the benefit of younger workers that even in those not too distant days it was customary to purify or repurify nearly all reagents--an A.R. grade was frequently unavailable. Reference sugars were often prepared from natural sources or certainly by careful recrystallisation of commercial material. Good samples were carefully hoarded and exchanged for others among trusted colleagues. One consequence of this is shown in the paper; we claimed that glucosamine and galactosamine gave the same amount of colour, others have denied this. All I can say is that with the samples available to me results were within 4%. We investigated the stability of the colour formed not so much from chemical interest but when performing many tests daily it was necessary to pace 'setting up' and 'reading' times.

"The paper really came into being with the arrival of a 'proper' spectrophotometer. It was decided that a thorough investigation of the Elson-Morgan method should be made and thousands of tests were done. The suggestion that the results should be examined statistically was met with mild concern by my senior colleague who felt that the results should be obvious and not require mathematical investigation or manipulation. However, investigated they were, but not manipulated! I think the effort was worthwhile because it showed the test to test variation to be expected in the method described. With hindsight I feel that the variation could have been lessened by the use of alkaline buffer rather than unbuffered sodium carbonate and indeed some later workers have done this.

"Why it has been so often cited I do not know. Perhaps because it reviews the earlier literature, gives precise instructions for a test which works and indicates the sort of accuracy to be expected from it. Possibly, too, it appeared at a time when interest in hexosamines was increasing, an interest which has been maintained to the present day.

"I can only say it was a pleasure to do the work, and a privilege to have behind me the support and guidance of my distinguished senior colleague."

# This Week's Citation Classic

CC/NUMBER 28
JULY 9, 1979

Reissig J L, Strominger J L & Leloir L F. A modified colorimetric method for the estimation of N-acetylamino sugars. *J. Biol. Chem.* 217:959-66, 1955.
[Instituto de Investigaciones Bioquimicas, Buenos Aires, Argentina & National Institute of Arthritis & Metabolic Disease, NIH, Bethesda, MD]

The authors describe a modification of the method of Aminoff et al.[1] for the estimation of N-acetylamino sugars, which is less time-consuming and affords enhanced sensitivity, more stringent specificity, and less susceptibility to factors which might interfere with color development. [The *SCI*® indicates that this paper has been cited over 730 times since 1961.]

Jose L. Reissig
Department of Biology
C.W. Post Center
Long Island University
Greenvale, NY 11548

January 17, 1977

"While doing a 'post-doctoral' in Leloir's laboratory in Buenos Aires I got involved in the study of phosphoacetylglucosamine mutase. At the time, the current method for acetylhexosamine determination was that of Aminoff, Morgan, and Watkins; but it turned out to be unsatisfactory for my enzyme assay because of buffering problems. I checked with 'el Dire' (i.e., the Director, as Leloir is affectionately called by his associates), and he mentioned that he had noticed a definite improvement in the test when substituting borate for carbonate buffer. This was the basis for the modification developed in Buenos Aires. Shortly afterwards Leloir visited Strominger in Bethesda, and found that he too was working on an improved method. It was decided that we should put out a joint paper including the observations gathered in both hemispheres. The more recent popularity of the method has to do with the growing interest in carbohydrate containing cell surface molecules.

"As to having made the 'hit parade' of the 'top 500' most quoted papers, whatever complacency it generates in me is overshadowed by other more pertinent considerations. I like to think that I have made contributions to science of greater originality and import than that one. After all, sooner or later, the method of Aminoff et al. would have been improved to reach the required level of reliability. What disturbs me most, however, is the establishment of this kind of scientific hall of fame utilizing a criterion which is only feebly correlated with scientifically significant parameters, but very much a transplant of quantitative marketing techniques from the world of business to that of science. Some of my colleagues feel that this sort of transplant is a price that we must pay for affluence. It is ironic, in this respect, that the research in question was performed during one year (1954-5) in which I worked without a salary. Of course, I welcome the organization of science that permits me now to make a living while pursuing my avocation. But, is there no choice other than the extremes of insolvency, or dancing unquestioningly to the tune of the prevailing business methods? At a time when many are having second thoughts about allowing even the world of economics to run unchecked, guided by the proverbial invisible hand and by management techniques which take for granted that economic expansion can go on forever, scientists should exercise special care about what is happening to their discipline. Unlike economists, they have known for a long time that 'small science is beautiful'; and just as the quest for *Appropriate Technology* has recently captured the imagination of many individuals, let me make the parallel and radical proposal that we should at all times pursue not just the fastest-growing, most-quoted kind of science, but simply an *Appropriate Science*.

"Should you choose to paraphrase, rather than transcribe verbatim, my above comments, I trust that in doing so you will include not only the anecdotal, but also the crux of my argument. If for reasons of space you find the abridgement is necessary, I would like to collaborate in the task, to make sure that the gist of my comments does not suffer in the process." [For obvious reasons, this commentary was published with no editing — E.G.]

---

1. **Aminoff D, Morgan W I J & Watkins W M.** The action of dilute alkali on the N-acetyl hexosamines and the specific blood-group mucoids. *Biochem. J.* 51:379-89, 1952.

# Citation Classics

Number 6 — February 6, 1978

Morris D L. Quantitative determination of carbohydrates with Dreywood's anthrone reagent. *Science* 107:254-5, 1948.

A solution of anthrone in 95% sulfuric acid produces a characteristic blue color when added to twice its volume of a water solution of carbohydrates. The depth of color can be used for quantitative determination of sugars and polysaccharides even when these are chemically combined. The effective range is from 20-500 micrograms. Prior hydrolysis to convert sugars to the free state is not needed; thus the reagent can be used for the quick determination of total carbohydrates in a mixture in terms of their glucose equivalent. Glycogen, starch, sucrose and other glucosides have been accurately measured. [The *SCI®* indicates that this paper was cited 577 times in the period 1961-1976.]

Dr. Daniel Luzon Morris
1202 8th Avenue West
Seattle, Washington 98119

February 2, 1977

"It would be a gross understatement to say I was astounded by your request for information on my 'classic' paper on the anthrone method. I have done no research work in biochemistry since the early fifties, and so was unaware that the method was even remembered, to say nothing of being used.

"This paper arose out of work on the glycogen which my wife and I had found to occur in large amounts in sweet corn.[1] I was trying to find out if the glycogen would give useful, prolonged elevation of blood sugar when injected intravenously. Blood sugar in rabbits, measured in conventional ways, did remain elevated for a number of hours, but there was still the question of the nature of this sugar, and also of how much of the injected carbohydrate was excreted in the urine. At the time, all methods of carbohydrate determination required first the removal of extraneous materials such as proteins, followed by hydrolysis of polysaccharides, and finally measurement of the reducing power of the simple sugar. When Dreywood's paper appeared[2] it looked as though it might provide a method for the measurement of this total excreted carbohydrate.

"In a way other than that implied in the title of this series, this work was a classic, in that everything clicked as the work was being done. All guesses turned out to be good ones, and all discrepancies could be quickly accounted for. Thus, as I remember it, the greater part of the investigation was completed in a matter of two or three weeks, with only a few weeks more needed for tying up loose ends.

"In the course of the work, I found that the anthrone method permitted shortcuts that had hitherto been unacceptable, and so greatly simplified sugar determination in biological fluids. As I have said, protein-containing solutions had to be freed of protein before conventional sugar methods could be applied. But with anthrone, lactose in milk could be accurately determined after simply diluting the whole milk. If whole blood were diluted, and then the anthrone method applied, the value for sugar that resulted was 30 to 60 mg % above that found after deproteinization. This higher value was presumably caused by carbohydrates attached to the precipitated protein. But the method could be used, and I believe has been used, for screening tests for hyperglycemia.

1. Morris D L & Morris C T. Glycogen in the seed of *Zea mays* (variety golden bantam). *J. Biol. Chem.* 130:535-44, 1939.
2. Dreywood R. Qualitative test for carbohydrate material. *Ind. Eng. Chem.* (Anal. ed.) 18:499, 1946.

# This Week's Citation Classic

CC/NUMBER 31
AUGUST 4, 1980

Brown A H. Determination of pentose in the presence of large quantities of glucose.
*Arch. Biochem.* 11:269-78, 1946. [Dept. Chemistry (Fels Fund), Univ. Chicago, Chicago, IL]

Glucose interferes with the orcinol reaction for colorimetric measurement of pentose. By measuring light absorption with a Klett colorimeter at two different spectral regions the errors of pentose assay were less than ±5% even when a 15-fold excess of glucose was present. A graphic procedure simplifies calculation of assay data. [The *SCI*® indicates that this paper has been cited over 535 times since 1961.]

Allan H. Brown
Department of Biology
University of Pennsylvania
Philadelphia, PA 19104

July 8, 1980

"I was surprised to learn that one of the least interesting (to me) papers I ever published has become a 'Citation Classic.' The work was done while I was on a postdoctoral appointment in chemistry at the University of Chicago where I was getting biologically reoriented after some years of World War II research. As a plant physiologist, I had joined a team of chemists to use for photosynthetic research the $^{14}C$ that the Atomic Energy Commission produced in an atomic reactor and was about to release for civilian use as a biochemical tracer.

"At Chicago I enjoyed one of the most intellectually stimulating periods I have known. We were introduced to photochemistry by James Franck, took lessons on building mica window Geiger counters (then not commercially available) from Willard Libby, and learned about photosynthesis by listening to Eugene Rabinowitch (and from reading the first of his three volume magnum opus, just published).[1] For the first time in years I could talk plant biology with people who 'spoke the language': E.W. Fager, an organic chemist (later to become a marine ecologist), and Hans Gaffron, a biochemist, research colleague, teacher, and friend. While waiting for the paperwork to run its course and provide us with $^{14}C$ (then at $400/mc!), I picked a minor problem I could pursue without tracer—to inventory the products of 'photoreduction' ($CO_2$ + $2H_2$ + $CH_2O$ + $H_2O$) in the same way that James H.C. Smith had done for photosynthesis some years earlier.[2] Pentose measurement posed a special problem due to interference from glucose during colorimetric assay. I solved it by reading with a Klett colorimeter at two different spectral regions to correct for absorption by hexose chromogen. After validating the method on known mixtures and getting respectable agreement with theory, I used it as part of an overall assay of carbohydrate constituents of *Scenedesmus* before and after some hours of photoreduction. Without tracer, the long time needed to accumulate in the organic fractions increments significant by gross analysis unfortunately insured that I could not possibly have demonstrated transient accumulation of a 'first product' of C-assimilation in any particular fraction.

"When I published those not very interesting results I decided to make the pentose assay method a separate paper just in case some pentose researcher might see it and save himself the trouble of reinventing it.[3] Buried in a physiological contribution it probably would have gone unnoticed. In the year after publication I received more reprint requests for that paper than I ever have since for papers that I believe are of substantially greater scientific interest. Since the paper has been quoted frequently, it seems to have served its intended purpose. Perhaps for some the moral of this story may be: if you value being widely quoted and if your scientific accomplishments fail to attract the attention you think they deserve, publish method papers instead!"

1. **Rabinowitch E I.** *Photosynthesis and related processes.* New York: Interscience, 1945. Vol. I. 599 p.
2. **Smith J H C.** Molecular equivalence of carbohydrates to carbon dioxide in photosynthesis. *Plant Physiol.* 18:207-23, 1943.
3. **Brown A H.** The carbohydrate constituents of *Scenedesmus* in relation to the assimilation of carbon by photoreduction. *Plant Physiol.* 23:331-7, 1948.

# Citation Classics

**Number 36** — September 5, 1977

Warren, Leonard. The thiobarbituric acid assay of sialic acids. *Journal of Biological Chemistry* 234:1971-5, 1959.

A colorimetric assay was developed for the measurement of sialic acids in which the products of periodate oxidation are coupled with thiobarbituric acid to form a red chromophore. Since only free sialic acids are measured the reaction could be used for the detection and measurement of neuraminidase (sialidase) which hydrolytically releases sialic acids from their bound form. [The *SCI*® indicates that this paper was cited 2,656 times in the period 1961-1975.]

Dr. Leonard Warren
The Wistar Institute
Thirty-Sixth Street at Spruce
Philadelphia, Pennsylvania 19104

March 1, 1977

"In the mid-fifties the biosynthesis of small molecules was regarded as a praiseworthy subject for study. I had the good fortune to work on the biosynthesis of the purine ring in an exciting laboratory (under J.M. Buchanan, in the Department of Biology, Division of Biochemistry, at M.I.T.) and after obtaining my doctorate it was only natural that I apply my knowledge and skills to determine how other compounds of biological interest were made. Not wanting to remain in the purine field, forever one of Buchanan's students, I decided to look for new sets of compounds to investigate. The structure for sialic acid was formulated at that time by A. Gottschalk and after some reading I decided to look into its mode of biosynthesis. On paper there were a few obvious pathways for its synthesis and these I investigated after arrival at the N.I.H. in Bethesda. Little progress was made, mainly because the existing assays (Direct Erlich, resorcinol and others) were not specific or sensitive enough. In retrospect I must have observed some of the desired *in vitro* synthesis but the backgrounds were so high and variable that I had little faith in the results.

"At that time, A. Weissbach and J. Hurwitz at the N.I.H. had just completed a study on 2-keto, 3-deoxygluconic acid (KDG) metabolism in bacteria and had used the thiobarbituric acid reaction to assay the compound.[1,2] There were striking analogies between the structures of KDG and sialic acid (N-acetylneuraminic acid) except for one difference that theoretically ruled out the application of the thiobarbituric acid test to the measurement of sialic acid; the amino group on sialic acids was always blocked with an acetyl or glycolyl group. This would prevent the formation of a fragment ($\beta$-formylpyruvic acid) from the first four carbons of sialic acid upon oxidation with periodate. Never letting mere theory interfere with my experiments, I used the Weissbach-Hurwitz procedure, after attempting to remove the amino blocking group. I obtained small and variable amounts of color. By systematically varying every parameter that could be altered I nudged the molar extinction co-efficient to about 57,000 in a reproducible way, tested biological materials, and devised methods for corrections. The entire exercise took about 2 months and, in fact, the resulting publication reflects a lack of extensive experience with the method. Exploiting this assay, it was not long before the precise pathway of biosynthesis of sialic acid was determined. The assay was found to be convenient in the measurement of neuraminidase activity because it measured only free sialic acid. Its widespread use over the years attests not only to its usefulness, reliability, and sensitivity but to the fact that it was devised at a time when membranes, cell surfaces, and glycoproteins were becoming very popular with biochemists, virologists, immunologists, and those studying the process of malignancy.

"It pleases me to be identified with this assay. At the same time, the published work has taken on a separate life of its own so that when I consult the paper it is as if someone else had written it. I suppose someone else did, a young aspiring biochemist."

1. **Weissbach A & Hurwitz J.** The formation of 2-keto-3-deoxyheptonic acid in extracts of E. Coli B.I. Identification. *Journal of Biological Chemistry* 234: 705-9, 1959.
2. **Hurwitz J & Weissbach A.** The formation of 2-keto-3-deoxyheptonic acid in extracts of E. Coli B. II. Enzymic studies. *Journal of Biological Chemistry* 234:709-12, 1959.

# This Week's Citation Classic

CC/NUMBER 26
JUNE 30, 1980

Aminoff D. Methods for the quantitative estimation of N-acetylneuraminic acid and their application to hydrolysates of sialomucoids. *Biochemical J.* **81**:384-92, 1961.
[Public Health Research Institute of the City of New York, Inc., New York, NY]

Two colorimetric procedures, the thiobarbituric acid (TBA) and alkali-Ehrlich, have been developed for the determination of free sialic acid in the presence of the bound sugar. Complementing each other, and those for the determination of total sialic acid in sialoglycoconjugates, it is now possible to follow the chemical and enzymatic release of sialic acid from many biologically important compounds. [The *SCI®* indicates that this paper has been cited over 1,050 times since 1961.]

David Aminoff
Departments of Biological Chemistry
and Internal Medicine
University of Michigan
Ann Arbor, MI 48109

May 16, 1980

"G.K. Hirst, whom I joined in 1957 at the Public Health Research Institute of the City of New York, Inc., requested that I develop an assay for neuraminidase. This enzyme is found in many bacterial extracts and also in the myxo group of viruses.

"My approach was determined by my previous experience. I had carefully re-examined the Morgan-Elson assay (MEA) for N-acetylhexosamines while working on my PhD thesis with W.T.J. Morgan in England. A. Gottschalk was visiting our labs at about that time. He had come from Australia, where he was working on the identification of the product released by the action of influenza virus on ovomucin. He found anomalies in the behavior of the product in the Morgan-Elson assay, and came to us for advice.

"The cause of the anomalies was not resolved until a few years later, when the material released by the virus was identified as sialic acid. The similarity of the colors given by sialic acid and N-acetylhexosamines in MEA convinced me that this assay could also be adapted to determine sialic acid. The remaining requirement, to determine *free* in the presence of the *bound* sialic acid, was ultimately achieved. The alkali-Ehrlich assay, however, is not very sensitive, and is of value only in certain circumstances.

"In search of a more sensitive technique to determine *free* sialic acid in the presence of the *bound* compound, my thoughts kept reverting back to my previous experience with the use of periodate in Conway units to determine the acetaldehyde released from *free* fucose. However, the oxidation products of sialic acid give no volatile material other than formaldehyde. In this, it does not differ from the other common hexoses found in glycoconjugates. The method would thus lack specificity.

"However, one day, while visiting with E. Racker at the same Institute, I observed Dan Levine, one of his graduate students, set up a colorimetric assay for deoxy-ribose. The method involved oxidation with periodate and the interaction of the oxidation product, malonaldehyde, with TBA to give a pink color. The implications hit me like a thunderbolt, for is not sialic acid also a deoxy sugar?! Rushing back to my own lab, I quickly established that free sialic acid readily reacted, while sialoglycoproteins did not. Far from ideal at the outset, it was laboriously refined and readied for use.

"The reasons for the article having been cited so much are attributable to: a) Sialoglycoconjugates play an important role in many biological systems. b) There is a great need for the determination of *free* in the presence of *bound* sialic acid. c) The TBA assay is extremely sensitive, specific, and reproducible. The only other assay that meets these requirements is Warren's TBA assay,[1] which was developed independently."

1. Warren L. The thiobarbituric acid assay of sialic acids. *J. Biol. Chem.* **234**:1971-5, 1959.

Number 6     **Citation Classics**     February 7, 1977

Trevelyan W E, Procter D P & Harrison J S. Detection of sugars on paper chromatograms. *Nature* 166:444-45, 1950.

A developed chromatogram is dried and drawn through a solution of silver nitrate in aqueous acetone. The solvent is allowed to evaporate, and the paper sprayed with NaOH in aqueous ethanol. Reducing sugars produce dense black spots of silver at room temperature, unreacted silver hydroxide being then removed by immersing the paper in ammonia solution. [The *SCI*® indicates that this paper was cited 2,248 times in the period 1961-1975.]

---

Dr. William E. Trevelyan
Tropical Products Institute
56-62 Grays Inn Road
London WC1X 8 LU, England

November 24, 1976

"In 1949 I was working on the composition of spent molasses from a distillery, in a newly-established water pollution research laboratory which had the task of disposing of the waste by anaerobic digestion. Equipment was minimal, and when I decided to look at the sugars present using the then new technique of paper chromatography, my chromatographic tank was constructed from the classic drain-pipe. No suitable oven was available, and after some exasperating failures I decided to develop a method of locating sugars on paper chromatograms which did not require the application of heat. A spot test described by Feigl seemed promising, but when a chromatogram was sprayed with silver nitrate solution followed by NaOH solution, the sugars ran badly. I mentioned to a colleague, famous for his gnomic utterances, that water caused trouble: 'Why have water, then?' said he, looking insufferably smug. Some time later it struck me that this oracular pronouncement perhaps had some sense in it.

"A search through the handbooks suggested that acetone, in which common sugars are not very soluble, but in which silver nitrate dissolved to some extent, fitted the requirements. It immediately occurred to me that it ought to be possible simply to dip the paper in an acetone solution of silver nitrate, more especially as spraying such a volatile and inflammable liquid would be difficult, not to say dangerous. This was, as far as I know, the first example of the dipping technique for applying reagents to paper chromatograms, which later became very popular. For some reason which I have forgotten I preferred to spray the silver nitrate-impregnated chromatogram with the alcoholic NaOH solution. Subsequently, other workers applied the dipping technique here, too, and suggested thiosulphate instead of the unpleasant ammonia solution for fixing the chromatogram.

"Dr. S.M. Partridge of the Low Temperature Research Station, Cambridge, [England] was kind enough to try the technique. He suggested I publish a description of it, and this I did in 1950 after joining the Distillers Company Ltd. yeast research unit at Epsom [Surrey, England], where a similar method was developed using triphenyltetrazolium chloride with practical assistance from D.P. Procter, a technician, and encouragement from Dr. J.S. Harrison, then head of the biochemical section at Epsom.

"Identification of sugars is a constantly recurring problem, especially in industrial laboratories, and I am not surprised that the method is widely used, as it is simple and requires no particular skill. It is also quite dramatic; many people, I find, experience a kind of aesthetic satisfaction in watching the spots come up. The silver nitrate dipping technique for visualising sugar spots is not suited to thin-layer chromatography (TLC), but my impression is that TLC of sugars in general has no advantage over paper chromatography.

"As to why the original paper has been so frequently cited, I have no idea. I wonder how many people who include it in their reference list have actually read it. No matter: I find a quiet satisfaction that the paper has become a Citation Classic, regarding it, as I do, as a flag stuck in the map for the small man, the lone worker long on enthusiasm but short on equipment, who does his best with what there is at hand. And I like to think that many found the method particularly useful."

# This Week's Citation Classic™

CC/NUMBER 13
MARCH 26, 1984

Gordon H T, Thornburg W & Werum L N. Rapid paper chromatography of carbohydrates and related compounds. *Anal. Chem.* 28:849-55, 1956.
[Dept. Entomology and Parasitology, Univ. California, Berkeley, and California Packing Corp., Emeryville, CA]

This paper presents mobility data for 71 carbohydrates and derivatives in 8:8:4:1 isopropanol:pyridine:water:acetic acid solvent, at loads from 0.005 to 1.0 micromole. Spots are detected by four color reagents differing in selectivity. Ways of eliminating interference caused by ionic interactions are described. [The *SCI*® indicates that this paper has been cited in over 395 publications since 1956.]

---

Harold T. Gordon
Department of Entomological Sciences
University of California
Berkeley, CA 94720

January 26, 1984

"This was the first of six research papers in a ten-year collaboration by me (a biochemically oriented entomologist) and two analytical chemists working for CalPak (now the Del Monte Corporation, a subsidiary of RJR industries). Our friendship started with a mutual interest in pesticides. Sugars entered into the picture when CalPak found that pineapple from the Philippines often developed brown spots during the canning process, making it unacceptable to consumers. They assigned this problem to Thornburg, who enlisted me (among others) as a consultant. The browning turned out to be thermal degradation of 2,4-diketogluconic acid, a bacterial oxidation product of glucose formed during ripening of the pineapple. The techniques worked out in our paper were helpful in solving this puzzle.

"The starting point was a method I had developed earlier, with the assistance of a premedical student (now a prominent radiologist) for chromatographic separation of cations in insect blood.[1] The solvent worked surprisingly well for sugars, polyols, and various organic acids and bases in crude biological extracts, since it handles high solute loads and minimizes ionic interactions that cause streaking of spots. We tried it on everything we could quickly lay our hands on, hoping to find something chemically similar to the mystery substance in pineapples. This required trying (and sometimes improving) many color reagents that had been used in prior carbohydrate work for the detection and characterization of spots. We ended up with a lot of data that we could see might be useful in later research by ourselves and others, so we decided to 'do it up right' for publication. This evolved into an unusually thorough study of the $R_f$ and spot dimensions of varying loads of many compounds—in effect, measuring capacity, separability, and detection sensitivity and selectivity. The CalPak chemists (both of them perfectionists!) deserve full credit for this. It led to recognition of their research skills by their company (Thornburg is now director of a large research group at Del Monte). Although I did much of the exploratory work, my ultimate role was in guiding and defining their experiments, and in writing the paper.

"We were surprised to learn that the paper has been frequently cited. It was perhaps the first 'quick-and-dirty' technique for learning something of the composition of biological extracts, and we later showed that it was applicable to an extraordinary variety of inorganic compounds.[2] The early appearance of a technique of general utility (and also simple, fast, and inexpensive) in a large-circulation analytical journal probably initiated a 'chain reaction' of citations, in multifarious contexts. Possibly only a few components of the method (such as the excellent color reagents, which did not require spraying) were widely adopted and so acknowledged.

"Modern chromatographic techniques use micron-size particles (of free or derivatized silica, synthetic polymers, etc.) instead of the matted cellulose fibers of paper, and have much higher resolving power. The plethora of TLC, GLC, and HPLC methods is reviewed by Churms,[3] who does not cite our paper. Nevertheless, our method may still prove useful for the preliminary stages of some investigations, or in teaching laboratories that cannot afford costly state-of-the-art instrumentation."

---

1. Gordon H T & Hewell C A. Paper chromatography of alkali and alkaline earth cations. *Anal. Chem.* 27:1471-4, 1955.
2. Gordon H T, Thornburg W W & Werum L N. Rapid paper chromatographic fractionation of complex mixtures of water-soluble substances. *J. Chromatography* 9:44-59, 1962.
3. Churms S C. Carbohydrates. (Heftmann E, ed.) *Chromatography*. New York: Elsevier, 1983. p. 223-86.

Number 43 **Citation Classics** October 24, 1977

Sweeley C C, Bentley R, Makita M & Wells W W. Gas-liquid chromatography of trimethylsilyl derivatives of sugars and related substances. *Journal of the American Chemical Society* 85:2497-507, 1963.

---

The paper describes the separation and estimation of carbohydrates and related polyhydroxy compounds by gas-liquid chromatography of trimethylsilyl (TMS) derivatives. [The *SCI®* indicates that this paper was cited 1258 times in the period 1961-1975.]

Professor Charles C. Sweeley
Department of Biochemistry
Michigan State University
East Lansing, Michigan 48824

February 28, 1977

"This paper on the gas-liquid chromatography of carbohydrates came about because the right people were in the right place at the right time. It involved a collaboration among three friends holding faculty appointments in two different departments at the University of Pittsburgh and a Postdoctoral Fellow from Japan who was working with one of us. The starting point was the development by Makita and Wells of a simple, rapid, and quantitative method for the preparation of trimethylsilyl derivatives of bile acids which were subsequently used in gas chromatography.[1] The method used pyridine as a solvent and derivatization with hexamethyldisilazane in the presence of trimethylchlorosilane. The possibility of using this method for polyhydroxy compounds was later discussed by Wells and Sweeley (who often drove to work together from the suburbs) and then with Bentley (who had a stock of carbohydrates). Preliminary experiments quickly confirmed the utility of the technique for carbohydrates and the next few months involved much harder but exciting developmental work. Most of the work reported in the paper was accomplished between October, 1962 and March of 1963.

"The trimethylsilyl group was becoming recognized as a valuable protecting agent for hydroxyl groups but the key observation was the development of the rapid preparative method by Makita and Wells. Prior to our work on carbohydrates, several investigators had prepared trimethylsilyl derivatives of carbohydrates and had subjected them to gas chromatography. However, the preparation methods were cumbersome and gave rise to mixtures and, furthermore, the column packings used were very inefficient. Our achievement was to put together a simple and reliable method for derivatization along with up-to-date gas chromatographic procedures and to apply them to a large number of compounds.

"The standard mixture that we developed for trimethylsilylation has been widely used and, indeed, is marketed under a variety of trademarks. It never occurred to any of us that so simple a mixture could have commercial possibilities. Furthermore, each of the three senior authors developed various other areas independently, following this work, and published a number of papers dealing with applications of the basic method. The problem of ordering the authors for publication was solved by a rotation process on the various papers in which we all collaborated and we have always regarded the three Pittsburgh authors as coequals in the achievement.[2,3] Although only Bentley remains at the University of Pittsburgh, Sweeley and Wells having moved to Michigan State University and Makita back to Japan (Faculty of Pharmaceutical Sciences, Okayama University), we all have remained the best of friends.

"It is perhaps noteworthy that we never had any direct grant support from any source for this research. In retrospect, it is remarkable that such a highly cited paper should not ever have received direct grant support from any federal agency."

1. Makita M & Wells W W. The quantitative analysis of fecal bile acids by gas-liquid chromatography. *Analytical Biochemistry* 5:523-30, 1963.
2. Bentley R, Sweeley C C, Makita M & Wells W W. Gas chromatography of sugars and other polyhydroxy compounds. *Biochemical and Biophysical Research Communications* 11:14-8, 1963.
3. Wells W W, Sweeley C C & Bentley R. Gas chromatography of carbohydrates, in *Biomedical applications of gas chromatography* (Szymanski H A, ed.) New York: Plenum Press, 1964, pp. 169-223.

Chapter

# 2

# Lipids and Related Compounds

Although advances in nucleic acid and protein chemistry may appear to overshadow progress in other branches of biochemistry, this opinion is based largely on the impact of nucleic acid and protein biochemistry on genetics, transplantation, cancer, and other subjects more prominent in the public eye than, for example, the structure of the cell membrane, where knowledge of lipids and related compounds has been intimately connected with our understanding of this important cell structure. If the Citation Classics on atherosclerosis were more frequently published in the life sciences rather than clinical practice, the importance of lipids would be more apparent. These Classics from the field of lipid chemistry, loosely defined, again tend to emphasize methodology, but quantitative techniques are a common forerunner of studies on synthesis, metabolism, and structure–function relationships regardless of the biochemical field being considered.

To introduce this chapter, a commentary on the widely applied mixture of chloroform and methanol used by Bligh and Dyer for the extraction of lipids is appropriate. By now this Classic has probably been cited at least 3,000 times. This method is still useful for its original purpose, and only after a total extraction do quantitative determinations become logical. Fortunately, titrimetry for this purpose (see Trout's rejected Classic), a tedious and subjective method, soon yielded to colorimetric and ionization procedures coupled to gas–liquid chromatography for quantitative purposes following thin-layer chromatography (TLC) for the separation of fatty acids, glycerides, and related compounds.

Helmut Mangold stands out as a stellar contributor to TLC of lipids. His association with three Classics is unparalleled in this Life Sciences volume. Although one of his articles was twice rejected, it was later published, all three papers ultimately being published in *Journal of the American Oil Chemists' Society*. The popularity of his 1961 paper stems from its utility as a manual plus its survey of TLC in lipid chemistry.

The next to the last paragraph of Van Handel's "This Week's Citation Classic" (TWCC) should not be missed, particularly if you seek "an original method, easily taught to technicians," as his triglyceride method obviously was. As Carson states in the next TWCC, automated triglyceride and cholesterol determinations later became absolutely required in clinical laboratories to meet the demand that the diagnosis and treatment of atherosclerosis posed. New cholesterol methods published in 1950, 1952, 1953, and 1954 all achieved Classic status. Moreover, the first three of these have been cited more than 1,000 times. This sequential infidelity of cholesterol "quantifiers" emphasizes the search scientists make for the perfect, perhaps even unmodifiable, method. This continual search exposed one amusing yet practical sidelight—that nighttime traffic signals can be useful timing devices (Abell's TWCC).

The close structural similarity of steroid molecules has plagued those in need of a quantitative determination of the individual components in a mixture. Once again, suitable extraction and fractionation procedures were necessary preludes. Bush describes his initial failings in the development of a paper chromatographic method but passes on Feldberg's advice that they both probably ignored: "remember ... when you are in research, you need plenty of holidays." Commentaries on steroid quantitation follow. Allen candidly admits to the pleasure his method and especially its mathematical proof gave him. Protein binding of steroids as a micromethod for their measurement was utilized by Murphy and later by Johansson. This approach is a less specific parallel to radioimmunoassay and enzyme-linked immunoassay methods based on the highly specific antibody binding of the steroid. Beverley Murphy writes very well; see her TWCC in Chapter 7 on Chemical Analysis and Preparative Methods.

Phosphate esters of lipids and lipoproteins are the subjects for the next set of Classics, five of which may be considered reviews. The five-part essay on fat transport in lipoproteins described by Levy has 3,973 citations to its credit. Several of these narratives are of interest and include the serendipitous production of ozone in an ultraviolet-illuminated room as the key to Bandurski and Axelrod's success in the chromatographic identification of phosphate esters. The mysterious presence of ethylformate in Dawson's chemical storeroom and its "solution of the solvent problem" might have, as he hints, some relationship to its role as an artificial rum flavor and the abundance of rum at the Christmas parties. The emergence of lipoproteins as important functional carriers of lipids is emphasized in Levy's Classic related to fat transport, in Green's Classic on the role of proteins in the lipoprotein bilayer cell membrane, and in the Framingham study (p. 39) on coronary heart disease. Also of medical importance is the British report of a committee headed by Inman that oral contraceptives were, because of then excessive dosage, responsible for thromboembolic disease. Fritz Lipmann's famed hydroxamic method for measuring acyl phosphates published in 1945 with Tuttle is certainly another important article. This method was used widely by those studying the terminal oxidative steps in carbohydrate metabolism, and it became the standard for measuring acetyl-coenzyme A. Its use to evaluate the transfer of two-carbon esters in the citric acid cycle gave way to the measure of activated acyl tRNA as the most frequent application of this method.

Of the six TWCCs describing methods for the quantitation of fatty acids and related compounds, Barker's will evoke some nostalgia, no doubt. Do you remember when you first used "Parafilm"? Barker found it useful to protect his lactic acid determinations from his own sweat. Since Barker and Summerson's article has been cited more than 2,500 times, many scientists have obviously suffered through those titrations with "0.002 N iodine to the faintest blue-gray endpoint with starch indicator." Having done this, it is difficult to identify with the colleague who complains that the scintillation counter can't be serviced until tomorrow. In these TWCCs the emergence of gas chromatography (and ionization detectors) as the replacements for titrimetry is described by James and Metcalfe.

The alternative lipoxygenase or cyclooxygenase pathways of arachidonic acid transformation each yield metabolites of extreme biological significance to the inflammatory state. The discovery of the thromboxanes as products of the second pathway is the basis for Hamberg's description of his work with Svensson and Samuelsson. Thromboxane $A_2$, though very unstable, is a powerful platelet aggregating agent. Another thromboxane, $B_2$, is a strong chemoattractant for leukocytes. This Classic is a part of the formidable research on arachidonic acid metabolism that enabled Samuelsson to share the 1982 Nobel Prize in Physiology and Medicine with Bergström and Vane.

# Citation Classics

Number 52 — December 25, 1978

**Bligh E G & Dyer W J.** A rapid method of total lipid extraction and purification. *Can.J. Biochem. Physiol.* **37:** 911-917, 1959.

Lipid decomposition studies in frozen fish have led to the development of a *simple, rapid,* and *reproducible* method for the extraction and purification of lipids from biological materials. The method has been applied to fish muscle and may easily be adapted to use with other tissues. [The *SCI*® indicates that this paper was cited 2,554 times in the period 1961-1977.]

---

E. Graham Bligh
Canada Dept. of Fisheries & Environment
Fisheries & Marine Service, Halifax Lab.
1707 Lower Water Street
Halifax, Nova Scotia, Canada B3J 2S7

April 3, 1978

"The Bligh and Dyer method has certainly been widely used, and it is most gratifying to have it identified as a Citation Classic. It all began back in the 1950s at the Halifax Laboratory of the Fisheries Research Board of Canada, where Dr. William J. Dyer and his group were investigating the deterioration of fish muscle proteins during frozen storage. The group's pioneering work attracted international attention and stimulated a great deal of research in similar laboratories around the world. They had found that protein denaturation in frozen fish muscle contributed substantially to consumer rejection of frozen seafoods and that it was accompanied by lipid deterioration, particularly hydrolysis. When I returned to the group in 1956, following graduate work at McGill University, my task was to investigate the role of lipids in the continuing research program on deterioration in frozen-stored fish.

"On initiating this lipid study, the well-known need for a gentle and reliable method for extraction and purification of total lipid from biological tissues became acute. We were working with cod muscle, where the highly unsaturated lipids rarely exceeded 1% wet weight and consisted primarily of protein-bound phospholipid. Existing procedures were unsatisfactory and attention was focused on the use of mixtures of chloroform and methanol to isolate the lipids from moist biological materials.

"Examination of the chloroform-methanol-water phase diagram led to the hypothesis that 'optimum lipid extraction should result when the tissue is homogenized with a mixture of chloroform and methanol which, when mixed with the water in the tissue, would yield a monophasic solution. The resulting homogenate could then be diluted with water and/or chloroform to produce a biphasic system, the chloroform layer of which should contain the lipids and the methanol—water layer the non-lipids.' The hypothesis was readily confirmed by experimentation and the method was proven to be very effective. It was enthusiastically accepted by the scientific community, and after 19 years, it is still being used extensively in research laboratories throughout the world, on a host of biological materials.

"I would mention that as a small government research laboratory, we were not encouraged to do fundamental research but in the course of providing a research and development service related to the problems of the Canadian seafood industry, we were continualy faced with situations where basic scientific knowledge was inadequate. This is even more true today as the world directs greater attention toward the oceans and their resources. Although Dr. Dyer is now retired, his colleagues continue to pursue the challenges in fisheries science at the Halifax Laboratory where I am now Director."

Number 40 **Citation Classics** October 3, 1977

Trout D L, Estes E H Jr. & Friedberg S J. Titration of free fatty acids of plasma: a study of current methods and a new modification.
*Journal of Lipid Research* 1:199-202, 1960.

This article identifies substances that interfere with Dole's titrimetric method for measuring plasma free fatty acids, and adds a step to remove them from the fatty acid extract. [The SCI® indicates that this paper was cited 922 times in the period 1961-1975.]

Dr. David L. Trout
Carbohydrate Nutrition Laboratory
Beltsville Agricultural Research Center
Beltsville, Maryland 20705

July 11, 1977

"Our widely used method of measuring free fatty acids (FFA) in blood plasma was devised, tested and written up during my first 18 months after completing graduate work at Duke University. My dissertation had been in pharmacology, chiefly on the interaction between lysergic acid diethylamide (LSD) and serotonin in Siamese fighting fish. I then started research in lipid metabolism at the Veterans Administration hospital nearby.

"My boss was E. Harvey Estes, a clinician who also taught at Duke University (Medicine). He had written extensively on cardiology and was greatly interested in the possibility that plasma FFA played a role in atherosclerosis. My main cohort was Samuel J. Friedberg, who was also doing research during his residency training.

"Interest in plasma FFA was new when we started working on FFA methodology in 1958. Previously, biochemists thought plasma FFA were largely artifacts due to the lipolytic activity in blood. Studies in 1956 showed that plasma FFA exist in vivo in widely varying concentrations. It was soon clear that plasma FFA are largely derived from adipose tissues, released into the blood at rapid, highly controlled rates, and promptly extracted by various tissues. FFA proved to be a major transport form of lipid.

"Unfortunately, early methods of determining plasma FFA were either inaccurate or extremely laborious. Dr. V.P. Dole's method,[1] however, looked promising. FFA were extracted by a solvent mixture. When water and hydrocarbon were added, the system formed two phases. An aliquot of the relatively non-polar upper layer, which contained most of the FFA, was finally titrated. The method was admirably direct and simple. However, we observed that the upper layer also contained a fraction of the lactic acid present and some titratable phospholipid. We then found that, by shaking the FFA extract with very dilute sulfuric acid and centrifuging, the interfering substances were satisfactorily removed. This constituted our modification of the Dole procedure. Henry Kamin, who had taught me biochemistry, urged us to determine FFA values on the same blood plasmas by several methods. The validity of our modification was supported by our finding that it gave lower values than the original Dole method and values similar to those obtained by R.S. Gordon's standard method.

"Our paper was rejected as being too long by a first journal but was published promptly by the *Journal of Lipid Research*. Within 6 months, Dole and Hans Meinertz published an alternate method of washing out interfering substances from the Dole extract of plasma FFA.[2] Subsequently, colorimetric and automated procedures for FFA analysis have been developed, and some of these are far more sensitive than Dole's or our procedure. At present, the chief problem in determining the plasma FFA in a person or animal is to collect the blood without first stimulating lipolysis in adipocytes through physical or mental stress.

"Many scientists still use our FFA procedure, perhaps largely out of habit. The fact that we modified, in a simple way, a fairly satisfactory and highly ingenious method, does not detract from the usefulness of the resulting procedure. Of course, we were extremely lucky that the Dole-Meinertz paper or a similar one from another laboratory did not reach print ahead of ours."

1. Dole V P. A relation between nonesterified fatty acids in plasma and the metabolism of glucose. *Journal of Clinical Investigation* 35:150-4, 1956.
2. Dole V P & Meinertz H. Microdetermination of long-chain fatty acids in plasma and tissues. *Journal of Biological Chemistry* 235:2595-9, 1960.

# This Week's Citation Classic

CC/NUMBER 15
APRIL 14, 1980

Malins D C & Mangold H K. Analysis of complex lipid mixtures by thin-layer chromatography and complementary methods.
J. Amer. Oil Chem. Soc. 37:576-8, 1960.
[Technological Lab., Bur. Commercial Fisheries, US Fish and Wildlife Serv., Seattle, WA and Univ. Minn., Hormel Inst., Austin, MN]

Studies in 1959 revealed that the standardized techniques of thin-layer chromatography according to Stahl[1] opened up new and exciting possibilities for the resolution of complex mixtures of lipids. Ultimately, these microtechniques largely replaced paper and column chromatographic systems for the routine analysis of industrial and biological mixtures. [The SCI® indicates that this paper has been cited over 275 times since 1961.]

Donald C. Malins
Northwest and Alaska Fisheries Center
Environmental Conservation Division
Seattle, WA 98112

March 27, 1980

"My recollections of 1959 include seemingly endless hours spent in a converted chicken coop at the Hormel Institute in Austin, Minnesota. Hermann Schlenck and his associates, Donald Sand and Joanne Gellerman, turned out some fine work on the synthetic chemistry of lipids in this celebrated coop (others worked in a converted stable, remnants of the estate of Jay C. Hormel).

"I met Helmut Mangold (now professor and director, Institut für Biochemie und Technologie, H.P. Kaufmann-Institut, Münster, FRG) in this unusual environment.

"Mangold extolled the virtues of the then largely unknown technique of thin-layer chromatography (TLC), but few listened. He suggested that I use Stahl's TLC procedure to analyze the acetoglycerides my colleagues and I had synthesized from marine oils. I mildly berated him for spending so much time with sophomoric issues, such as fiddling around with powders on glass plates.

"To our surprise, the 'Stahl technique' worked wonders, and my colleagues and I published a paper which included a thin-layer chromatogram of herring oil acetoglycerides.[2] We stated: 'The work indicated that the method (TLC) may have wide use for the separation and identification of other lipid classes as well.' On reflection, it was an extreme understatement.

"Mangold asked me to join him in a blitzkrieg effort to look into the potential of TLC for the separation of complex mixtures of lipids. We scurried around obtaining as many types of mixtures as possible, which were placed on thin-layer plates. Remarkable separations were obtained in less than an hour. We advocated the virtues of TLC over paper and column chromatography, but not everyone was convinced.

"My time at Hormel had run out, so the work continued between Seattle and Austin. We only had one Stahl applicator (perhaps the only one in the country), so we shipped it back and forth. I attempted a thin-layer separation of 'glycerides' from dogfish liver. Two large spots appeared where only one (triacylglycerols) usually occurred. Had we completely separated triacylglycerols from diacylglycerol ethers for the first time and in less than an hour? We had indeed.[3] But credibility is hard to establish. Some time went by before some colleagues accepted this finding. After all, no one had apparently obtained a complete separation by column or paper chromatography — the yardsticks of that time. But once the doubters used the technique in their own work, the method was quietly sanctified.

"In this and subsequent papers, we demonstrated how Stahl's standardized plates could be used for preparative purposes and for reverse-phase chromatography.[4,5] We also showed that resolved lipids could be removed from plates and analyzed in a complementary way by other techniques, such as gas-liquid chromatography.[4,5] A number of papers (authored by us separately and together) demonstrated the great power and versatility of TLC. Those were great times.

"Relatively simple and rapid thin-layer chromatographic techniques, used in conjunction with complementary methods (e.g., gas chromatography), are described for the analysis of complex synthetic and natural mixtures of lipids. This would account for the frequent citation of the paper."

1. Stahl E. Thin-layer chromatography: a laboratory handbook. New York: Springer-Verlag, 1969. 1,041 p.
2. Greger E H, Jr., Malins D C & Gauglitz E J, Jr. Glycerolysis of marine oils and the preparation of acetylated monoglycerides. J. Amer. Oil Chem. Soc. 37:214-7, 1960.
3. Mangold H K & Malins D C. Fractionation of fats, oils, and waxes on thin-layers of silicic acid. J. Amer. Oil Chem. Soc. 37:383-5, 1960.
4. Malins D C & Mangold H K. Thin-layer chromatography. (F J Welcher, ed.) Standard methods of chemical analysis. New York: D. Van Nostrand, 1966. Vol. 3. p. 738-80.
5. Malins D C. Recent developments in the thin-layer chromatography of lipids. (Holman R T, ed.) Progress in the chemistry of fats and other lipids. Oxford: Pergamon Press, 1966. Vol. 8, Pt. 3. p. 303-58.

# This Week's Citation Classic

CC/NUMBER 12
MARCH 22, 1982

Mangold H K & Malins D C. Fractionation of fats, oils, and waxes on thin layers of silicic acid. *J. Amer. Oil Chem. Soc.* 37:383-5, 1960.
[Univ. Minnesota, Hormel Inst., Austin, MN and Technological Lab., Bureau of Commercial Fisheries, US Fish and Wildlife Serv., Seattle, WA]

The fractionation of lipids by thin-layer chromatography (TLC) is described. Adsorption chromatography on thin layers of silicic acid permits the rapid resolution of complex lipid mixtures into classes of compounds ranging in polarity from hydrocarbons to phospholipids. [The *SCI®* indicates that this paper has been cited over 260 times since 1961.]

Helmut K. Mangold
Institute for Biochemistry and Technology
H.P. Kaufmann Institute
Federal Center for Lipid Research
D-4400 Münster
Federal Republic of Germany

November 30, 1981

"In September 1958, the US Public Health Service, National Institutes of Health (NIH), informed me that my application for a research grant had been approved for a period of three years. The grant for work on the 'Isolation of unusual lipids...' covered my salary and included a few hundred dollars a year for equipment, glassware, and chemicals. With this starting capital I purchased the equipment E. Stahl had developed for the coating of glass plates with a thin layer of an adsorbent.

"The first publication I credited to my NIH grant described the use of adsorption-TLC for the fractionation of radioactively labeled lipid derivatives into classes of compounds having the same type and number of functional groups, and the further resolution of each of these lipid classes by reversed-phase paper chromatography.[1]

"Early in 1959, I began cooperating with Donald C. Malins, a guest who had spent several months at our institute. We agreed to try the fractionation of complex natural lipid mixtures, such as vegetable oils and animal fats and oils. Our enthusiasm grew day by day as we recognized that TLC on silicic acid allowed us to distinguish olive oil, essentially a mixture of triacylglycerols, from jojoba oil, which consists of a mixture of wax esters. Even more surprising and impressive was the resolution of such structurally related lipids as the triacylglycerols and alkyldiacylglycerols of shark liver oils.

"In the fall of 1959, we submitted our first manuscript to a journal, but it was rejected. We sent the manuscript to another journal, which didn't want it either. Well, eventually, the editor of the *Journal of the American Oil Chemists' Society* accepted our manuscript for publication, and we sent him a few more papers.

"The response to our publications was enormous. The TLC technique caught on like wildfire, not only in the lipid field, but also in other areas of natural products chemistry and in biochemistry. In 1960 we reported on our work at the meeting of the American Oil Chemists' Society in New York City, and we won the Bond Award. An invitation to the Gordon Conference on Lipid Metabolism and numerous invitations by universities and professional societies followed. I became a consultant to the Oak Ridge Institute of Nuclear Studies, Oak Ridge, Tennessee, an appointment I cherished for over ten years.

"My NIH grant was renewed year after year. I received additional support, and when I left the University of Minnesota in 1969, I 'bequeathed' to my successors two postdoctoral positions that were financed by the NIH. At Münster, I continued work on unusual lipids, such as ether lipids,[2] pheromones, and cyclopentenyl fatty acids,[3] and I investigated the lipids in plant cell cultures.[4] In 1977 I won the Heinrich Wieland Prize for studies on the synthesis and biosynthesis of alkoxylipids.[2]

"I am sure my friend Malins will agree with me that our paper has been quoted so frequently because it was the first to describe the fractionation of natural fats and oils and waxes, and not of model mixtures of synthetic lipid compounds. We had made a point of developing a procedure for the analysis of 'greases,' something hardly anybody wanted to deal with. It worked, and it still does!"[5]

1. **Mangold H K.** Zur Analyse von Lipiden mit Hilfe der Radioreagenz-Methode.
   *Fette Seifen Anstrichm.* 61:877-81, 1959.
2. ──────, Synthesis and biosynthesis of alkoxylipids. *Angew. Chem. Int. Ed.* 18:493-503, 1979.
3. **Mangold H K & Spener F.** The cyclopentenyl fatty acids. (Tevini M & Lichtenthaler H K, eds.)
   *Lipids and lipid polymers in higher plants.* Berlin: Springer-Verlag, 1977. p. 85-101.
4. **Radwan S S & Mangold H K.** Biochemistry of lipids in plant cell cultures. (Fiechter A, ed.)
   *Advances in biochemical engineering.* Berlin: Springer-Verlag, 1980. Vol. 16. p. 109-33.
5. **Mangold H K & Mukherjee K D.** New methods of quantitation in thin-layer chromatography: tubular thin-layer chromatography. *J. Chromatogr. Sci.* 13:398-402, 1975.

# Citation Classics

Number 5     January 30, 1978

**Mangold H K.** Thin-layer chromatography of lipids.
*J. Amer. Oil Chem. Soc.* 38:708-27, 1961.

The author describes the technique of thin-layer chromatography (TLC) and its applicability in the lipid field, stressing the simplicity, sensitivity, capacity, versatility, and efficiency of the method. [The *SCI®* indicates that this paper was cited 560 times in the period 1961-1976.]

Professor K. Mangold
Institute for Biochemistry and Technology
H.P. Kaufmann-Institute
Federal Center for Lipid Research
D-4400 Munster, Germany

February 28, 1977

"This paper was certainly not written with an aim to make it a 'Citation Classic'; in fact, it was almost not written at all. Early in 1961, while I was working at The Hormel Institute, a research unit of the University of Minnesota, in Austin, Minnesota, the American Oil Chemists' Society invited me to give a talk on thin-layer chromatography during a short course the society was planning to conduct at the University of Rochester, in July of that year. This presentation was to be published in *Journal of the American Oil Chemists' Society*. Although I had in mind to enter matrimony shortly after I had received this invitation, I considered it easily possible to get married in April or May, write the paper on our wedding trip—a voyage across the Atlantic—have it typed by my wife, Anne, and give the talk a few weeks later. How foolish I was! I did give the talk at Rochester, N.Y., but the deadline for submission of the manuscript had long passed when I started writing.

"Actually, I was in a perfect situation for writing: Professor W.O. Lundberg, who headed The Hormel Institute, had given me a spacious, though empty, laboratory which I tried to equip, at least with glassware and chemicals. One of the first items I purchased was the basic equipment for thin-layer chromatography which Professor E. Stahl had developed.[1]

"In writing the article for *Journal of the American Oil Chemists' Society* I certainly was at a great advantage, inasmuch as this was to be the first survey of the method to be published. Moreover, although TLC is useful in the analysis of almost all non-volatile organic substances, I was convinced that adsorption-TLC would become the method of choice for the fractionation of complex lipid mixtures—and this holds true, even today. I demonstrated that not only adsorption but also other principles of chromatography, such as reversed-phase partition, are applicable in TLC and that the method is valuable also as a semi-preparative tool. Still, I did not try to make it appear to be the panacea for all problems but, instead, I emphasized that TLC is fully exploited only when it is used in conjunction with other techniques. I showed how the method could be combined with gas chromatography, and I tried to integrate it in a system of lipid analysis.

"Since, at that time, I wasn't burdened with any administrative duties nor too many speaking obligations, I could manage to try out various aspects of TLC at the lab bench between the hours I spent writing. Thus, I could reveal numerous tips and hints and personal experiences I had gained with the new method. In addition, I included several experimental procedures in much detail. At the time I wrote this manuscript, I had published four papers on the use of TLC in the lipid field, two of them with D.C. Malins, who is now Director of the Environmental Conservation Division at the Northwest Fisheries Center in Seattle, Washington; two more publications were in press.

"As time went on I developed more and more enthusiasm, not only for the method but also for the job of writing about it. What was meant to be a review became a bit of a 'manual' on the use of TLC with emphasis on the method's application in the lipid field.

"For over fifteen years now, and even today, I am getting reprint requests for the article I published in 1961. Of course, I am very pleased to see that people still find it useful."

REFERENCE

1. Pelick N, Bolliger H R & Mangold H K. *Advances in chromatography.* (Giddings J C & Keller R A, eds.) New York: Marcel Dekker, Inc. Vol. 3, pp. 85-118, 1966.

# Citation Classics

Number 16 — April 18, 1977

Van Handel E & Zilversmit D B. Micromethod for the direct determination of serum triglycerides. *Journal of Laboratory Clinical Medicine* 50:152-57, 1957.

A microprocedure for the direct determination of triglyceride concentrations in biologic specimens is presented. The method depends on the quantitative removal of phosphatides from the sample and the subsequent determination of esterified glycerol. The procedure has been tested on whole blood and plasma. [The *SCI®* indicates that this paper was cited 1,440 times in the period 1961-1975.]

---

Dr. Emile Van Handel
Department of Health
and Rehabilitative Services
Florida Medical Entomology Laboratory
P.O. Box 520
Vera Beach, Florida 32960

December 8, 1976

"In the early fifties, Don Zilversmit was working on the turnover of phospholipids at the University of Tennessee, while I was working on the chemistry of phospholipids at the University of Amsterdam. This shared interest led to my appointment at the University of Tennessee. Had Memphis offered an ocean beach, it might have lasted longer than the three years it did. Ironically, the team became best known for removing from serum, in one swoop, the phospholipids to which we had both dedicated our lives.

"To study the mechanism of absorption in dogs, we used the then available I 131 labeled fat. In order to test whether the label would stick during absorption, we needed a way to measure serum triglycerides. Determinations then in use were done by difference, extracting and weighing the total lipids, and subtracting whatever value was determined for cholesterol, cholesterol esters, and phospholipids. Since triglycerides constitute only a small fraction of this total, these determinations often yielded negative values. We therefore designed a 'micromethod for the direct determination of serum triglycerides,' based on the glycerol rather than on the fatty acid moiety of the molecule. Had we been able to mix I 131 with C 14 labeled triglycerides, the non-validity of the I 131 labeling would have been immediately apparent, and the chemical method unnecessary. The I 131 method sank, the triglyceride method rose.

"To make the paper acceptable to a clinical journal, we determined 'normal values' in 12 students. Twenty years and millions of determinations later, the average value in the population is still the same.

"In 1957, the term 'triglyceride' could not be found in the index of any clinical textbook. Within a few years, 'high triglycerides' became a common household scare-word along with high cholesterol and high glucose.

"The need for automation in clinical chemistry required a modification which combines a specific lipase with a glycerol dehydrogenase and avoids solvent extraction and heating. However, the 1957 method is still widely used, in spite of my 'improvement' published in 1961.[1] This 1961 cluster (information science jargon for a buster) is yearly siphoning off enough citations to knock the 1957 mother paper just off the list of 50 most cited papers.

"Don Zilversmit, now at Cornell, is still working on problems related to mammalian lipids and lipoproteins. I moved to the Entomological Research Center in Vero Beach, Florida, and have been working on insect metabolism ever since. I have published several other methods which are frequently cited. There has always been a much greater demand for my method papers than for my biological results, except when I started the title of an article with the word 'sex.'[2] Perhaps the readers expected again an original method, easily taught to technicians.

"Serum triglycerides seem to be a permanent feature on the clinical scene. What is the significance of this determination? I don't know. Check with your doctor."

1. **Van Handel E.** Suggested modifications of the micro determination of triglycerides. *Clinical Chemistry* 7:249-51, 1961.
2. ——————. Sex as regulator of triglyceride metabolism in the mosquito. *Science* 134:1979-80, 1961.

# This Week's Citation Classic
NUMBER 2
JANUARY 8, 1979

Carlson L A. Determination of serum triglycerides.
J. Atheroscler. Res. 3: 334-6, 1963.

This paper summarizes 10 years' work in 3 pages. It describes the final version of my method for determination of serum triglycerides with major emphasis on practicality. Water-soluble interfering substances are removed; the method is specific for triglycerides, and it can easily be used for tissues. [The SCI® indicates that this paper was cited 407 times in the period 1963-1977.]

---

Lars A. Carlson
Department of Internal Medicine
King Gustaf V Research Institute
S-10401 Stockholm, Sweden

February 23, 1978

"My job is that of a physician in an academic department of internal medicine. I have no biochemical training but I am a somewhat unusual (read crazy) physician in being fascinated in and obsessed by biochemical methods. When the day's bedside and administrative work is done, I have spent and do spend the evenings and nights in the laboratory engaged with various aspects on methodology. I do not understand, but am deeply grateful for the patience of Kerstin, Bjorn, Mats, and Pia which made/makes my night work possible.

"This paper is number four and the last of my work on the method for determination of triglycerides in blood serum. The first was published in 1956 by myself and L. B. Wadstrom. My interest in blood lipids and lipoproteins was aroused as a medical student by the article in Science, 1950 by Gofman and co-workers on the role of lipids and lipoproteins in atherosclerosis. I was then given a lab bench by Nanna Svartz in the institute of which I am now head, and started my work on serum lipids. At that time triglycerides of serum were unheard of in the clinic. Serum cholesterol was determined, but only used as an aid in the diagnosis of diseases such as nephrosis, myxedema, etc. I soon realized that there was urgent need in the lipid field for a method for determining triglycerides.

"The method has three essential steps. In principle these were available, but they had not been combined before: (1) Extraction of lipids by chloroform-methanol, (2) removal of phospholipids by silicic acid, and (3) determination of the triglyceride-glycerol by the chromotropic acid reaction. I believe that this paper became a 'Citation Classic' because we succeeded in greatly simplifying steps 1 and 2. We are still using these two steps daily. Particularly the extraction procedure for blood and tissues, with its practical simplicity, still charms visitors and ourselves.

"We had used the modified method since 1960 and I had not thought of publishing it until M. F. Oliver of Edinburgh, at the first meeting on the lipid-lowering compound clofibrate in 1963 stated that he—then editor of the *Journal of Atherosclerosis Research*—would immediately print any good paper on the difficult task of 'Determination of serum triglycerides.' He had this paper 10 days later, had it printed in 4 months—and still later—became one of my best international friends.

"In 1950 no direct serum triglyceride analysis could be done; in 1963 one technician could do 20-30 per day. Looking in my laboratory today where the analyzing machine feeds 100 completed analyses per hour into the computer, one can talk about progress. But why determine serum triglycerides? Simply because they are so closely related to atherosclerotic diseases."

# This Week's Citation Classic

CC/NUMBER 22
MAY 30, 1983

Sperry W M & Webb M. A revision of the Schoenheimer-Sperry method for cholesterol determination. *J. Biol. Chem.* **187**:97-106, 1950.
[Depts. Biochemistry, New York State Psychiatric Inst., and Coll. Physicians & Surgeons, Columbia Univ., New York, NY]

A revision of the Schoenheimer-Sperry method for the determination of total and free cholesterol in blood serum is described. [The *SCI®* indicates that this paper has been cited in over 1,650 publications since 1961.]

Warren M. Sperry
151 Edgewood Avenue
Yonkers, NY 10704

April 15, 1983

"In 1930, when I moved from the University of Rochester to the Babies' Hospital at the Columbia Medical Center, nothing was known about free and esterified cholesterol in children's blood serum because the only available method, gravimetry of the digitonide, required large amounts of blood. I undertook an attempt to adapt this method to a microscale.

"In 1932, Rudolf Schoenheimer of Freiburg, knowing my interest in a micromethod for cholesterol, invited me to collaborate in the development of a method based on his brilliant idea of applying the widely used colorimetric method to the digitonide. He realized that a procedure which works in one laboratory may fail in another.

"There followed a voluminous correspondence in which Rudolf in long letters described his procedures and the good results they gave, and I in equally long letters described my utter failure to obtain such results. The impasse was ended by Rudolf's move from Freiburg to the department of biochemistry at Columbia University. He came to my laboratory to show me where I had gone wrong. He failed as miserably as I had to reproduce his Freiburg results. Although this was a severe setback to our plans, I must confess that it did much to restore my ego, which was at a low ebb. Rudolf was already deeply immersed in plans for his pioneering work on the use of isotopes as metabolic indicators, and he turned the further development of the method over to me.

"I went back to square one and began a study in which I varied solvents, reagents and their concentrations, temperatures and times at various steps, mechanical procedures, etc. That was about 50 years ago, and I have forgotten the details. Finally, a method which gave accurate results on 0.2 ml. of serum was achieved.[1]

"In several studies, hundreds of samples of blood serum were analyzed. No trouble was encountered until our supply of digitonin was interrupted by the war. Aqueous solutions of domestic digitonin were unstable, but solutions in 50 percent ethanol were found to be stable. This and several other changes which facilitated the technique were incorporated in a revision of the method, which was published in the paper cited above.

"I was surprised by the number of citations. Because the method requires considerable time and effort, it has been used in few, if any, clinical laboratories. It has been used as a standard for calibration of rapid colorimetric methods which determine only total cholesterol. It made possible the finding that the percentage of esterified in total cholesterol in healthy humans is relatively constant, varying only from about 70 to 75 percent.[2] In contrast, the level of total cholesterol in the same subjects varied by nearly 200 percent. Studies of the mechanism by which the ratio between the cholesterol fractions is governed and its significance in high- and low-density lipoproteins should be rewarding."

1. Schoenheimer R & Sperry W M. A micromethod for the determination of free and combined cholesterol.
   *J. Biol. Chem.* **106**:745-60, 1934.
   [The *SCI* indicates that this paper has been cited in over 355 publications since 1961.]
2. Sperry W M. The relationship between total and free cholesterol in human blood serum.
   *J. Biol. Chem.* **114**:125-33, 1936.

23

## This Week's Citation Classic

CC/NUMBER 34
AUGUST 20, 1979

Abell L L, Levy B B, Brodie B B & Kendall F E. A simplified method for the estimation of total cholesterol in serum and demonstration of its specificity.
*J. Biol. Chem.* **195**:357-66, 1952. [Columbia Research Service, Goldwater Memorial Hospital, Roosevelt (formerly Welfare) Island, NY]

This paper describes a simple, rapid, and specific colorimetric method for the determination of total cholesterol in serum. [The *SCI*® indicates that this paper has been cited over 1,410 times since 1961.]

Liese L. Abell
7 Peter Cooper Road
New York, NY 10010

March 23, 1978

"During the late forties our laboratory was deeply involved in research concerned with arteriosclerosis. When it became apparent that cholesterol was implicated in the etiology of this disease, much time was spent determining serum cholesterol levels in various groups. Existing methods were precise but laborious and time-consuming; therefore we decided to try and find a shorter procedure, which was just as accurate and specific. A starting point was provided by the method of W.M. Sperry and F.C. Brand,[1] and we proceeded to adapt this method to the needs of a laboratory where a very large number of samples had to be analyzed daily.

"By simplifying the extraction and saponification steps and by using a modified Liebermann-Burchard reagent we were able to eliminate several manipulations; this resulted in a considerable saving of time and omitted many sources of error.

"This new analytical procedure had to fulfill the following crucial conditions: (1) it must be specific, i.e., the product measured had to be proven to be cholesterol and nothing else; (2) it must be reproducible; (3) the results must be validated by comparison with existing and established methods; and (4) it must be simple enough to be performed by a competent technician.

"We were very fortunate that Dr. Bernard B. Brodie was willing to cooperate. He and his group undertook the exacting task of providing the proof of specificity by countercurrent distribution. His laboratory was located some distance from ours, at the end of a long corridor, and many miles were covered carrying the little bottles containing the final extract to his laboratory, and innumerable separatory funnels migrated back and forth between our research divisions.

"The proof of reproducibility and the validation of the method were the most time-consuming parts of the work. This demanded infinite patience; it was a repetitive and exacting chore and involved coding the samples to exclude subjective judgment. My compulsive personality helped. Much of this work had to be done in the evenings because our only spectrophotometer was used for routine work during the day. Since my laboratory afforded a beautiful view of the lights of Manhattan across the East River, the monotony of the precise timing of the Liebermann-Burchard reaction was relieved when I discovered that I could time myself accurately by the change of appropriate traffic lights.

"As the procedure is simple and accurate it was chosen by the laboratories which participated in the first Cooperative Study of Lipoproteins and Atherosclerosis in 1951.[2] It is still used as a standard method in many laboratories around the world."

1. **Sperry W M & Brand F C.** The colorimetric determination of cholesterol.
   *J. Biol. Chem.* **150**:315-24, 1943.
2. **Gofman J F.** Evaluation of serum lipoprotein and cholesterol measurements as predictors of clinical complications of atherosclerosis. Report of a cooperative study of lipoproteins and atherosclerosis.
   *Circulation* **14**:691-742, 1956.

## This Week's Citation Classic

CC/NUMBER 12
MARCH 23, 1981

Zlatkis A, Zak B & Boyle A J. A new method for the direct determination of serum cholesterol. *J. Lab. Clin. Med.* **41**:486-92, 1953.
[Depts. Chem., Pathol., and Med., Wayne Univ., and City of Detroit, Receiving Hosp., Detroit, MI]

The authors present a new, sensitive, and stable color reaction for the simple determination of serum cholesterol by direct treatment of the serum with a reagent composed of ferric chloride dissolved in a glacial acetic acid-sulfuric acid mixture. [The *SCI*® indicates that this paper has been cited over 1,110 times since 1961.]

Bennie Zak
Department of Pathology
Wayne State University School of Medicine
Detroit, MI 48201
and
Albert Zlatkis
Department of Chemistry
Houston University
Houston, TX 77004

January 27, 1981

"This procedure developed out of a need for both simplicity in handling micro samples as well as sensitivity of signal achieved in a project in which many serum samples drawn from treated and control rabbits had to be evaluated for their cholesterol concentrations for the purpose of a special investigation into the formation and reversal of atherosclerotic plaques.[1] The idea of generating these plaques in a susceptible animal such as the rabbit was not a new one; however, the attempt to cause the reversal of those atherosclerotic lesions by some potentially appropriate chemical treatment with a chelating agent was a newer one. In a sense, the then somewhat unusual direct approach was made possible by the use of a specially purified glacial acetic acid which was glyoxal-free and thereby did not result in an interfering Hopkins-Cole reaction of the matrix reagent with either the tryptophane of the still intact proteins or the free tryptophane of the serums. In the ensuing years many modifications appeared, a few from ourselves, many from others. Then, with the advent of continuous flow mechanization and its adaptation to the automated handling of many serum samples, it became an often used reaction in the modified cholesterol procedures of many routine chemistry laboratories. Commercial organizations were quick to cash in on its virtues for easy salability, and modified procedures ensured profits for those enterprising enough to package them in kit forms for both manual and automated systems.[2,3] In more recent years the Bureau of Standards felt the reaction itself was important enough to study the elucidation of the mechanism.[4]

"Since the introduction of enzymes as reagents into cholesterol methodologies, a new simplicity of technology without the need for a viscous strong acid matrix of reaction has begun to replace the latter and a waning popularity is the result even though comparisons have shown a similarity in results between the new and the old methodologies. Any future for strong acid reactions now could only reside in the comparative economics of such a simple reagent over that of the more complex sequenced enzyme reagent systems that have begun their displacement process, for the old procedures may only serve as cheaper screening devices in the future for this important and popular determination.

"It is always difficult to understand why a simple procedure is referred to enough to make it a *Citation Classic*. One could take the optimistic view that it filled an immediate need for investigators who could use a simple, sensitive, and stable color reaction as compared to a less sensitive, unstable, and more complicated reaction system of the past. The method was, in spite of its simplicity, easily amenable to alteration, so between application and modification it appeared in the literature many times. We are still studying its reaction characteristics and how they are affected by matrix modification. Hopefully, a few more references are still to appear."[5]

1. Uhl H S, Brown H H, Zlatkis A, Zak B, Myers G B & Boyle A J. Effect of ethylenediamine tetraacetic acid on cholesterol metabolism in rabbits. *Amer. J. Clin. Pathol.* **23**:1226-33, 1953.
2. Wybenga D R, Pileggi V J, Dirstine P H & DiGiorgio J. Direct manual determination of serum total cholesterol with a single stable reagent. *Clin. Chem.* **16**:980-4, 1970.
3. Levine J B & Zak B. Automated determination of serum total cholesterol. *Clin. Chim. Acta* **10**:381-4, 1964.
4. Burke R W, Diamondstone B I, Velapoldi R A & Menis O. Mechanisms of the Liebermann-Burchard and Zak color reactions for cholesterol. *Clin. Chem.* **20**:794-801, 1974.
5. Zak B. Cholesterol methodologies: a review. *Clin. Chem.* **23**:1201-14, 1977.

## This Week's Citation Classic

CC/NUMBER 18
MAY 4, 1981

Zak B, Dickenman R C, White E G, Burnett H & Cherney P J. Rapid estimation of free and total cholesterol. *Amer. J. Clin. Pathol.* 24:1307-15, 1954.
[Depts. Pathology, Wayne Univ. College of Medicine, Detroit Receiving Hosp., and Detroit Memorial Hosp., Detroit, MI]

The paper describes the determination of extracted total cholesterol and digitonide precipitated free cholesterol by ferric iron reaction in a glacial acetic-sulfuric acid milieu and discusses the several variables affecting the equilibrium reaction. [The *SCI®* indicates that this paper has been cited over 415 times since 1961.]

Bennie Zak
Department of Pathology
Wayne State University
School of Medicine
Detroit, MI 48201

January 27, 1981

"This paper came into being because a previously described direct reaction for serum cholesterol,[1] also a *Citation Classic*, resulted in interference problems with Hopkins-Cole reactants owing to glyoxal impurities in the acetic acid and tryptophane in the sample as the indicted offending compounds. In the original work of 1953, care was taken to inform readers that a high purity solvent was required, but some analysts ignored or overlooked the warning and unfortunately encountered the side reaction which resulted in overlapping spectral interference. A partial cleanup by means of simple extraction along with digitonide purification lengthened the procedure as an analytical modification. However, this then newer technique ensured that the results obtained would be more in harmony with true values. Later studies by others[2] on the nature of the milieu in which the reaction occurs proved that interferences could at least be partially or mostly eliminated by means of this analytical consideration. Subsequently, I have had the opportunity to emphasize the importance of the reaction medium more fully.[3] Several observations on the effect of the medium on the reaction as an optimization phenomenon were investigated and discussed here.

"Even though this procedure was more involved technically than the totally direct approach, it was still simple enough to popularize it somewhat over the extant use of techniques at least as complicated, ending up with variations of the much less sensitive and stable Liebermann-Burchard reaction as the concluding step of those procedures.

"Perhaps the most important realization that I gleaned from the application of an acceptable new reaction in the face of entrenched technology is the understanding that the older way is not necessarily the only way, and that when given the seed of an idea, curious investigators will modify and improve what appears in an attempt to hone it to perfection, while more innovative investigators, dissatisfied now with both the old as well as the new analytical devices, will visualize still newer reactions because such new reactions now seem to be a real possibility. The overpowering influence of a totally accepted equilibrium reaction which virtually everyone uses, partly from the belief that nothing else is available or partly from the belief that nothing else could be available, sometimes seems to program a slowing of the progression of inventiveness. In addition, total acceptance by most of the workers in an area even seems to cause resentment about what appears to them as a drastic change, and this in turn may impede acceptance of what might be a useful replacement of older technology, at least for the moment."

1. Zlatkis A, Zak B & Boyle A J. A new method for the direct determination of serum cholesterol.
    *J. Lab. Clin. Med.* 41:468-92, 1953.
    [Citation Classic. *Current Contents/Life Sciences* (12):20, 23 March 1981.]
2. Wybenga D R, Pileggi V J, Dirstine P H & DiGiorgio J. Direct manual determination of serum total cholesterol with a single stable reagent. *Clin. Chem.* 16:980-4, 1970.
3. Zak B. Cholesterol methodologies: a review. *Clin. Chem.* 23:1201-14, 1977.

## This Week's Citation Classic™

CC/NUMBER 3
JANUARY 16, 1984

Bush I E. **Methods of paper chromatography of steroids applicable to the study of steroids in mammalian blood and tissues.** *Biochemical J.* 50:370-8, 1952.
[National Institute for Medical Research, Mill Hill, London, England]

Using aqueous methanol as stationary phase, the separation of steroids by paper partition chromatography is achieved without special treatment of the paper. A family of solvent systems is described capable of separating the full range of biologically active steroid hormones and their metabolites, as are a new fluorescence reaction for $\Delta^4$-3-ketosteroids and the 'wick' and 'running-up' techniques for complete transfer of extracts to paper chromatograms. [The *SCI*® indicates that this paper has been cited in over 1,580 publications since 1955.]

---

Ian E. Bush
Research Service
Veterans Administration
Medical and Regional Office Center
White River Junction, VT 05001

October 18, 1983

"My main goal as a graduate student at the University of Cambridge was to identify the hormone(s) secreted by the adrenal cortex. Vogt's discovery of the high biological activity of adrenal venous blood[1] made this an ambitious but feasible task. But no one knew which, if any, of the steroids isolated from gland extracts was secreted into the blood. The big risk was that the activity measured by Vogt was due to an unidentified steroid in the 'amorphous fraction' of adrenal extract, whose biological activity was so high that it would be chemically undetectable in adrenal venous blood.

"In 1949, paper partition chromatography had been successfully applied to most of the important classes of hydrophilic biochemicals. Lipids, however, were another matter. In theory, the adrenal steroids should have been easily separable with simple benzene/water systems, but they weren't! Impatient with theory, I wasted a year with alumina-impregnated paper[2] proving that my first supervisor, P.R. Lewis, had been right; adsorption chromatography gave pretty pictures with pure steroids, but was almost useless with blood extracts. Duly humbled, I returned to theory suspecting that my first failures with partition systems had been due to the low solubilities of steroids, their adsorption by cellulose, and the low temperature in Cambridge. At the National Institute for Medical Research (London), with its resources (including radiators!), I decided to try systems based on hydrocarbon/methanol/water — horribly volatile in all senses compared with the stolid reliability of butanol, phenol, and collidine. In the first experiment (30 percent methanol/benzene at 35°C), adsorption was still evident but the major active adrenal steroids were separated as short 'comets.' Two days later, perfect 'spots' were obtained with 40 percent methanol as stationary phase. In a few more weeks, I had devised a 'family' of solvent systems based on aqueous methanol capable of separating the less polar androgens, estrogens, progesterone, and their metabolites. Attempting to get the Zimmerman reaction to work with $\Delta^4$-3-ketosteroids, I noticed a dim orange fluorescence which led to the extremely sensitive soda fluorescence reaction for this group.

"The frequent citation is probably due to the very wide applicability of this rather simple family of solvent systems and later variants.[3] More convenient than Zaffaroni's alternatives[4] (impregnation of paper with formamide or propylene glycol), they played a major role in the discovery of aldosterone (Simpson, Tait, and Reichstein[5]), hitherto unknown estrogens (G.F. Marrian[6]) and the adrenal steroids in the blood of many vertebrate species.

"My overwhelming memory is of the role of luck and friendliness in making this work possible. Luck in having Lewis, R.K. Callow, and W. Feldberg as supervisors; and the generous provision of reference compounds and advice by Reichstein, Vogt, C.S. Hanes, A.J.P. Martin, and others to a somewhat rambunctious graduate student who would not be allowed to attempt such a project today. The spirit of the times is best illustrated by Feldberg's first advice to me after a disquisition on the virtues of chloralose as an anesthetic: 'And remember, Bush, when you are in research, you need plenty of holidays!'"

---

1. Vogt M. The output of cortical hormone by the mammalian suprarenal. *J. Physiology* 102:341-56, 1943.
2. Datta S P, Overell B G & Stack-Dunne M. Chromatography on alumina-impregnated filter paper. *Nature* 164:673-4, 1949.
3. Sherma J & Zweig G. Steroids, bile acids, and cardiac glycosides. *Paper chromatography and electrophoresis.* Volume II. *Paper chromatography.* New York: Academic Press, 1971. p. 200-47.
4. Zaffaroni A, Burton R B & Keutmann E H. Adrenal cortical hormones: analysis by paper partition chromatography and occurrence in the urine of normal persons. *Science* 111:6-8, 1950. (Cited 250 times since 1955.)
5. Simpson S A, Tait J F, Wettstein A, Neher R, von Euw J, Schindler O & Reichstein T. Aldosteron. Isolierung und Eigenschaften. Über Bestandteile der Nebennierenrinde und verwandte Stoffe. *Helv. Chim. Acta* 37:1163-200, 1954. (Cited 180 times since 1955.)
6. Loke K H, Marrian G F & Watson E J D. The isolation of a sixth Kober chromogen from the urine of pregnant women and its identification as 18-hydroxyoestrone. *Biochemical J.* 71:43-8, 1959. (Cited 35 times since 1959.)

# This Week's Citation Classic

Allen W M. **A simple method for analyzing complicated absorption curves, of use in the colorimetric determination of urinary steroids.**
*J. Clin. Endocrinol. Metab.* 10:71-83, 1950.
[Dept. Obstet. & Gynecol., Washington Univ. Sch. Med., St. Louis, MO]

This paper gives the rigorous mathematical proof for a simple formula which makes possible the analysis of contaminated absorption curves in both the visible and ultraviolet regions of the spectrum. The only requisite is that the 'background' absorption be linear in the region analyzed. [The *SCI®* indicates that this paper has been cited over 485 times since 1961.]

---

Willard M. Allen
Department of Obstetrics
and Gynecology
University of Maryland
Baltimore, MD 21201

March 29, 1978

"My first exposure to absorption spectra came in 1933 when Oskar Wintersteiner reported to me that the pure corpus luteum hormone which I had isolated had an absorption maximum at 240 mu. This fact, together with other analytic data he had obtained, virtually established the structural formula of progesterone.

"My next exposure came in 1941 when I was using a Coleman photoelectric spectrophotometer to measure estriol in extracts of human pregnancy urine. This instrument gave excellent absorption spectra of the colors produced by the Kober Reagent (mostly concentrated sulfuric acid). Good results were obtained when the estriol content was high, but abnormally high results were always obtained when the estriol content was low. By 1942, I was fully aware of the problem. I had looked at hundreds of absorption curves and had seen the 'estriol hump' sitting, as it were, on a high baseline due to contaminating brown colors.

"These background colors plagued everyone doing steroid analyses for years and no one, including me, was able to achieve a perfect blank so that the colorimeter would automatically subtract the background color. Consequently, I turned to mathematics. Sometime during 1942 I succumbed to the important conclusion that a mathematical solution was possible if the slight curve of the absorption spectrum of the background was considered to be a straight line.

"By 1943, I had a workable formula which used the optical densities at 420-450-480 mu, and two constants which were derived from the separate absorption spectra of pure estriol and estriol-free 'background.' Apparently, I was not satisfied with this one because I soon developed a closely related formula which was devoid of arbitrary constants. Actually, I think this new one came from the top of my head without mathematical proof.

"In 1947, after we had discovered the Allen-Blue Test for adrenal tumors I decided to develop the mathematical proof for this formula which we had used for about three years. This proved to be an ordeal in logic and analytical geometry which consumed me for several months. Success finally came when I saw that the addition of a simple expression to both sides of the equation would break the impasse. Only then did my cherished formula emerge.

"This formula soon became known as 'The Allen Correction' in steroid laboratories around the world. The beauty of the formula lies in its simplicity. Nothing I have ever done has given me more pleasure and satisfaction."

# This Week's Citation Classic

CC/NUMBER 2
JANUARY 10, 1983

Brown J B. A chemical method for the determination of oestriol, oestrone and oestradiol in human urine. *Biochemical J.* **60**:185-93, 1955.
[Clinical Endocrinology Research Unit, Medical Research Council, Univ. Edinburgh, Scotland]

This paper described the first chemical method to be developed for the separate measurement of the three classical estrogens: estriol, estrone, and estradiol in the urine of men and nonpregnant women. The specificity, sensitivity, reproducibility, accuracy, convenience, and application of the method were discussed. [The *SCI*® indicates that this paper has been cited in over 980 publications since 1961.]

---

James B. Brown
Department of Obstetrics and Gynaecology
University of Melbourne
Parkville, Victoria 3052
Australia

November 30, 1982

"Estrogen is the most important female hormone. Its measurement provides the key to many studies in human female reproduction including the monitoring of ovarian function in normal and infertile women, the identification of the times of fertility and ovulation during the cycle, the action of contraceptives, the use of fertility drugs, and the achievement of test-tube babies.

"Estrogen assay is also important in the study of cancers of the breast, endometrium, and ovary. The pressing need for a quantitative assay of estrogens in nonpregnant women was well recognized in 1950 and many other groups were engaged on the problem. The method cited was the first to be developed. It thus ranks as a classic biochemical procedure which helped to open up a whole new and important field of research.

"The method measured the then three known estrogen metabolites in human urine. It was developed in the Clinical Endocrinology Research Unit, MRC, University of Edinburgh, with the expert assistance of H.A.F. Blair. The work qualified me for a PhD degree and was supervised by G.F. Marrian. I was also involved in the development of methods for measuring the other important female hormones and establishing their patterns of production.[1]

"The first step in developing the estrogen method involved a thorough study of the highly specific but notoriously unstable Kober color reaction. Four interdependent variables were identified and optimization of these for the two stages of the reaction and for each of the three estrogens provided an exceptionally stable system which became the accepted method for estrogen measurement for many years. Later, Ittrich[2] introduced a solvent extraction step which, with fluorimetry, increased the sensitivity and specificity 10,000-fold.

"The new color method was applied to the development of optimum extraction and purification procedures from urine. A novel phase change procedure was included involving methylation of the estrogens. The methyl ethers were ideally suited to alumina chromatography by which they were separated from one another and further purified. Success was due to meticulous optimization of every step; the elimination of unnecessary manipulations; the recent availability of ground-glass joints, clean solvents, and a modern spectrophotometer; and Marrian's support. In collaboration with others, the method was further validated against bioassay, isotope methods, and gas-liquid chromatography. Workers involved included Bauld, Bulbrook, Greenwood, Diczfalusy, Gallagher, Fishman, Preedy, and Kellie. The method with modifications was widely applied,[3] and it was the only one which was clinically viable until the development of radioimmunoassays for plasma estradiol in the 1970s.[4] For myself, I started as an organic chemist in New Zealand, became a hospital biochemist, and then a reproductive endocrinologist. I am now a professor of obstetrics and gynecology, without ever having delivered a baby. The paper cited is my most important and the measurement of estrogens in body fluids is now widely used in still increasing numbers in the study and treatment of human infertility."

---

1. **Brown J B, Klopper A & Loraine J A.** The urinary excretion of oestrogens, pregnanediol and gonadotrophins during the menstrual cycle. *J. Endocrinology* **17**:401-10, 1958.
2. **Ittrich G.** Eine neue Methode zur chemischen Bestimmung der oestrogenen Hormone im Harn. *Hoppe-Seylers Z. Physiol. Chem.* **312**:1-14, 1958.
3. **Brown J B & Beischer N A.** Current status of estrogen assay in gynecology and obstetrics. *Obstet. Gynecol. Survey* **27**:205-35, 1972.
4. **Abraham G E, Odell W D, Swerdloff R S & Hopper K.** Simultaneous radioimmunoassay of plasma FSH, LH, progesterone, 17-hydroxyprogesterone, and estradiol-17β during the menstrual cycle. *J. Clin. Endocrinol. Metab.* **34**:312-18, 1972.

# This Week's Citation Classic

CC/NUMBER 3
JANUARY 19, 1981

Murphy B E P. Some studies of the protein-binding of steroids and their application to the routine micro and ultramicro measurement of various steroids in body fluids by competitive protein-binding radioassay.
J. Clin. Endocrinol. Metab. 27:973-90, 1967. [Dept. Invest. Med., McGill Univ., and Clin. Invest. Unit, Queen Mary Vet. Hosp., Montreal, Canada]

This article extended and refined the methodology employed in our original competitive protein-binding assay for plasma corticoids,[1] increasing the sensitivity 100-fold. It showed how the basic method could be applied to the measurement of a number of different steroids in plasma, urine, and cerebrospinal fluid. [The SCI® indicates that this paper has been cited over 1,690 times since 1967.]

Beverley E. Pearson Murphy
Montreal General Hospital
Montreal H3G 1A4
Canada

October 17, 1980

"This paper was an outgrowth of my PhD thesis.[2] The original assay developed accidentally during the final year of my postgraduate training in internal medicine (July 1961-June 1962) as a research fellow under Chauncey Pattee. Since my predecessor, Richard Gillies, had found low 17-hydroxycorticoid levels in a few patients with advanced cirrhosis of the liver and had suggested these might be due to low levels of transcortin, I decided to embark on a study of protein-binding in plasma of cirrhotic subjects. To gain some experience with the technique of dialysis, I repeated a study by W.R. Slaunwhite and A.A. Sandberg in 1959 in which they had looked at the fall in cortisol binding with increasing levels of cortisol in diluted plasma. Plotted as unbound cortisol vs. cortisol added, the results resembled a 'standard curve' for cortisol and our technical assistant William Engelberg and I considered the possibility that this relationship might indeed be used to measure cortisol. Fortunately our knowledge of steroid chemistry was too limited to be discouraging and with the help and blessing of Pattee we had an assay working within a few weeks. In June 1962, I presented my work — my first scientific presentation — to the Quebec Society of Clinical Chemists. Though rejected after four months by Endocrinology, the manuscript was accepted promptly by its sister journal. Impressed by the similarity of our assay to those employing antibodies (later 'radioimmunoassays') I also wrote a general paper suggesting that many proteins, even intracellular ones, could probably be used in this fashion.[3] We had already applied the technique to thyroxine.

"From 1963 to 1966 I looked at ways to streamline and extend our assays, and these efforts for steroids culminated in the paper cited. These included the use of various species of transcortin and the investigation of different adsorbents to separate bound and unbound steroid moieties — initially done by trying out every particulate reagent on the shelf from Lloyd's reagent to charcoal, measured out with a cocktail spoon. Sensitivity was also increased by substituting tritium for $^{14}C$ as a tracer. With the resulting techniques it was easy to develop methods for a number of different steroids and a large mass of data soon accumulated. Since it was all related I attempted to combine it into one long paper — which took months to review. The reviewers could not agree as to its acceptability and a third reviewer was involved. All the reviewers had many suggestions for revision and all in all it took well over a year to appear in print. This delay was a source of mild embarrassment since some colleagues to whom I had given the methods pre-publication published earlier.

"I suppose the fact that this paper describes a whole group of new methods accounts for its popularity. It intrigues me that, when I do see it quoted, the quoters rarely specify just which method they are referring to, so that one is left to guess. I suspect that it is quoted much more often than read."

1. Murphy B E P, Engelberg W & Pattee C J. A simple method for the determination of plasma corticoids. J. Clin. Endocrinol. Metab. 23:293-300, 1963.
2. Murphy B E P. Some aspects of the protein-binding of corticosteroids and thyroxine in human blood. Thesis, McGill University, 1964. 205 p.
3. ——————. The application of the property of protein-binding to the assay of minute quantities of hormones and other substances. Nature 201:679-82, 1964.

# This Week's Citation Classic

NUMBER 42
OCTOBER 15, 1979

Johansson E D B. Progesterone levels in peripheral plasma during the luteal phase of the normal human menstrual cycle measured by a rapid competitive protein binding technique. *Acta Endocrinol.* 61:592-606, 1969.
[Department of Obstetrics & Gynecology, University of Uppsala, Uppsala, Sweden]

A rapid method for the estimation of progesterone in plasma is described. After a simple extraction with petroleum ether, progesterone is quantified by competitive protein binding. The catch was that the batch of petroleum ether had to have certain properties to avoid crossreacting steroids. [The *SCI®* indicates that this paper has been cited over 265 times since 1961.]

Elof D. B. Johansson
Department of Obstetrics
& Gynecology
University Hospital
S-750 14 Uppsala 14, Sweden

August 3, 1979

"In the summer of 1967 Jimmy Neill, Ernst Knobil, J.K. Datta and myself had published a method for progesterone based on the pioneer work on competitive protein binding by Beverly Murphy.[1,2] I had the privilege to come into this work as a postdoctoral fellow at the department of physiology in Pittsburgh (Ernst Knobil).

"When I got back to the department of obstetrics and gynecology at the University of Uppsala, Sweden, I immediately started to assemble equipment and to train personnel to get the method started. I had lots of projects, and my chief, Carl Gemzell, gave me constant encouragement.

"The method contained a tedious thin layer chromatography step. This gave me the biggest problem and I often had to do the chromatography myself. On the morning of February 2nd I was very tired and very happy. My daughter had been born at 4 o'clock in the morning. I went down to the lab and started to work with the assay. I had always tried to get rid of the thin layer step. I had tried it in Pittsburgh with rhesus monkey plasma but it did not work. This morning I decided on another try to omit the thin layer step. I was simply too tired to do it. When the assay was done I regretted it. What a waste of time, I thought. The next morning when I calculated the results my heartbeat accelerated. The pool samples were almost identical to the results from the thin layer assays. Why? It turned out that the lot of petroleum ether that I used extracted progesterone well, but only minute amounts of cortisol and 17$\alpha$-hydroxyprogesterone, the main crossreacting steroids in the quantification step.

"I spent the following months trying to prove that progesterone in women could be measured with this rapid technique. The method worked very well in our hands. With the same manpower as before we could describe the pattern of progesterone during the menstrual cycle and normal pregnancy and start mode of action studies on hormonal contraceptives. Family planning was and is my main interest. That was why I started to do research.

"Nowadays the assay methods using competitive protein binding have been replaced by radioimmunoassays, which are even more rapid, sensitive, and specific.

"There are several reasons for the many citations of this paper. First of all it describes a new method. The method was so simple that it attracted the interest of people who usually at that time would not think of steroid measurements. The paper also presented daily plasma levels of progesterone from 20 ovulatory menstrual cycles. This was a quantity of steroid measurements never seen before. It attracted gynecologists, veterinarians, and physiologists who saw new possibilities for studies in their own field. The third explanation for its many citations is my own frequent citation of the paper."

1. Neill J D, Johansson E D B, Datta J K & Knobil E. Relationship between the plasma levels of luteinizing hormone and progesterone during the normal menstrual cycle. *J. Clin. Endocrinol. Metab.* 27:1167-73, 1967.
2. Murphy B E P. Some studies of the protein-binding of steroids and their application to the routine micro and ultramicro measurement of various steroids in body fluids by competitive protein-binding radioassay. *J. Clin. Endocrinol. Metab.* 27:973-90, 1967.

## This Week's Citation Classic

CC/NUMBER 48
NOVEMBER 30, 1981

Bandurski R S & Axelrod B. The chromatographic identification of some biologically important phosphate esters. *J. Biol. Chem.* 193:405-10, 1951.
[Kerckhoff Labs. Biol., Calif. Inst. Technol., Pasadena, and Enzyme Res. Div., Bur. Agr. and Indust. Chem., Agr. Res. Admin., US Dept. Agr., Albany, CA]

The objective of this research was to provide a means for separating and identifying phosphate esters involved in glycolysis in higher plants. At the time we didn't know whether plants had a glycolytic system or whether they did oxidative phosphorylation and made ATP. These ideas were only then being developed in animal and microbial systems. Our procedure was based on two-dimensional paper chromatography of plant extracts with successive development in an acidic and in a basic solvent. The solvents finally selected gave the best overall resolution of the intermediates involved in plant glycolysis. [The *SCI®* indicates that this paper has been cited over 475 times since 1961.]

Robert S. Bandurski
Department of Botany and
Plant Pathology
Michigan State University
East Lansing, MI 48824

October 8, 1981

"The frequent citation of this paper reflects not a classic quality in the paper but simply the needs of biochemistry for methods. When the work was done, paper chromatography was only then being developed as a sensitive and rapid means of detecting chemical compounds in complex biological mixtures. Although paper chromatography was already quite useful for separating sugars, amino acids, and organic acids, it could not readily be applied to phosphorylated compounds since the phosphorus so dominated the chemical properties of the compounds that they were never sufficiently separated. The solvents Axelrod and I described in the paper gave reasonably adequate resolution of the ten or so major phosphorylated compounds occurring in biological extracts.

"A further difficulty with chromatography of phosphorylated compounds on paper was that methods of visualizing the compounds caused almost total destruction of the paper. Here is where chance entered in. Axelrod and I had sprayed a paper chromatogram with the then used Hanes-Isherwood reagent. This reagent contained strong acids and had the paper been heated to hydrolyze phosphorus-containing compounds, the paper would have disintegrated. There was insufficient phosphorus for detection so we examined the sprayed sheet in a dark room under ultraviolet illumination. To our astonishment, a blue spot gradually became visible. Apparently, the ozone generated by the UV lamp, in the presence of acid and molybdate from the spray, easily degraded even resistant esters to inorganic phosphate which then reacted to form the blue colored phosphomolybdate product. Thus, accidental use of an ultraviolet lamp led to a way for degrading organophosphates to inorganic phosphorus.

"Now, 30 years later, we realize how primitive our methods were—but we did have the correct idea and that was to try to spread out all the compounds being made and then in a gross way to try to decide what major metabolic products were being made and what major perturbations the experimental treatments were causing.

"Also, the work illustrates the principle of serendipity. One starts out with a reasonable project and a humble and simple question and all else follows automatically. The methods are developed as one needs them and then from the methods come answer after answer, each one leading to a new level of understanding. Accident plays a role, of course, as it did in our case, but accident only provides the opportunity; it is the determination to answer a question that provides the progress.

"The reference cited illustrates the dwarfing complexity of the work done since our primitive publication."[1]

1. **Stahl E,** ed. *Thin-layer chromatography: a laboratory handbook.*
    New York: Springer-Verlag, 1977. 1041 p.

# This Week's Citation Classic

NUMBER 8
FEBRUARY 19, 1979

Dawson R M C. A hydrolytic procedure for the identification and estimation of individual phospholipids in biological samples.
*Biochem. J.* 75:45-53, 1960.

The author describes a simplified, single-assay method using paper chromatography after mild alkaline hydrolysis for the rapid quantitative determination of the individual phospholipids in a small sample of biological material. [The *SCI®* indicates that this paper has been cited 225 times since 1961.]

R.M.C. Dawson
Agricultural Research Council
Institute of Animal Physiology
Babraham, Cambridge, CB2 4AT
England

December 13, 1977

"It is surprising that this paper should have proved of such value since it reports modifications, albeit substantial, of a method for investigating phospholipid structure published in principle six years earlier.[1] At that time there existed no method of resolving the individual phospholipids present in a small sample of tissue. Although the solvent fractionation techniques evolved by the pioneering work of J. Folch had indicated the complex nature of the kephalins, this procedure could not be scaled down to yield a worthwhile separation when only a few milligrams of phospholipid phosphorus was available.

"We began to think of the possibility of isolating identifiable fragments of the phospholipids after they had been subjected to various degradative procedures. The final breakthrough came as a result of wasting time. In my laboratory in the Biochemistry Department at Oxford, I was, at that period, helped by an attractive female assistant who acted as a magnet for the other young scientists around. Consequently, at midmorning an extensive break was taken, with coffee drinking and the exchange of much gossip and scientific chit-chat. During one such session, we discussed the observations of the late Professor Baer and Dr. Kates showing that the deacylation of synthetic phosphatidylcholine by methanolic alkali was complete long before any liberation of choline. Would this work for the other deacylated phosphoglycerides and produce recognizable fragments? Experiments quickly showed this was so and that the deacylated parent structure was left essentially intact producing 'glycerylphosphoryl' derivatives, which could be adequately separated by paperchromatographic techniques.

"This principle proved extremely useful for accurately measuring the specific radioactivities of the diacylphosphoglycerides. Eventually, it was developed into a complete analytical technique, and extended so that those phospholipids not rendered water-soluble by the alkali treatment could be examined. This involved much trial and error research, although I believe the principle of using a simple, low molecular weight ester for rapid neutralization of the alkaline digests before examining the deacylated plasmalogens was novel. By far the most effective of these esters was ethylformate. How this came into our chemical stores is a mystery, but it is perhaps not without significance that ethylformate is used in the manufacture of artificial rum and that Christmas parties were an annual event in the laboratory.

"After five years it was decided to write the accumulated experience in collaboration with my colleagues Norma Hemington and James Davenport, who had played a substantial part in the development of the method. Subsequently, the technique has been modified by many workers, but it is gratifying to see many papers appearing even today in which unidentified phospholipids and glycolipids resolved as spots by T.L.C. have had their structures investigated using the same principle of successive selective degradations."

## REFERENCE

1. Dawson R M C. The measurement of $^{32}P$ labeling of individual kephalins and lecithin in a small sample of tissue. *Biochim. Biophys. Acta* 14:374-9, 1954.

| Number 1 | **Citation Classics** | January 2, 1978 |

Skipski V P, Peterson R F & Barclay M. Quantitative analysis of phospholipids by thin-layer chromatography. *Biochem. J.* 90:374-8, 1964.

The authors describe the conversion of their previously-developed procedure for qualitative separation of phospholipids by thin-layer chromatography to a quantitative analysis which permitted the determination of the main known phospholipids in animal tissues. [The SCI® indicates that this paper was cited 696 times in the period 1964-1976.]

Dr. Vladimir P. Skipski
Sloan-Kettering Institute for Cancer Research
Rye, New York 10580

February 28, 1977

"In the beginning of the sixties, we were faced with the problem of characterizing the phospholipid profiles in different human serum lipoprotein classes, as well as the phospholipid composition of animal cell surface membranes. The attempt to use silicic acid column chromatography along with paper chromatography for phospholipid determination had only moderate success at our hands. At that time, the first papers concerning the separation of phospholipids by thin-layer chromatography began to appear in different journals. Wagner et al.[1] described the separation of phospholipids by a one-dimensional thin-layer chromatographic procedure. However, this system did not give reliable separation of phosphatidylserine and phosphatidylinositol and some other non-nitrogen containing phospholipids. We tried to modify the procedure of Wagner et al. for the goal of separating all phospholipid classes. However, our attempt never became reality.

"In our first attempt to modify the Wagner et al. system, experiments revealed that the position of phosphatidylserine on chromatoplates prepared from silica gel G (containing calcium sulfate as a binder) depended not only upon the type of compound (phosphatidylserine) but also upon the amount. This 'load effect' was due to the presence of calcium sulfate and was eliminated empirically by using the basic silica gel G chromatoplates. However, this sysem still did not permit separation of even the most common phospholipids present in animal tissues. Therefore, as soon as silica gel without calcium sulfate appeared on the market, we worked out another system of thin-layer chromatography for separation of phospholipids using this type of adsorbent and converted it to a quantitative procedure. This one-dimensional chromatographic system permitted the separation of most common phospholipids present in mammalian tissues including phosphatidylserine and phosphatidylinositol.

The limitations of this procedure, some realized only retrospectively, are as follows:

1. Phosphatidylglycerol has a mobility similar to phosphatidylethanolamine.

2. With changes in both air humidity and batch number of silica gel, it was necessary to alter the proportion of ingredients in the developing solvent; e.g., if the phosphatidylethanolamine spot has an Rf value lower than 0.70, it is advisable to increase the content of water or acetic acid in the developing solvent and *vice versa*. The amount of methanol in the developing solvent apparently determines the distance between phosphatidylinositol and phosphatidylcholine.

3. Some non-nitrogen-containing phospholipids did not separate in the system described in the Abstract.

Problems 1 and 3 were solved by developing additional thin-layer chromatographic systems which permitted the separation of phosphatidic acid, phosphatidylglycerol, cardiolipin and phosphatidyl ethanolamine. All these systems are described in *The Methods in Enzymology*.[2]

The authors are very happy that the procedure worked out by them found a wide application in many laboratories.

## REFERENCES

1. Wagner H, Horhammer L & Wolff P. Dünnschichtchromatographie von phosphatiden und glykolipiden. (Thin-layer chromatography of phosphatides and glycolipids.) *Biochem. Z.* 334:175-84, 1961.
2. Lowenstein J M, ed. *The methods in enzymology.* Vol. 14. *Lipids.* New York: Academic Press, 1969. 771 pp.

## This Week's Citation Classic

CC/NUMBER 20
MAY 19, 1980

Marinetti G V. Chromatographic separation, identification, and analysis of phosphatides. *J. Lipid Res.* 3:1-20, 1962.
[Dept. Biochem., Univ. Rochester Sch. Medicine and Dentistry, Rochester, NY]

This review article deals with chromatographic methods for the qualitative analysis of intact phosphatides and their hydrolysis products. Chromatography of phosphatides was carried out on silicic acid-impregnated paper. Identification of the different phosphatides was carried out by partial chemical hydrolysis and by enzymatic cleavage. [The *SCI®* indicates that this paper has been cited over 665 times since 1962.]

G.V. Marinetti
Department of Biochemistry
School of Medicine and Dentistry
University of Rochester
Rochester, NY 14642

April 25, 1980

"When I began my PhD research in 1951, I became interested in the structure, metabolism, and function of phosphatides and glycolipids. The isolation from brain tissue of certain phosphatides, such as sphingomyelin, or of glycolipids, such as cerebrosides, in gram quantities for structural studies was relatively easy. However, when I became interested in metabolic studies on phosphatides, the awesome truth of the limited technology for the separation, identification, and analysis of small quantities of these lipids made it clear that these metabolic studies were not feasible. Therefore, I turned my efforts toward the development of paper chromatographic methods for the analysis of these lipids. This work was done over a period of several years in collaboration with George Rouser (now at the City of Hope Medical Center, Duarte, California) and James Berry (now at the University of Minnesota). We systematically tested various solvents of different polarity, using ordinary Whatman paper. Our first results were very discouraging since only smears or streaks were obtained. We tried hard to improve these systems but to no avail. It was apparent that some type of modification of the filter paper would be required to resolve this problem.

"Column fractionation of lipids was a major effort of the PhD thesis of Dorothy Rathmann of the biochemistry department of the University of Rochester.[1] She found that silicic acid gave a partial separation of some phosphatides by column fractionation. I thought that if it were possible to impregnate filter paper with silicic acid, I might achieve a better system for phosphatide separation. This was accomplished, and I will never forget the day when I developed the phosphatide chromatogram with Rhodamine 6G to detect the lipids. There before my eyes was a beautiful sight to behold—a clear separation of the various phosphatides as discrete spots. This technique opened the doors to the solution of numerous problems dealing with the analysis, biosynthesis, and enzyme degradation of phosphatides in microgram amounts. Indeed, this chromatographic breakthrough (which was developed independently by Lea, Rhodes, and Stoll in Cambridge, England[2]) led to the elucidation of how phosphatides are biosynthesized and degraded in cellular and sub-cellular systems. This led to a flurry of activity in the period 1960-1972. The time was right for this area of research to proceed with all due speed since phosphatide and glycolipid biochemistry had too long lagged behind protein, carbohydrate, and nucleic acid biochemistry.

"My review article apparently found wide appeal and served as a springboard for the next phase of technology, namely thin-layer chromatography utilizing silicic acid-coated glass plates or plastic sheets. The article was timely for the area of membrane biochemistry since, after the burst of research in the 1960s dealing with the elucidation of the manner in which phosphatides are biosynthesized, attention turned to the role of phosphatides in cell membranes. This may account for the article's frequent citation. I was fortunate to have written the review article at a propitious time. I hope it helped to promote research in these exciting frontier areas of biochemistry. I published a more recent review in this field in 1976."[3]

1. **Rathmann** D M. *Adsorption analysis of phospholipids*. Unpublished PhD thesis, University of Rochester, Rochester, NY, 1944. 156 p.
2. **Lea** C H, **Rhodes** D N & **Stoll** R D. Phospholipids. 3. On the chromatographic separation of glycerophospholipids. *Biochemical J.* **60**:353-63, 1955.
3. **Marinetti** G V, ed. *Lipid chromatographic analysis*. New York: Marcel Dekker, 1976. Vol. 2.

# This Week's Citation Classic

NUMBER 12
MARCH 19, 1979

Green D E & Fleischer S. The role of lipids in mitochondrial electron transfer and oxidative phosphorylation. *Biochim. Biophys. Acta.* 70:554-82, 1963.
[Institute for Enzyme Research, University of Wisconsin, Madison, WI]

The interactions of phospholipids with membrane complexes or membrane proteins are of two types—electrostatic and hydrophobic. The electron transfer function of lipid-depleted complexes can be reconstituted by adding back phospholipid (hydrophobic interaction). Cytochrome c forms complexes with phospholipid that are soluble in organic solvents (electrostatic interaction). Intrinsic membrane proteins such as structural protein combine hydrophobically with phospholipids to form stable complexes with unique properties. The structure and function of membranes are largely determined by lipid-protein interactions. (The *SCI*® indicates that this paper has been cited over 285 times since 1963.]

David E. Green
University of Wisconsin
Institute for Enzyme Research
Madison, WI 53706

December 22, 1977

"Historically, this article marked a turning point in the membrane field—traditionally the preserve of anatomists and electron microscopists. The emphasis exclusively on the lipid bilayer dominated thinking about the structure of membranes until the 1960s. The protein was relegated to a minor role, if any, in the determination of membrane structure. After a long period of exploration, the technology was finally developed in our laboratory for resolving the mitochondrial energy coupling system into its component elements—the electron transfer complexes (Y. Hatefi, D. Ziegler), the headpiece-stalk sector (T. Oda, P. Blair, H. Fernandez-Moran) and the membrane proteins (R. Criddle).

"At this stage we became aware that the function as well as the structure of membrane systems (complexes as well as proteins) depended in an absolute way on interactions between protein and phospholipid. S. Fleischer, G. Brierley, R. Lester, and F. Crane were the principal investigators in these pioneer studies. It then became necessary to specify the nature of lipid-protein interactions and to determine why these interactions were crucial for the exercise of membrane function.

"The *BBA* article was the first introduction to this realm of lipid-protein interactions—opening the door to the study of membrane structure via function. Strange to realize that in the 1960s ours was the only laboratory in the world to study both membrane structure and function at the same time and in the same place. The *BBA* article was widely read, as the *Science Citation Index*® indicates, but did it modify the prevailing thinking? In part, yes; in part, no. The concept of intrinsic membrane proteins, as well as the reality of lipid-protein interactions, was universally accepted. But there was no relaxation of the insistence on the paramount position of the lipid bilayer in the determination of membrane structure. Proteins became the raisins in the pudding instead of the icing on the cake. That was the extent of progress induced by this article.

"It is interesting that developments in the study of mitochondrial function have traditionally compelled new assessments in the membrane field. The recent evidence that the transducing function of the mitochondrial inner membrane is localized in a ribbon continuum, and that the protein domain of this membrane is a continuous structure, may provide the extra leverage that the original article lacked. It may finally set the proper course for the membrane field."[1]

REFERENCE

1. Haworth R A, Komai H, Green D E & Vail W J. The ribbon structure of the mitochondrial inner membrane. *J. Bioenerg. Biomembrane* 9: 151-70, 1977.

# This Week's Citation Classic
NUMBER 45
NOVEMBER 5, 1979

Vandenheuvel F A. Study of biological structure at the molecular level with stereomodel projections. I. The lipids in the myelin sheath of nerve.
*J. Amer. Oil Chem. Soc.* 40:455-71, 1963. [Animal Research Institute, Canada Department of Agriculture, Ottawa, Ontario, Canada]

Data on myelin obtained from analytical, X-ray, and other studies have been integrated by using exact molecular parameters and force calculations. A plausible tri-dimension arrangement of molecules in the lipid bilayer of myelin was obtained. [The *SCI*® indicates that this paper has been cited over 200 times since 1963.]

---

Frantz A. Vandenheuvel
287 Clemow
Ottawa, Ontario K1S 2B7
Canada

August 1, 1979

"Late in 1961, I came across an article proposing an arrangement of lipid chains in membranes. The figure describing this arrangement was undoubtedly the most detailed thus far published on the subject. However, molecular parameters and spacings were inaccurate to the point that the lipid density estimated from this drawing was 3 times that expected! The thought occurred to me that while a more realistic model could undoubtedly be produced, such a model would be more interesting, and probably more useful, if it were based on data for a specific membrane system, such as myelin. Thus the above paper, soon to be followed by a study of the whole myelin membrane,[1] came into being.

"This work met with immediate acceptance in part, I believe, because the DNA model of Watson and Crick had convinced many scientists that accurate molecular models are important research tools in biology where structure and function are indissolubly linked. Indeed, this paper showed how properties of myelin could be predicted from a realistic tri-dimensional model of this structure. As in the case of DNA, the basic data were of diverse disciplinary origin, and the methods used for their integration relied on exact molecular parameters and properties. No doubt the myelin paper gave a further boost, not only to the use of molecular models, but also to the multidisciplinary team approach to membrane research.

"The myelin paper displays several models of lipid molecules which later were used by others, either to represent lipid components of biological structures, in some cases other than membranes, or to back up various arguments and theories. Hence, these models have been reproduced in many papers, review articles, and books. Another feature of the myelin paper to which reference is often made is a detailed discussion of inter- and intramolecular forces as they specifically apply to membrane components. It is probable that the paper is still used as a reliable reference because none of several increasingly sophisticated X-ray investigations made over the last 15 years has seriously challenged its conclusions regarding the structure of myelin.

"It is evident that this paper involved considerable graphical work of a painstaking, very precise nature. Less obvious is the time and effort required by countless trial and error molecular force estimations made without the benefit of a computer. With the deadline imposed by the symposium at which the paper was to be delivered, the work had all the ingredients of a grueling task. Perhaps it was. All I can remember, however, is a feeling of elation, as, one by one, the pieces of the puzzle fell into place."

---

1. **Vandenheuvel F A.** Structural studies of biological membranes: the structure of myelin. *Ann. NY Acad. Sci.* 122:57-76, 1965.

# This Week's Citation Classic

CC/NUMBER 6
FEBRUARY 8, 1982

Huang C. Studies on phosphatidylcholine vesicles. Formation and physical characteristics. *Biochemistry* 8:344-51, 1969.
[Department of Biochemistry, University of Virginia, School of Medicine, Charlottesville, VA]

This paper describes a method to prepare spherical unilamellar phospholipid bilayer vesicles and the application of various physical techniques to characterize these vesicles as model bilayer membranes. [The *SCI®* indicates that this paper has been cited over 595 times since 1969.]

---

Ching-hsien Huang
Department of Biochemistry
University of Virginia
School of Medicine
Charlottesville, VA 22908

October 16, 1981

"After working on planar bilayer membranes at the Johns Hopkins Medical School with Thomas E. Thompson for three years, I joined Manfred Eigen's group in 1965 as a fresh postdoctoral fellow in Göttingen, Federal Republic of Germany. I planned to learn the chemical relaxation technique and to apply it to studying the dynamic properties of phospholipid bilayers. It became apparent very quickly that the chemical relaxation technique cannot practically be applied to the planar bilayer membrane, since any rapid perturbation such as temperature jump will break the planar bilayer at once. Consequently, I decided to prepare a new phospholipid model membrane system which could survive the rapid perturbation. As the only lipid biochemist at Göttingen then, and the lowest on the totem pole among all the eminent scientists there, I had trouble assembling the necessary chemicals and supplies to begin the long investigation. In fact, by the end of my two-year stay at Göttingen, I did not have enough data to write even a brief communication. Nevertheless, I knew that it was only a matter of time before I finished the work and I had a good, albeit slow, start. Looking back, it was quite a sharp contrast that scientists at Göttingen at the time were making many significant contributions to our understanding of allosteric behavior of multisubunit enzymes by chemical relaxation technique, while I was slowly plowing the furrow.

"I spent my last year as a research fellow of the Helen Hay Whitney Foundation in the biochemistry department, University of Virginia School of Medicine, Charlottesville, Virginia. I knew from my ultracentrifugation work at Göttingen that small unilamellar phospholipid vesicles can be generated by subjecting the phospholipids in aqueous solution to ultrasonic irradiation. I was not satisfied with the ultrasonic irradiation, however, because the vesicles prepared are extremely heterogeneous both in size and in slope. Sepharose-4B, a gel-filtration column material for separating large macromolecules, became available commercially in 1967-1968. Quickly, I used the column material and subjected the ultrasonic irradiated lipid dispersion to the Sepharose-4B column. It was quite a day at Charlottesville when the clear separation of large liposomes from the small vesicles was seen. The elution pattern of the gel-filtration step became the first figure used in the 1969 publication. I consider this gel-filtration step as the most crucial step for the preparation of homogeneous phospholipid vesicles.

"I believe that my first independent publication after graduate school became a *Citation Classic* due to the nature of the work. It describes a simple method for the preparation of a new model membrane system and, in addition, it contains a large number of physical characteristics of the system. Our current understanding of the lipid dynamics in bilayer membranes is, to some degree, promoted by the availability of various membrane model systems.[1] This publication can be credited for providing a simple model system. A more complete description of the small unilamellar vesicle was published much later in *Proceedings of the National Academy of Sciences of the USA*.[2] This paper was communicated by Eigen in October 1977, ten years after the original idea was conceived in his laboratory in Göttingen."

1. Thompson T E & Huang C. Dynamics of lipids in biomembranes. (Andreoli T E, Hoffman J F & Fanestil D D, eds.) *Physiology of membrane disorder*. New York: Plenum, 1978. p. 27-48.
2. Huang C & Mason J T. Geometric packing constraints in egg phosphatidylcholine vesicles. *Proc. Nat. Acad. Sci. US* 75:308-10, 1978.

# This Week's Citation Classic

Kannel W B, Castelli W P, Gordon T & McNamara P M. Serum cholesterol, lipoproteins, and the risk of coronary heart disease. The Framingham Study.
*Ann. Intern. Med.* 74:1-12, 1971.
[Heart Disease Epidemiology Study, Framingham, MA and Natl. Heart & Lung Inst., Natl. Insts. Health, Bethesda, MD]

The 14-year risk of coronary heart disease (CHD) is described according to cholesterol, Sf 0-20, and Sf 20-400 lipoproteins. Sf 20-400 was not an independent risk factor taking total cholesterol and other risk factors into account, except possibly in women over 50. Sf 0-20 lipoproteins showed a linear independent association with risk but added nothing to the estimate of risk achieved by the total cholesterol alone. [The *SCI®* indicates that this paper has been cited in over 645 publications since 1971.]

---

William B. Kannel
Heart Disease Epidemiology Study
National Heart Institute
National Institutes of Health
118 Lincoln Street
Framingham, MA 01701

May 11, 1983

"The 1971 report from Framingham was an attempt to bridge a growing gap in the rekindled interest in lipoproteins and our further experience with simple lipid measures such as the total cholesterol. To place this report in perspective one needs to recall that interest in the lipoproteins was stimulated in the late-1940s by John Gofman[1] and his associates at University of California, Berkeley. They correctly reasoned that knowledge of how the fats in our blood are transported might provide important insights into how blood lipids are related to cardiovascular disease. Out of their earlier work there came a series of lipoprotein determinations in the analytical ultracentrifuge which culminated in a postulated 'atherogenic index.' This index was a composite of Sf 0-20 beta (low density lipoproteins or LDLs) and Sf 20-400 prebeta (very low density lipoproteins or VLDLs), and emphasized the VLDLs which seemed to them to have a stronger association with coronary heart disease (CHD). The Gofman group's insistence that this was a better test than the serum cholesterol alone provoked a reaction from some of the cardiologists and epidemiologists interested in lipid atherogenesis who were skeptical as to how much better such an index was compared to the simple total cholesterol. They wanted to justify the added cost and difficulty of the ultracentrifuge lipoprotein analysis. This led to a multicenter trial in the mid-1950s to compare the efficiency of total serum cholesterol versus the atherogenic index in separating coronary cases from controls.[2] The report of this study was clouded by disagreement among the principal investigators who could not reach a consensus and the results were summarized in two versions, one written by each protagonist. The cholesterol proponents claimed a single cholesterol test did just as well; the lipoprotein advocates stuck to their original claim of their superiority.

"In actuality, the simple cholesterol test group won as interest in doing lipoproteins waned in the US until Fredrickson, Levy, and Lees[3] revived such interest in the mid-1960s using their lipoprotein typing system, a much simpler procedure. However, definitive 'typing' also requires preparatory ultracentrifuge analysis. Their lipid studies, correlated with careful family studies, revealed powerful genetic relationships which stimulated great interest. Controversy reemerged as the value of these more detailed lipoprotein analyses were questioned in relation to the simple cholesterol test. Our 1971 paper was an effort to examine some of the trade-offs. It was one of the largest bodies of data showing the impact of cholesterols and lipoproteins on risk using prospective data. The report is now out of date since the revival of interest in high density lipoproteins (HDLs) in the late-1970s. In our most recent paper, in *Circulation*,[4] we examine the new concepts and the evidence seems to clearly indicate that lipoprotein studies add greatly to our estimation of cardiovascular risk, particularly in people over 50, and that most of this added knowledge is contained in the LDL and HDL measures.

"Thus, Gofman's original contention has proved correct. Knowledge of the lipoprotein transport does enhance risk assessment since the serum total cholesterol reflects chiefly the atherogenic LDL-cholesterol component but fails to take into account the protective HDL-cholesterol fraction reflecting removal of cholesterol. His emphasis on VLDL, however, still awaits confirmation."

---

1. **Gofman J W, Lindgren F, Elliott H, Mantz W, Hewitt J, Strisower B & Herring V.** The role of lipids and lipoproteins in atherosclerosis. *Science* 111:166-71; 186, 1950.
2. **Committee on Lipoproteins and Atherosclerosis, National Advisory Heart Council.** Evaluation of serum lipoproteins and cholesterol measurements as predictors of clinical complications of atherosclerosis: report of a cooperative study of lipoproteins and atherosclerosis. *Circulation* 14:691-725, 1956.
3. **Fredrickson D S, Levy R I & Lees R S.** Fat transport in lipoproteins: an integrated approach to mechanisms and disorders. *N. Engl. J. Med.* 276:34-44; 94-103; 148-56; 215-25; 273-81, 1967.
   [Citation Classic. *Current Contents* (3):11, 16 January 1978.]
4. **Castelli W P, Abbott R D & McNamara P M.** Summary estimates of cholesterol used to predict coronary heart disease. *Circulation* 67:730-4, 1983.

# This Week's Citation Classic

CC/NUMBER 12
MARCH 21, 1983

Carlson L A & Böttiger L E. Ischaemic heart-disease in relation to fasting values of plasma triglycerides and cholesterol: Stockholm Prospective Study.
*Lancet* 1:865-8, 1972.
[King Gustaf V Res. Inst., Karolinska Hosp., Stockholm, and Dept. Geriatrics, Univ. Uppsala, Sweden]

This is a report about a prospective study of risk factors for coronary heart disease (CHD), the first to include measurements of fasting concentrations of plasma triglycerides. The paper describes the first, nine-year follow-up of 3,168 men of whom 71 had developed CHD. The rate of CHD increased linearly with increasing concentrations of both plasma cholesterol and plasma triglycerides. Plasma cholesterol and triglycerides were risk factors for CHD independent of each other. [The *SCI*® indicates that this paper has been cited in over 605 publications since 1972.]

---

Lars A. Carlson
King Gustaf V Research Institute
Karolinska Institute
S-104 01 Stockholm
Sweden

December 23, 1982

"In my previous *Citation Classic*,[1] I described the evolution of my method for determining the concentration of plasma triglycerides. Armed with that method, I set out to analyze a hitherto unstudied plasma lipid component. It was indeed a virgin field from which to harvest. Being a clinician it was obvious that I should study the normal as well as various diseased states.

"My first and still most striking observation was the presence of a very high frequency of hypertriglyceridaemia in survivors of myocardial infarction.[2] Until that time, one had only discussed plasma cholesterol in that context. The question that immediately arose was whether the hypertriglyceridaemia preceded the infarct, i.e., according to present terminology was a risk factor, or was caused by it. The answer could only be obtained by a prospective study. Sven Lindstedt, now professor of clinical chemistry in Gothenburg, joined me and we performed the initial survey, the Stockholm Prospective Study (SPS), on 6,464 healthy men and women during 1961-1962, including for the first time in a prospective study fasting values for serum triglycerides. The study could never have been carried through if we had not had the enthusiastic support of Sune Bergström, Karolinska Institute.

"It was then necessary to wait for time to pass. However, after nine years, I was too impatient to wait any longer. Fortunately, my scientific playmate from the 1950s, Lars Erik Böttiger, now professor of medicine at Karolinska, had for many years dealt with the use of computers in medicine and he then joined me for this first—as well as for all ensuing—follow-ups of the SPS. Altogether, 71 men had developed an infarction during the nine years which had elapsed since the initial examination. When the results had been processed and the first figures had been drawn, M.F. Oliver, now professor of cardiology in Edinburgh, visited me. He looked at the figures and said: 'I think this ought to go to *Lancet*,' and so it did. In my previous *Citation Classic*, I described how Oliver, remarkably enough, also influenced its publication.[1] He is welcome a third time!

"The main message in the *Lancet* paper was that serum triglycerides indeed are risk factors for myocardial infarction. This has been confirmed in the 15-year follow-up[3] and also in the 20-year follow-up.[4]

"I believe the paper has been cited frequently because this is the first prospective study in which, from the lipid point of view, not only plasma cholesterol but also fasting plasma triglycerides were measured.

"No, we did not measure HDL cholesterol! No, we did not measure apolipoprotein AI! The fact is that the fuse to the deep freeze where we had all 6,464 plasma samples broke after two years and we did not find that out until 1966. What a mess!"

---

1. Carlson L A. Citation Classic. Commentary on *J. Atheroscler. Res.* 3:334-6, 1963.
   *Current Contents/Life Sciences* 22(2):10, 8 January 1979.
2. ─────. Serum lipids in men with myocardial infarction. *Acta Med. Scand.* 167:399-413, 1960.
3. Carlson L A, Böttiger L E & Åhfeldt P E. Risk factors for myocardial infarction in the Stockholm Prospective Study.
   *Acta Med. Scand.* 206:351-60, 1979.
4. Böttiger L E & Carlson L A. Unpublished results.

## This Week's Citation Classic

CC/NUMBER 15
APRIL 13, 1981

Miller G J & Miller N E. Plasma-high-density-lipoprotein concentration and development of ischaemic heart-disease. *Lancet* 1:16-19, 1975.
[MRC Pneumoconiosis Unit, Llandough Hosp., Penarth, South Wales, and Dept. Cardiology and Lipid Res. Lab., Royal Infirmary, Edinburgh, Scotland]

A low plasma high-density lipoprotein (HDL) cholesterol concentration in subjects at increased risk for coronary-heart disease, and the discovery of an inverse relation between HDL concentration and body cholesterol-pool size, suggested that this lipoprotein is important for the removal of cholesterol from tissues and the retardation of atherosclerosis. [The *SCI®* indicates that this paper has been cited over 510 times since 1975.]

---

George J. Miller
UK Medical Research Council
External Scientific Staff
Caribbean Epidemiology Centre (PAHO)
P.O. Box 164
Port-of-Spain, Trinidad
West Indies

March 27, 1981

"This paper has its origins among the hill-farmers of rural Jamaica. Hypertension, diabetes mellitus, and high cholesterol concentrations are not uncommon in these people, yet few develop coronary-heart disease (CHD). This puzzle eventually prompted me to review the literature in search for ideas.

"The vast bibliography on the subject was highly repetitive and seemingly obsessed with the belief that hyperlipidaemia was responsible for coronary atherosclerosis. The relative tolerance of a high cholesterol concentration among men in rural Jamaica, and among women in general, suggested that the problem was not that simple.

"One observation, confirmed many times, appeared not to have aroused interest. Subjects at relatively high risk for CHD have a low plasma high-density lipoprotein (HDL) concentration, no matter how risk is identified (e.g., obesity, hypertriglyceridaemia, masculinity). Curious, I searched for more information about HDL, and John Glomset's[1] studies on lecithin-cholesterol acyltransferase appeared highly relevant. Suddenly the penny dropped! Perhaps HDL is required for the transport of cholesterol from the arterial wall to the liver for catabolism. If so, then a low HDL concentration might indicate inadequate clearance of cholesterol and potentiation of the atherosclerotic process. Was hypercholesterolaemia in rural Jamaica benign because of an association with a high HDL concentration?

"The hypothesis predicted an inverse association between the amount of cholesterol in the body and plasma HDL concentration, but I could find no reference to this topic. I sent my ideas to my brother Norman, then involved in lipid metabolism, and by a remarkable coincidence he possessed data with which to explore this possibility. These formed baseline observations for published studies of drug effects undertaken with Paul Nestel and colleagues.[2] Norman's discovery of the relation was exciting, and I hurried to join him in Edinburgh. He strengthened the argument considerably and we completed the manuscript over a weekend.

"Several weeks later I sat at my desk facing the rejected manuscript—returned with apologies. Somewhat bewildered, I was urged by colleagues to seek the editor's advice. To my delight, he recalled the paper, re-considered his decision, and accepted in little more than 24 hours!

"The article owes its success partly to the promise of new pastures for CHD research, and partly to the support quickly forthcoming from epidemiologists, notably in Framingham and Tromso. A recent review is available,[3] and the Jamaican hill-farmer does indeed have a high HDL cholesterol concentration!"[4]

1. Glomset J A. Physiological role of lecithin-cholesterol acyltransferase. *Amer. J. Clin. Nutr.* 23:1129-36, 1970.
2. Miller N E, Clifton-Bligh P & Nestel P J. Effects of colestipol, a new bile-acid-sequestering resin, on cholesterol metabolism in man. *J. Lab. Clin. Med.* 82:876-90, 1973.
3. Miller G J. High density lipoproteins and atherosclerosis. *Annu. Rev. Med.* 31:97-108, 1980.
4. Miller G J, Miller N E & Ashcroft M T. Inverse relationship in Jamaica between plasma high-density lipoprotein cholesterol concentration and coronary-disease risk as predicted by multiple risk-factor status. *Clin. Sci. Mol. Med.* 51:475-82, 1976.

# This Week's Citation Classic

CC/NUMBER 46
NOVEMBER 17, 1980

Lipmann F & Tuttle L C. A specific micromethod for the determination of acyl phosphates. *J. Biol. Chem.* 159:21-8, 1945.
[Biochemical Research Laboratory, Massachusetts General Hospital, Boston, MA]

The discovery of acetyl phosphate as a metabolic intermediary made a simple, sensitive method for its determination desirable. Acetyl phosphate is an acid anhydride, and a described qualitative method for determination of acetyl chloride with hydroxylamine converts it to the hydroxamic acid which gives a bright purple color on addition of ferric chloride. There was no difficulty converting this into a quantitative method applicable also to the determination of other reactive acyl derivatives under a variety of conditions. [The *SCI*® indicates that this paper has been cited over 850 times since 1961.]

---

Fritz Lipmann
Rockefeller University
New York, NY 10021

October 9, 1980

"I was somewhat surprised to learn that during the period 1961-1975 this colorimetric method still seems to have been extensively used, because radioactive compounds came into use during those years. Now that we are able to assay very small amounts so easily by the use of radioactivity, this often seems preferable to colorimetry. However, we subsequently found this method very useful for detecting activated carboxylic acid derivatives of various kinds, and it has become valuable for detecting them as metabolic intermediaries.

"To give some examples, we found that the determination of acyl phosphates required a concentration of hydroxylamine of only 0.02 M. However, in pork liver extracts, it appeared that due to the presence of an esterase, the salts of fatty acids in general could be converted to hydroxamate, but concentrations of 2-3 M hydroxylamine were needed. This indicated that by reacting with the esterase, an activated acyl intermediary formed which would react with the high concentration of hydroxylamine to form hydroxamate. Furthermore, it was known that at strongly alkaline reaction, any ester would give hydroxamates. However, unusually reactive esters yielded hydroxamates at the much lower pH of 5.5.

"An hydroxamate formation at low pH became of considerable interest in the case of amino acid activation. This was found to be biphasic: the primary reaction with ATP leads to an aminoacyl adenylate, followed by aminoacyl transfer on the same enzyme to the 3'-terminal of a low molecular ribonucleic acid called the transfer RNA (tRNA). The aminoacyl adenylate and the 3'-(or 2'-) amino acid ester with tRNA are in reversible equilibrium. Therefore, the ester link between the carboxyl group of the amino acid and the tRNA had to be of the energy-rich type, comparable to the phosphoanhydride in aminoacyl adenylate. Accordingly, we found this aminoacyl-tRNA reactive with hydroxylamine at a pH of 5.5 where ordinary aminoacyl esters are completely nonreactive.

"It is my impression that the popularity of the hydroxamate method is due to the versatility of the hydroxamate reactivity for various acyl derivatives, using different pH or different concentrations of hydroxylamine as indicators of the type of reaction product being studied."

# This Week's Citation Classic

CC/NUMBER 19
MAY 9, 1983

Inman W H W, Vessey M P, Westerholm B & Engelund A. Thromboembolic disease and the steroidal content of oral contraceptives: a report to the Committee on Safety of Drugs. *Brit. Med. J.* 2:203-9, 1970. [Comm. on Safety of Drugs, Queen Anne's Gate, London; Dept. Med., Radcliffe Infirmary, Oxford, England; Swedish Adverse Drug Reactions Comm., Stockholm, Sweden; and Adverse Reactions Board, Danish Natl. Health Serv., Brønshøj, Denmark]

A study of reports of thromboembolic disease in women using oral contraceptives revealed that the greatest risk was in those using high doses of oestrogen. The demonstration of a dose-response effect confirmed that the Pill probably was a cause of thrombosis and led to the safer 'Mini Pill.' [The *SCI*® indicates that this paper has been cited in over 360 publications since 1970.]

---

W.H.W. Inman
Drug Surveillance Research Unit
University of Southampton
Botley, Hampshire SO3 2BX
England

March 1, 1983

"This study provided the first proof of a causal relationship between oral contraceptives and thrombosis. In 1964, I was invited by the late Sir Derrick Dunlop, the founder and chairman of the Committee on Safety of Drugs, to become its first Medical Assessor of Adverse Reactions. I was responsible to the Subcommittee on Adverse Reactions until 1980, when I became director of an independent Drug Surveillance Research Unit at the University of Southampton.

"From the very early days, reports of thromboembolism in young women suggested a possible association with oral contraceptives, and in 1965, I designed what was to become the first epidemiologic study to provide statistical proof of such an association.[1] This study involved all deaths of women between the ages of 15 and 44 who died from pulmonary embolism, or coronary or cerebral thrombosis, in England and Wales during 1966. By August of that year, sufficient material had been accumulated to demonstrate a positive association with oral contraceptives. We persuaded the Medical Research Council (MRC) and the College of General Practitioners to start additional studies late in 1966, and when these had demonstrated a similar association, we published a joint report to the MRC in April 1967.[2]

"Epidemiological studies prove associations but do not prove causation. Early in 1966, I had noticed among the spontaneous reports received on 'yellow cards' an unusual distribution of reports in relation to the oestrogen that had actually been taken by the patient. At that time, the market for oral contraceptives containing mestranol, or ethinyloestradiol, was divided in the proportion of 52 percent to 48 percent respectively; so were the reports of nearly all types of suspected adverse reactions, with the single exception of reports of venous thromboembolism. Here the distribution was in the ratio of 72 percent to 28 percent, a significant difference which I attributed to a possible thrombogenic effect of mestranol, since there was no other reason why thrombosis should have been reported selectively in women using this particular oestrogen.

"Somewhat distracted by doubts whether this was an intrinsic effect of the chemical or a dose-response effect, I withheld publication until many more reports had accumulated. By late-1969, it was clear that the dose rather than the nature of the oestrogen was the important factor. My results were reported to the Department of Health with the reservation that, at that time, I had not yet looked closely at the progestogen component. Unfortunately, but perhaps predictably, the material leaked to the press before any communication could be prepared informing prescribers of the added risk of high-dose oral contraceptives.

"Early in 1970, I flew to Denmark and Sweden, returning home with data on their reports and sales figures which showed exactly the same relationship.

"Although I was also responsible for the first epidemiological study to show a significant association between the Pill and venous thrombosis,[1] it was my work on the oestrogen dose effect, which clinched causality, that earned me the title 'Father of the Mini Pill'! Interest in the latter probably accounts for the frequent reference to this work. For some time, in the dark winter of 1970, I wondered whether the low-dose pills which were rapidly introduced would prevent conception or whether I might, inadvertently, have fathered rather more than I had bargained for!"

---

1. **Inman W H W & Vessey M P.** Investigation of deaths from pulmonary, coronary, and cerebral thrombosis and embolism in women of child-bearing age. *Brit. Med. J.* 2:193-9, 1968.
   [Citation Classic. *Current Contents/Clinical Practice* 8(44):12, 3 November 1980.]
2. **Medical Research Council.** Risk of thromboembolic disease in women taking oral contraceptives. A preliminary communication to the Medical Research Council by a subcommittee. *Brit. Med. J.* 2:355-9, 1967.

## This Week's Citation Classic

CC/NUMBER 46
NOVEMBER 14, 1983

Barker S B & Summerson W H. The colorimetric determination of lactic acid in biological material. *J. Biol. Chem.* **138**:535-54, 1941.
[New York Hospital, Depts. Medicine and Biochemistry, Cornell Univ. Medical College, New York, NY]

This procedure achieved the sensitive and specific quantitation of lactate in biological fluids, eliminating the previous need for oxidation, distillation, and titration. Heating in concentrated sulfuric acid produced acetaldehyde which was directly determined by the purple color formed with p-hydroxydiphenyl in the presence of cupric copper. [The *SCI*® indicates that this paper has been explicitly cited in over 2,500 publications since 1961. Of these, 70 were in 1980, 70 in 1981, and 55 in 1982.]

---

S.B. Barker
University of Alabama in Birmingham
University Station
Birmingham, AL 35294

August 23, 1983

"At the present time, it is difficult to appreciate the importance attached to lactic acid in carbohydrate metabolism during the 1930s and 1940s. Measurement of this stabilized form of pyruvate was an essential aspect of many studies, at both *in vivo* and *in vitro* levels—exercise and Warburg tissue slice experiments being good examples of each. Determination of changes in glycogen, glucose, and lactate levels was essential in those days. Of the three, lactate was by far the most laborious, since the Friedemann, Cotonio, and Shaffer[1] (F-C-S) procedure was the standard and involved an incredibly tricky distillation apparatus in which the acetaldehyde produced by permanganate oxidation of lactate was received in an excess of bisulfite. The final eye-straining titration required addition of 0.002N iodine solution to the faintest possible blue-grey endpoint with starch indicator.

"Many laboratory groups which viewed the complicated array of glassware with mixed devotion and loathing tried one or another of the colorimetric procedures proposed for lactate, but found them unreliable. Summerson's primary interest was the development of colorimetric methods for his photoelectric instrument and I was anxious to have a less temperamental procedure than the F-C-S. The effectiveness of the collaboration was probably enhanced by the fact that we worked in separate laboratories. As individually independent investigators, we routinely subjected each other's results to careful scrutiny and reconciled any discrepancies. The relatively long-lived success of the published Barker-Summerson procedure can undoubtedly be explained by the combination of our rigorous control of each step plus the elimination of a complicated distillation and painstaking titration. Simplicity was achieved by carrying out both stoichiometric acetaldehyde production and its reaction with p-hydroxydiphenyl in the same solution contained in a single appropriate test tube.

"Although this was a technologically primitive era, as viewed nowadays, nonetheless several advances were crucial to the success of the method. Quantitative dispensing of highly purified $H_2SO_4$ required the use of an all-glass system, including a grease-free, accurately ground standard taper stopcock. Thorough but contamination-free mixing of copper-calcium hydroxide reagents with biological solutions being analyzed was greatly facilitated by the opportune invention of 'Parafilm,' sometimes mistranslated as paraffin. Above all, making a long series of color readings in very concentrated sulfuric acid was practical only by the advent of photoelectric colorimeters which accepted closely standardized test tubes. Early variable results, as well as one laboratory's claim that lead was necessary, caused us to test various metallic ions, with the important discovery of a considerable enhancement of color development by added copper or by a mixture of ferrous and ferric iron.

"This procedure, for its day, was remarkably sensitive, one of the first to be accurate in the microgram range. In fact, its very sensitivity was the source of most of the contamination problems as other laboratories adopted it. Lactate is present in sweat and saliva at easily detectable levels. Rigorous manipulative standards had to be adopted, e.g., avoiding handling of pipette tips or sneezing over open test tubes. After some two decades of wide acceptance, it was eventually superseded by spectrophotometric analysis utilizing lactic dehydrogenase.

"For a report of recent work in the field, see reference 2."

---

1. **Friedemann T E, Cotonio M & Shaffer P A.** The determination of lactic acid. *J. Biol. Chem.* **73**:335-58, 1927.
2. **Byers F M.** Automated enzymic determination of L(+)- and D(-)-lactic acid. *Meth. Enzymology* **89**:29-34, 1982.

# This Week's Citation Classic

CC/NUMBER 33
AUGUST 13, 1979

James A T & Martin A J P. Gas-liquid partition chromatography: the separation and microestimation of volatile fatty acids from formic acid to dodecanoic acid.
*Biochem. J.* 50:679-90, 1952.
[National Institute for Medical Research, Mill Hill, London, England]

Gas-liquid chromatography was suggested by Martin and Synge in their classical paper on the liquid-liquid chromatogram in 1941. Our development of the new chromatogram in 1951[1] gave birth to a new, very powerful analytical and purification technique that is now found in almost every laboratory in the world. The first paper outlined the general principles, a specific application, and the general theory and produced an explosive development of a new subject. [The *SCI*® indicates that this paper has been cited over 525 times since 1961.]

A.T. James
Division of Biosciences
Unilever Research Laboratory
Colworth House
Sharnbrook, Bedfordshire, England

June 26, 1978

"The development of the technique was due to a series of fortunate accidents. In 1950, I was working at the Lister Institute of Preventive Medicine in London on the structure of Gramicidin S, initially with R.L.M. Synge. I was therefore acquainted with the new liquid chromatographic method of amino acid analysis developed by Martin and Synge.[2] W.T.J. Morgan was having problems with the paper chromatographic analysis of amino sugars and asked for my help. As a sideline, I therefore developed a liquid-liquid column chromatographic system for the separation of the 2:4 dinitrophenyl derivatives of the amino sugars. For this I needed an automatic fraction collector but none was available. Fortunately, A.J.P. Martin had just temporarily arrived at Lister before moving on to permanent facilities at the National Institute for Medical Research. Already interested in such devices, he was delighted to collaborate in the design of automatic mechanical fraction collectors.

"As we got on so well together, he invited me to join him as co-worker at the National Institute in order to attempt to develop a technique of continuous counter current crystallation. I spent some months at this, but we got very poor results. Seeing my dejection, Martin suggested that we attempt to turn the suggestion in the original paper on the liquid-liquid partition chromatogram[2] (for which they received the Nobel Prize) into reality. This we very quickly did after an initial setback and opened up this new field. The potential was obviously enormous and we spent a couple of years in developing a new general purpose detector—the gas-density balance[3]—separating a wide range of volatile compounds, and studying the relationship between solution effects and relative retention volumes.

"Our results with paraffin hydrocarbons quickly got the attention of workers in the petroleum industry and the subject took off into its present ramifications. Other British and also American workers took it up and most of the potential power of the technology was soon outlined.

"It was very pleasant for Martin and me for a few years to know more about a major development than anyone else in the world, but it did not last long. Our later development of the technique for the separation, identification, and quantitative analysis of long chain fatty acids, led me directly into fatty acid biochemistry, where I have remained ever since, except for an increasing involvement in scientific bureaucracy. However, in both spheres I have attempted to apply the fundamental precepts I learned from A.J.P. Martin: (1) Nothing is too much trouble provided someone else does it; (2) Never answer the first letter; if it's important they'll write again; and (3) If there are twelve ways of tackling a problem, they're all wrong."

1. **James A T, Martin A J P & Randall S S.** Automatic fraction collectors and a conductivity recorder. *Biochem. J.* 49:293-9, 1951.
2. **Martin A J P & Synge R L M.** A new form of chromatogram involving two liquid phases. *Biochem. J.* 35:1358-68, 1941.
3. **Martin A J P & James A T.** Gas-liquid chromatography: the gas-density meter, a new apparatus for the detection of vapours in flowing gas streams. *Biochem. J.* 63:138-43, 1956.

# This Week's Citation Classic

CC/NUMBER 2
JANUARY 12, 1981

Metcalfe L D & Schmitz A A. The rapid preparation of fatty acid esters for gas chromatographic analysis. *Anal. Chem.* 33:363-4, 1961.
[Armour Industrial Chemical Company, McCook, IL]

The routine application of gas chromatography to the determination of the composition of fatty acid samples made it essential to prepare methyl esters rapidly and simply. Boron trifluoride-methanol converts fatty acids to their methyl esters in about two minutes. The esters are comparable to those obtained by other procedures. [The *SCI®* indicates that this paper has been cited over 825 times since 1961.]

Lincoln D. Metcalfe
Research Laboratories
Armak Company
McCook, IL 60525

October 21, 1980

"We were pleased to learn that this publication has become a *Citation Classic*. My observations have led me to the conclusion that if you want to write a highly cited paper, it should be one concerned with methodology in an area of high general interest. Like most authors, I set out to solve a problem in my particular field of interest. The fact that our solution was of great interest to many people was a happy coincidence.

"The story of the paper begins with H. J. Harwood, the research director of Armour and Company, asking me to become involved with the National Heart Institute's Lipid Analysis Committee (LAC) in 1957. This committee was organized by Evan Horning, who was with the National Heart Institute at that time. Because of the suspected role that fat played in heart disease, rapid methods for the analysis of fats and fatty acids were urgently needed. The LAC was set up to solve the formidable analytical problems using gas chromatography. Scientists from various universities, government laboratories, and industry made up this rather loose organization.

"In meeting with the LAC, it quickly became apparent that using the primitive gas chromatograph available to me at that time, our chromatography contribution would be very limited. I decided that the chemical area would be the one where we could make a useful contribution. We did make such a contribution by supplying the Committee with various fats, fatty acids, and methyl esters which were readily available to us at Armour.

"One chemical problem that kept coming up at the LAC meetings in 1958 was the need for a rapid and convenient method for preparing methyl esters of fatty acids. The fatty acids are usually gas chromatographed as their methyl esters. Most of the proposed esterification techniques were cumbersome, slow, or used reagents that were considered hazardous. In any event, the methods were not readily applicable for use with large numbers of routine samples.

"After one of the committee meetings, it occurred to me that an esterification technique using a relatively large amount of a powerful Lewis acid catalyst in methanol might be feasible. As an analytical chemist, I was familiar with an earlier use of boron trifluoride catalyst in a number of analytical techniques involving Karl Fischer reagent.[1] I decided that I would investigate $BF_3$ as an esterification catalyst for fatty acid methyl ester preparation.

"On recently examining my research notebook of that period, I found my first entry on the use of $BF_3$-methanol reagent was on January 11, 1959. I was surprised to find that the procedure I used in this early experiment was essentially the same as the one that appeared in the final publication. I never did get to present the new esterification procedure before the LAC.

"Late in 1959, gas chromatography was introduced at Armour Industrial Chemical Company (now Armak Company) as a quality control tool. Preparation of methyl esters for this plant gas chromatography control work involved a two-hour reflux period. We quickly introduced the two minute $BF_3$-methanol esterification procedure into the plant control laboratory when we became aware of their difficulties.

"Shortly afterward, we submitted a paper for publication in *Analytical Chemistry*. We later published an expanded version that could be applied to most lipids."[2]

1. Mitchell J, Jr. & Smith D M. *Chemical analysis*. New York: Interscience, 1948. Vol. 5.
2. Metcalfe L D, Schmitz A A & Pelka J R. Rapid preparation of fatty acid esters from lipids for gas chromatographic analysis. *Anal. Chem.* 38:514-15, 1966.

# This Week's Citation Classic

CC/NUMBER 22
MAY 30, 1983

Williamson D H, Mellanby J & Krebs H A. Enzymic determination of D(—)-β-hydroxybutyric acid and acetoacetic acid in blood. *Biochemical J.* 82:90-6, 1962.
[Medical Research Council Unit for Research in Cell Metabolism, Dept. Biochemistry, Univ. Oxford, England]

A rapid and specific enzymic method for measuring ketone bodies (acetoacetate and 3-hydroxybutyrate) is described using a purified D-3-hydroxybutyrate dehydrogenase from *Rhodopseudomonas spheroides*. [The *SCI®* indicates that this paper has been cited in over 755 publications since 1962.]

---

Dermot H. Williamson
Metabolic Research Laboratory
Nuffield Department
of Clinical Medicine
Radcliffe Infirmary
Oxford OX2 6HE
England

March 22, 1983

"In 1960, I was invited by Sir Hans Krebs to return to the MRC Unit for Research in Cell Metabolism in Oxford, a group I had left some six years previously when it was still located in Sheffield. On discovering that in the interval I had acquired some expertise in the use of enzymes as analytical tools, it was suggested that I should develop an enzymic method for the determination of ketone bodies (acetoacetate and D-3-hydroxybutyrate) using the D-3-hydroxybutyrate dehydrogenase known to be active in rat liver mitochondria. The existing chemical methods for the determination of ketone bodies were time-consuming and unspecific.

"After three months of abortive attempts to solubilize this enzyme which is tightly bound to the inner mitochondrial membrane, the project seemed doomed to failure. A chance conversation between Sir Hans and an Australian microbiologist, June Lascelles, who was a lecturer in the department, provided the vital clue to further progress. Certain bacteria, including *Rhodopseudomonas spheroides*, are known to accumulate a polymer, poly-hydroxybutyrate, as a reserve fuel and they also contain soluble 3-hydroxybutyrate dehydrogenase to oxidize the monomer, D-3-hydroxybutyrate.[1] Within a few weeks I had grown sufficient amounts of *Rhodopseudomonas spheroides* to obtain a partially purified preparation of the enzyme and to show that it could indeed be used for the enzymic determination of acetoacetate and hydroxybutyrate. Perhaps more importantly, the enzyme appeared stable. It was at this stage that Jane Mellanby, a PhD student working on the metabolism of ketone bodies, joined me in the large-scale production of the enzyme. Soon we were growing 30 litre aspirators of the red-coloured bacteria and occasionally covering the floor of the 37°C room because of a too effective aeration system. At a later stage we encouraged H.U. Bergmeyer of Boehringer Mannheim to take up production of the enzyme and this resulted in a joint publication on the crystalline protein.[2]

"We realized the potential clinical value of the method and described its application to blood samples in the original paper. The method has been widely used for research on the regulation of ketone body production[3] and of ketone body utilization.[4] In particular, the method was used by Cahill and his colleagues[5] to demonstrate that ketone bodies are an important substrate for brain in starvation. Despite numerous minor modifications by other workers, the original method appears to be firmly established."[6]

---

1. Carr N G & Lascelles J. Some enzymic reactions concerned in the metabolism of acetoacetyl-coenzyme A in Athiorhodaceae. *Biochemical J.* 80:70-7, 1961.
2. Bergmeyer H U, Gawehn K, Klotzsch H, Krebs H A & Williamson D H. Purification and properties of crystalline 3-hydroxybutyrate dehydrogenase from *Rhodopseudomonas spheroides*. *Biochemical J.* 102:423-31, 1967.
3. McGarry J D & Foster D W. Regulation of hepatic fatty acid oxidation and ketone body production. *Annu. Rev. Biochem.* 49:395-420, 1980.
4. Robinson A M & Williamson D H. Physiological roles of ketone bodies as substrates and signals in mammalian tissues. *Physiol. Rev.* 60:143-87, 1980.
5. Owen O E, Morgan A P, Kemp H G, Sullivan J M, Herrera M G & Cahill G F. Brain metabolism during fasting. *J. Clin. Invest.* 50:1589-95, 1967.
6. Bach A, Métais P & Jaeger M A. Etude critique de la détermination enzymatique de l'acétoacétate et du D-(—)-béta-hydroxybutyrate par la méthode de Williamson: application au plasma et au tissu hépatique. *Ann. Biol. Clin. Paris* 29:39-50, 1971.

47

# This Week's Citation Classic

CC/NUMBER 36
SEPTEMBER 8, 1980

Duncombe W G. The colorimetric micro-determination of long-chain fatty acids.
*Biochem. J.* 88:7-10, 1963.
[Wellcome Research labs, Beckenham, Kent, England]

Long-chain fatty acids dissolved in chloroform will form chloroform-soluble copper soaps. The reaction of these with a chromogenic reagent for copper is the basis for the determination described. [The *SCI*® indicates that this paper has been cited over 705 times since 1963.]

---

W.G. Duncombe
Wellcome Research Laboratories
Langley Court
Beckenham Kent BR3 3BS
England

July 17, 1979

"The start of this work was the need for a sensitive quantitative analysis for radioactive long-chain fatty acids eluted from paper chromatograms, so that their specific activities could be determined. I found a reference to a method in which the copper soaps of such acids were analysed by spectroscopy of their solutions in chloroform, making use of their blue colour.[1] However, the sensitivity was not adequate for my purpose, so thoughts about increasing it naturally followed, and it needed no great intellectual feat to wonder whether using one of the chromogenic reagents for copper would have the desired effect. Further search in the literature failed to show that anyone had done this, so I pursued the idea and developed a sensitive and reproducible method for the analysis of solutions of fatty acids.

"Ironically, the application for which the method was devised proved to be unsatisfactory. Extracts from paper gave high and unreproducible blanks, which could not be made acceptable by pre-washing the paper and purifying the solvents. However the basic method seemed to be worth publishing, and soon reports started appearing in the literature in which it was being used in a variety of applications in which the concentration of long-chain fatty acids extracted from biological materials was required.

"A further development led to increased use. Non-esterified fatty acids in blood plasma had previously been determined by titration. An adaptation[2] of the colorimetric method now permitted this determination to be carried out directly on plasma. Although not entirely specific under these conditions, comparison trials showed that the colorimetric method gave results comparable with those obtained by titration, and it began to be adopted for clinical and other investigations in which changes in fatty acid levels are more important than absolute values.

"It is of course well known that the frequency of citation of a paper is no criterion of its usefulness or significance. Happily this does not seem to be the case with the paper in question, since citations mostly occur when the method or some variation on it is actually being used. Improvements and modifications are indeed suggested from time to time, but perhaps this too is a reflection of the basic value of the method."

1. Iwayama Y. New colorimetric determination of higher fatty acids. *Yakugaku Zasshi* 79:552-4, *Chem. Abstr.* 53:14819, 1959.
2. Duncombe W G. The colorimetric micro-determination of non-esterified fatty acids in plasma. *Clin. Chim. Acta* 9:122-5, 1964.

# This Week's Citation Classic

CC/NUMBER 19
MAY 7, 1984

Mauzerall D & Granick S. The occurrence and determination of δ-aminolevulinic acid and porphobilinogen in urine. *J. Biol. Chem.* 219:435-46, 1956.
[Rockefeller Institute for Medical Research, New York, NY]

Delta-aminolevulinic acid (ALA) and porphobilinogen (PBG) in biological fluids are separated by the use of ion exchange resins. A pyrrole is quantitatively formed by condensing ALA with acetyl acetone. Both the pyrrole and the PBG are determined colorimetrically with a modified Ehrlich's reagent. The factors influencing the optimal conditions for these reactions are discussed. [The *SCI*® indicates that this paper has been cited in over 960 publications since 1956.]

---

David Mauzerall
Rockefeller University
New York, NY 10021

March 28, 1984

"In 1955, Sam Granick realized the importance of a simple assay for the crucial metabolite δ-aminolevulinic acid (ALA). For a fresh graduate in physical organic chemistry, having just joined Granick at the Rockefeller Institute, this was a breeze. However, the adage on the time required to solve an 'easy' problem held true: multiply the guess by two and go to the next higher time scale. There existed an easy color reaction for the next intermediate in the biosynthetic pathway, porphobilinogen (PBG): the Ehrlich reaction. Why not convert ALA to a pyrrole and use the same reaction? One less reagent on the rack. But (and still a but for the few of us interested in the photosynthetic origin of life) ALA→PBG has a low yield 'in vitro.' Therefore, back to the Knorr pyrrole synthesis and use of β-dicarbonyl compounds. This is where the escalation to the higher time scale occurred. Suitable study showed that the conditions could be specified for the quantitative yield of pyrrole using ALA. Writing this reminds me that I never published that work, something about which Granick always chided me.

"Since we had a (nearly) quantitative reaction for ALA, the Ehrlich color reaction definitely needed improvement. It faded characteristically faster than the technician could insert the tube into the Beckman DU (1955). Rates were competing with equilibria, and it was a textbook case of kinetic analysis which was carried out on an analog computer built by C.C. Yang. It was great fun—I could simulate autocatalytic reactions just by the leakage in the capacitors. Unfortunately, all this was chopped out of the *Journal of Biological Chemistry* paper by a referee. Some things do not change that much.

"The molar extinction for the Ehrlich reaction was doubled, and the stability of the color much increased. This was accomplished by increasing the acidity of the solvent: perchloric acid in acetic acid *versus* hydrochloric acid in water. In a fit of worry just before publishing, I strongly heated some of the 'Special Ehrlich Reagent' (taking suitable precautions). It decomposed quite vigorously but did not detonate. Relieved, we sent the manuscript off with a warning that the reagent be disposed of periodically.

"The 'normals' in the test assay included Granick and myself as is usual. Only later[1] was it found that these 'normals' were actually high (possibly because of the presence of lead in the urban environment). Granick tried for many years to interest the drug companies in a 'packaged' column for the analysis. At that time there was no interest in what has now become an active enterprise. So it goes.

"I believe this article is quoted because it is a useful analytical method for an important metabolite. Its presence reflects the genetic defects in hereditary hepatic porphyrias or tyrosinemia and the effects of environmental hazards such as lead. Recent reviews summarize the use of this assay.[2-4] The choice of the particular problem was Granick's. It was a result of his constant probing and questioning. He let me indulge in theorizing to try to answer his questions. It was a good way to start."

---

1. Sassa S, Granick S & Kappas A. Effect of lead and genetic factors on heme biosynthesis in the human red cell. *Ann. NY Acad. Sci.* 244:419-40, 1975.
2. Lien L F & Beattie D S. Comparisons and modifications of the colorimetric assay for delta-aminolevulinic acid synthase. *Enzyme* 28:120-32, 1982.
3. Bishop D F, McBride L & Desnick R J. Fluorometric coupled-enzyme assay for delta-aminolevulinate synthase. *Enzyme* 28:94-107, 1982.
4. Sassa S. Delta-aminolevulinic acid dehydratase assay. *Enzyme* 28:133-43, 1982.

# This Week's Citation Classic™

Ferreira S H & Vane J R. Prostaglandins: their disappearance from and release into the circulation. *Nature* 216:868-73, 1967.
[Dept. Pharmacology, Inst. Basic Medical Sciences, Royal Coll. Surgeons, London, England]

The blood-bathed organ technique was used to detect and continuously estimate the concentration of prostaglandins $E_1$, $E_2$, and $F_{2\alpha}$ in the circulation, to determine their stability in the blood and their disappearance in different vascular beds. Prostaglandins are released into the splenic venous blood during spleen contractions. The pulmonary circulation provides an efficient protective mechanism for it removes almost all the prostaglandins before they reach the arterial circulation. [The *SCI®* indicates that this paper has been cited in over 745 publications since 1967.]

Sergio Henrique Ferreira
Department of Pharmacology
Faculty of Medicine of Ribeirão Preto
University of São Paulo
São Paulo
Brazil

November 17, 1983

"In previous papers, we investigated the disappearance of peptides and amines[1,2] in the circulation and realized the usefulness of the blood-bathed organ technique for measuring the metabolism of biologically active substances in blood or during a single passage through a vascular bed. The present study with prostaglandins started as a sequel to our paper on bradykinin.[1] We developed a set of bioassay tissues for measuring certain prostaglandins in the circulating blood and applied it for detecting this substance after endogenous release or after removal in various vascular beds. At that time, we had just received from Upjohn samples of the precious synthetic prostaglandins $E_1$, $E_2$, and $F_{2\alpha}$. I remember Vane asking me to investigate the fate of angiotensin I in the pulmonary circulation as well as the action of the peptide extracts from the venom of *Bothrops jararaca* which I had brought with me. This preparation, the bradykinin potentiating factor (BPF), was known to inhibit inactivation of bradykinin and potentiate its effect. I had been working mostly with peptides since my initial research training in Brazil with Rocha e Silva and I decided then to work with prostaglandins in order to get some experience with another class of substances. In consequence, I missed the opportunity of discovering the role of the lung in the conversion of angiotensin I and of the snake venom peptides as inhibitors of converting enzyme![3-5]

"I believe that our prostaglandins paper became a *Citation Classic*™ because the experiments led to the right results, being made at the right time and at the right place. In fact our results were soon confirmed by various groups using different techniques. The time was right because it coincided with the beginning of the explosive development of the prostaglandin field.

"Was Vane's laboratory the right place? While the experiments were being carried out we realized the importance of the lung as a metabolic organ. Placed at a very strategic position in the circulatory system, the lung could control the arterial level of circulating substances. A substance which escaped pulmonary clearance might be a systematic mediator, while those removed by the lungs might play a role only as local hormones. Thus, our descriptive paper led to a new concept and Vane geared his group to test this hypothesis. In a short time, the importance of the lungs as a filter was established allowing passage of some endogenous mediators but removing others.[4] The groups also found that with certain provoking stimuli, the lungs could release a novel substance (RCS) which was able to contract the isolated rabbit aorta.[6] The activity of RCS was later shown to be due to thromboxane $A_2$. It was by the inhibition of the generation of prostaglandins in lung homogenates and of the release of prostaglandins by spleen that the mode of action of aspirin-like drugs was established.[7-9]

"Some papers become 'instant' classics either due to their relevance, as a missing piece in the biological puzzle, or because they describe a new practical research tool. This paper fits in neither group in my opinion. But those first papers on the disappearance of bradykinin and prostaglandins played a strongly stimulating role in Vane's laboratory. However, bradykinin was losing momentum and prostaglandins were becoming the circus mistress. The fact that Vane and his colleagues successfully demonstrated that the lungs were an important organ for the metabolism of endogenous mediators was an important concept which kept a stream of references to our early work. Unhappily, many other important papers are now 'unknown classics' perhaps because they had no luck to be made at the right time or, more importantly, at the right place, i.e., in a laboratory able to maintain a continuous rate of citation for a long period of time."

1. Ferreira S H & Vane J R. The detection and estimation of bradykinin in the circulating blood. *Brit. J. Pharmacol. Chemother.* 29:367-77, 1967.
2. ──────────. Half-lives of peptides and amines in the circulation. *Nature* 215:1237-40, 1967.
3. Bakhle Y S. Conversion of angiotensin I to angiotensin II by cell-free extracts of dog lung. *Nature* 220:919-21, 1968.
4. Ng K K F & Vane J R. Fate of angiotensin I in the circulation. *Nature* 218:144-50, 1968.
5. Vane J R. The release and fate of vaso-active hormones in the circulation. *Brit. J. Pharmacol.* 35:209-42, 1969.
6. Piper P J & Vane J R. Release of additional factors in anaphylaxis and its antagonism by anti-inflammatory drugs. *Nature* 223:20-35, 1969.
7. Vane J R. Inhibition of prostaglandin synthesis as a mechanism of action for aspirin-like drugs. *Nature New Biol.* 231:232-5, 1971.
8. ──────────. *Citation Classic.* Commentary on *Nature New Biol.* 231:232-5, 1971. *Current Contents/Life Sciences* 23(42):12, 20 October 1980.
9. Ferreira S H, Moncada S & Vane J R. Indomethacin and aspirin abolish prostaglandin release from the spleen. *Nature New Biol.* 231:237-9, 1971.

# This Week's Citation Classic

CC/NUMBER 42
OCTOBER 20, 1980

Vane J R. Inhibition of prostaglandin synthesis as a mechanism of action for aspirin-like drugs. *Nature New Biol.* 231:232-5, 1971.
[Department of Pharmacology, Institute of Basic Medical Sciences, Royal College of Surgeons of England, Lincoln's Inn Fields, London, England]

The generation of prostaglandins by a cell-free enzyme preparation in vitro was measured by bioassay. Aspirin-like drugs inhibited the formation of prostaglandins. This enzyme inhibition was proposed as the mechanism of therapeutic action and side effects of aspirin-like drugs, implicating prostaglandins in inflammation. [The *SCI®* indicates that this paper has been cited over 1,915 times since 1971.]

John R. Vane
Wellcome Research Laboratories
Langley Court
Beckenham, Kent BR3 3BS
England

July 17, 1979

"The aspirin-like drugs are not only amongst the most widely used synthetic pharmaceuticals but also amongst the oldest. Even so, there was no widely accepted theory to explain their actions. The drugs' three salient properties—anti-pyresis, anti-inflammation, and analgesia—seemed unrelated to one another as did their shared side effects.

"The discovery that aspirin and similar compounds prevented the biosynthesis of prostaglandins (PGs) suggested that this enzyme inhibition accounted for their therapeutic effects. It came at a time when interest was burgeoning in prostaglandins, which had already been detected in inflammatory exudates and shown to mimic some of the signs and symptoms of inflammation.

"The finding was a consequence of our development of a method of parallel bioassay which allowed the immediate and continuous detection of the release of vasoactive substances, such as histamine, serotonin, epinephrine, bradykinin, and some of the PGs. The method had the advantage of simplicity and allowed the maximum opportunity for serendipity. We were using this technique to study the release from lungs of putative mediators of anaphylaxis, and found some unexpected ones — $PGE_2$, $PGF_{2\alpha}$ and an ephemeral substance which we called 'rabbit aorta contracting substance' or RCS. Aspirin abolished the output of RCS (nowadays known as thromboxane $A_2$). Later experiments using a different system showed that aspirin also abolished the release of PGs. Whilst working on a review over the weekend, I realized that stimulated tissues released more PGs than they contained, equating release with stimulation of biosynthesis. Could aspirin be preventing PG biosynthesis? On that Monday morning I said to my colleagues, 'I think I know how aspirin works.' I told them my hypothesis and set about experimenting to put it to the test. The most direct test of whether aspirin inhibited prostaglandin biosynthesis was to make an in vitro preparation of the enzyme involved. As a classical pharmacologist, I had always been taught to study whole tissues, rather than smash them into bits. However, I found a paper describing prostaglandin synthetase made from homogenised lungs[1] and on that same day made my very first enzyme preparation. It worked, and so did aspirin as an inhibitor!

"The forging of the aspirin-prostaglandin link was important for several reasons. First, it established a scientifically satisfactory mode of action for aspirin-like drugs, now generally accepted. Secondly, it provided a simple enzyme test for discovering new, and hopefully better, aspirins. Thirdly, it provided a tool by which biologists could study the importance of prostaglandins in the body's processes. Give an aspirin-like drug, see what it does to a particular organ or function and that tells you what a lack of prostaglandins does. Fourthly, from the elucidation of the role of prostaglandins in the body, it suggested new uses for aspirin-like drugs."

---

1. Ånggård E. & Samuelsson B. Biosynthesis of prostaglandins from arachidonic acid in guinea pig lung. *J. Biol. Chem.* 240:3518-21, 1965.

## This Week's Citation Classic

CC/NUMBER 36
SEPTEMBER 5, 1983

Flower R J. Drugs which inhibit prostaglandin biosynthesis.
*Pharmacol. Rev.* 26:33-67, 1974.
[Department of Pharmacology, Institute of Basic Medical Sciences, Royal College of Surgeons of England, London, England]

This was the first review to draw together all the data then available on the inhibition of prostaglandin biosynthesis by aspirin-like drugs and other compounds. It aimed to provide a useful guide to biological scientists who wished to use these drugs as pharmacological tools. Much of the subject matter has been reviewed several times since then. Reference 1 is a particularly comprehensive article. [The *SCI®* indicates that this paper has been cited in over 905 publications since 1974.]

Roderick J. Flower
Department of Prostaglandin Research
Wellcome Research Laboratories
Beckenham, Kent BR3 3BS
England

June 15, 1983

"When I was told that this paper had become a *Citation Classic*, I was both surprised and a little amused, for the review started out as a sort of private student project.

"I began my PhD studies with John Vane at the Royal College of Surgeons in 1971. He and his colleagues had just made their fundamental discovery that the aspirin-like drugs prevented prostaglandin generation by blocking the biosynthetic cyclo-oxygenase enzyme.[2-4] His department of pharmacology was a very busy and exciting place to work in those days, and the notion that inhibition of prostaglandin generation could explain the anti-inflammatory effects of aspirin was one of the hottest concepts of the time.

"I began my postgraduate work by examining the effects of many anti-inflammatory drugs on the prostaglandin synthesizing system and was fortunate to be able to contribute to several significant publications on the topic (e.g., reference 5). In my spare time, I began to collect together all our results and those published by other groups and arrange them into tables; I also collated a lot of data about inhibition of the enzyme by other agents such as fatty acid analogues and antioxidants.

"Apart from my enthusiasm for the subject, my motive for doing this was laziness. I thought that if I could sort out all this data a little at a time, it would save me a great deal of effort later when I came to write my PhD thesis! Eventually, however, the idea that my Sunday afternoon jottings might be put together into some sort of review began to claim my attention. With the help of a senior colleague, Mick Bakhle, I prepared a manuscript and later showed it to Vane. After he had read it and suggested various changes, he surprised me by suggesting that it might be suitable for *Pharmacological Reviews*. I was even more surprised when it was accepted almost immediately by that august journal.

"The literature on prostaglandins absolutely exploded in the early 1970s. Because prostaglandins are formed by practically every tissue in the body, one cannot resort to the techniques of classical endocrinology (i.e., removal or ablation of a particular gland or organ) to test for their involvement in physiological events. Neither were there any really reliable antagonists of prostaglandin action, and so the key to many experiments was the use of the aspirin-like drugs to prevent the cellular biosynthesis of these lipids. Naturally, many scientists wanted to know what dose of these drugs to use, how long they would last, or what the likely side effects were. Because of this demand for information my little review was warmly received and, it now appears, highly cited. Incidentally, I am not sure whether it is always strictly ethical to cite reviews: I often notice that people refer to this particular manuscript as if it were the source of the original information, instead of citing the author who first published the data. This use of comprehensive reviews as 'umbrella references' undoubtedly gives them a spurious citation score.

"The entire prostaglandin system has become much more complex than it was in 1974. Thromboxane and prostacyclin have been discovered and a plethora of new lipoxygenase enzymes have also made their appearance, adding substantially to the number of target enzymes in the cascade.

"Sometimes I think it would be a good idea to update the review to include inhibitors of all these new pathways. Rash impulses like this are easily checked: I stand by an open window in our library, take several deep breaths, and thumb through the latest *Current Contents®*. The sight of all those new papers on prostaglandins is enough to strike terror into the heart of the most seasoned reviewer!"

1. **Shen T Y.** Anti-inflammatory drugs. (Vane J R & Ferreira S H, eds.) *Handbook of experimental pharmacology. Vol. 50 (II).* Berlin: Springer-Verlag, 1979. p. 305-42.
2. **Vane J R.** Inhibition of prostaglandin synthesis as a mechanism of action for aspirin-like drugs. *Nature New Biol.* 231:232-5, 1971.
   [Citation Classic. *Current Contents/Life Sciences* 23(42):12, 20 October 1980.]
3. **Ferreira S H, Moncada S & Vane J R.** Indomethacin and aspirin abolish prostaglandin release from the spleen. *Nature New Biol.* 231:237-9, 1971.
4. **Smith J B & Willis A L.** Aspirin selectively inhibits prostaglandin production in human platelets. *Nature New Biol.* 231:235-7, 1971.
5. **Flower R, Gryglewski R, Herbaczyńska-Cedro K & Vane J R.** Effects of anti-inflammatory drugs on prostaglandin biosynthesis. *Nature New Biol.* 238:104-6, 1972.
   [The *SCI* indicates that this paper has been cited in over 330 publications since 1972.]

# This Week's Citation Classic

Winter C A, Risley E A & Nuss G W. Carrageenin-induced edema in hind paw of the rat as an assay for antiinflammatory drugs.
*Proc. Soc. Exp. Biol. Med.* 111:544-7, 1962.
[Merck Institute for Therapeutic Research, West Point, PA]

This paper presented a method for testing compounds for antiinflammatory activity. It was the first technique available which permitted an assay after single oral doses at nontoxic levels within a single day, and which yielded linear and parallel log dose responses. [The *SCI®* indicates that this paper has been cited in over 1,000 publications since 1962.]

---

Charles A. Winter
2915 W. Lake Sammamish Parkway NE
Redmond, WA 98052

January 4, 1983

"In the absence of full knowledge of the etiology of inflammatory diseases, antiinflammatory drugs seem to offer the best treatment for the relief of symptoms, but at the time of this research, the only generally useful nonsteroidal antiinflammatory drugs were aspirin and phenylbutazone. Medicinal chemists, eager to proceed with synthetic programs looking for safer and more effective compounds, were hampered in their study of structure-activity relationships by the failure of pharmacologists to provide adequate guidance. Assays of antiinflammatory activity based upon the cardinal signs of inflammation included the classic method of Meier and co-workers[1] which tested for the inhibition of granuloma formed around a cotton pellet inserted subcutaneously in the rat. The biological activity of indomethacin was discovered in this way.[2]

"Indomethacin soon established itself as the antiinflammatory drug by which others were judged, and we at the Merck Institute for Therapeutic Research felt increased urgency to improve our methods of testing new compounds. The usefulness of the granuloma inhibition assay was limited to compounds available in sufficient quantity to treat a group of animals daily for a week. We therefore sought a method responsive within hours and requiring only a single administration. A technique based upon swelling induced by acute inflammation seemed to offer a potential solution.

"Others had found that antiinflammatory drugs could inhibit the edema produced by a phlogistic agent,[3] but responses were. nonspecific, reacting to compounds of various classes, and often not to antiinflammatory compounds except in toxic doses. A more specific phlogistic agent was found in carrageenan. The finding was empirical; at first we had little knowledge of the properties of carrageenan but we soon found that it induced reproducible edema which responded in a fairly specific way to nontoxic doses of antiinflammatory drugs, yielding parallel linear log dose-response data. We were now able to test small samples of new compounds and report results within hours. Eventually, an expanded study of large series of newly synthesized agents became possible,[4] which earned the Directors' Award for scientific achievement given by the directors of Merck & Co., Inc.

"Our publication apparently filled a need, for it soon became the procedure most widely used for assay of antiinflammatory activity.[5] In our original publication, our preferred antiinflammatory compound, indomethacin, was not mentioned, although by that time, investigation of indomethacin was well under way. Not until the following year were the wraps taken off this compound.[2] It later developed that not all samples of carrageenan were the same, that edema induced by carrageenan was biphasic, and that antiinflammatory compounds inhibit only the second phase.[5] It wasn't until a decade later that Vane[6] proposed the hypothesis that antiinflammatory drugs owe their activity to inhibition of prostaglandin synthesis. At last it became clear that carrageenan releases prostaglandins, and that in using our method for seeking new drugs, what we are actually looking for are new inhibitors of prostaglandin synthesis. This offers a reasonable explanation for the fact that antiinflammatory drugs are useful in treating symptoms but do not attack the fundamental causes of rheumatic disorders; the latter requires a new approach, and such research is well under way."

1. **Meier R, Schuler W & Desaulles P.** Zur Frage des Mechanismus der Hemmung des Bindgewebswachstums durch Cortisone. *Experientia* 6:469-74, 1950.
2. **Winter C A, Risley E A & Nuss G W.** Anti-inflammatory and antipyretic activities of indomethacin, 1-(p-chlorobenzoyl)-5-methoxy-2-methyl-indole-3-acetic acid. *J. Pharmacol. Exp. Ther.* 141:369-76, 1963.
3. **Domenjoz R.** The pharmacology of phenylbutazone analogs. *Ann. NY Acad. Sci.* 86:263-91, 1960.
4. **Shen T Y & Winter C A.** Chemical and biological studies on indomethacin, sulindac and their analogs. *Advan. Drug Res.* 12:89-245, 1977.
5. **Swingle K F.** Evaluation for antiinflammatory activity. (Scherrer R A & Whitehouse M W, eds.) *Antiinflammatory agents, chemistry and pharmacology.* New York: Academic Press, 1974. Vol. 2. p. 33-122.
6. **Vane J R.** Inhibition of prostaglandin synthesis as a mechanism of action of aspirin-like drugs. *Nature New Biol.* 231:232-5, 1972.

[Citation Classic. *Current Contents/Life Sciences* 23(42):12, 20 October 1980.]

# This Week's Citation Classic

CC/NUMBER 41
OCTOBER 11, 1982

Moncada S, Higgs E A & Vane J R. Human arterial and venous tissues generate prostacyclin (prostaglandin X), a potent inhibitor of platelet aggregation.
*Lancet* 1:18-21, 1977.
[Wellcome Research Labs., Beckenham, Kent, England]

Human arterial and venous tissues generate prostacyclin, an unstable substance which potently inhibits platelet aggregation. Prostacyclin synthesis is inhibited by 15-hydroperoxy arachidonic acid, a lipid hydroperoxide. These suggest that prostacyclin plays a role in preventing platelet aggregation on the vessel wall. Inhibition of prostacyclin synthesis by lipid peroxides may contribute to the genesis of certain diseases. [The *SCI®* indicates that this paper has been cited in over 475 publications since 1977.]

S. Moncada
Department of Prostaglandin Research
Wellcome Research Laboratories
Beckenham, Kent BR3 3BS
England

June 21, 1982

"The discovery of prostacyclin in 1976[1] and a series of later papers, published both from our laboratory and in collaboration with the scientists of the Upjohn Company, created the background for the demonstration of the synthesis and release of prostacyclin by human vascular tissue. We had been trying to demonstrate the synthesis of thromboxane $A_2$ ($TXA_2$), a powerful vasoconstrictor and inducer of platelet aggregation by the vessel wall, following my hypothesis that its generation in the vasculature might synergise with platelet-derived $TXA_2$ in the formation of the haemostatic plug. We showed instead that the vessel wall converts arachidonic acid into a powerful vasodilator and inhibitor of platelet aggregation, prostacyclin. The demonstration of the existence of such a compound in human vessels was of obvious interest and excitement. Prostacyclin is the biological counterpart of $TXA_2$. We have suggested that a balance between these two compounds plays a role in the control of platelet aggregation *in vivo* and that some of the thromboresistant properties of the vascular endothelium might be related to prostacyclin generation. Some of our hypotheses are at present being studied and it will probably be some time before elucidation of the biological role of prostacyclin is made.

"Some pathological conditions have been associated with a decrease in prostacyclin formation. The most striking is the association between the vascular complications of diabetes and a reduced prostacyclin generation; such a reduction has also been implicated in atherosclerosis. As yet, research into the possible role of lipid peroxides (strong inhibitors of prostacyclin formation[2,3]) in the development of atherosclerosis is just beginning. This promises to be a fascinating area of research.

"While basic and clinical research continues, synthetic prostacyclin has been successfully used clinically in a number of conditions. These include situations in which the blood has to be exteriorized from the body, such as cardiopulmonary bypass operations, renal dialysis, and charcoal haemoperfusion. Other conditions such as peripheral vascular disease, Reynaud's syndrome, pre-eclampsia, and thrombotic thrombocytopenic purpura are being studied intensively.

"Prostacyclin has also provided a naturally occurring molecule which comprehensively inhibits platelet aggregation. It is highly likely that new inhibitors of platelet aggregation, based on the chemical structure of prostacyclin, will be developed, probably orally active and long lasting. These compounds will have a superior antiplatelet activity to those available at the moment and, therefore, will allow more comprehensive study and better treatment of thrombotic disease.

"Looking back at January 1977, I would say that this paper aroused so much interest because human tissue was used, giving a 'degree of respectability' to the discovery of prostacyclin. Five years later, I think prostacyclin has already established itself as an endogenous substance to be 'reckoned with' if one wants to understand platelet/vessel wall interactions. As very often happens, thinking back to that time, I find myself wondering how it happened that prostacyclin was there for so long but nobody saw it before us."

1. Moncada S, Gryglewski R J, Bunting S & Vane J R. An enzyme isolated from arteries transforms prostaglandin endoperoxides to an unstable substance that inhibits platelet aggregation. *Nature* 263:663-5, 1976.
2. ............................................................. A lipid peroxide inhibits the enzyme in blood vessel microsomes that generates from prostaglandin endoperoxides the substance (prostaglandin X) which prevents platelet aggregation. *Prostaglandins* 12:715-33, 1976.
3. Salmon J A, Smith D R, Flower R J, Moncada S & Vane J R. Some characteristics of the prostacyclin synthesizing enzyme in porcine aorta. *Biochim. Biophys. Acta* 523:250-62, 1978.

# This Week's Citation Classic

CC/NUMBER 2
JANUARY 10, 1983

Hamberg M, Svensson J & Samuelsson B. Thromboxanes: a new group of biologically active compounds derived from prostaglandin endoperoxides.
*Proc. Nat. Acad. Sci. US* 72:2994-8, 1975.
[Department of Chemistry, Karolinska Institutet, Stockholm, Sweden]

Stimulation of human blood platelets results in the formation of thromboxane $A_2$ from platelet arachidonic acid. Thromboxane $A_2$ in very low concentrations causes clumping of human platelets and has a mediator role in hemostasis and in the generation of vascular disease. [The *SCI*® indicates that this paper has been cited in over 1,305 publications since 1975.]

---

Mats Hamberg
Department of Chemistry
Karolinska Institutet
S-104 01 Stockholm 60
Sweden

November 10, 1982

"The discovery of the thromboxane family of compounds in 1974-1975 can be regarded as a logical outcome of our previous work on the mechanism of prostaglandin biosynthesis from certain polyunsaturated fatty acids carried out from 1965 to 1967.[1,2] This work led to the proposal of the existence of endoperoxide intermediate(s); however it was not until 1973 that such intermediates (prostaglandins $G_2$ and $H_2$) could be isolated.[3] The access to pure endoperoxides and the finding of their pro-aggregated activity on human blood platelets[3] necessitated a study on the metabolic fate of arachidonic acid in human platelets[4] and also made possible the work by Moncada and Vane which led to the discovery of prostaglandin $I_2$ (prostacyclin).[5]

"In platelets, two pathways of arachidonic acid metabolism were found. One was initiated by a novel lipoxygenase and resulted in the formation of 12-hydroxyeicosatetraenoic acid. This was the first example of a lipoxygenase-catalyzed reaction in animal tissue. Subsequently, several other lipoxygenases catalyzing dioxygenation of polyunsaturated fatty acids were found in animal tissue. Of special interest is arachidonic acid 5-lipoxygenase which catalyzes the first reaction in the formation of leukotrienes.[6]

"The second pathway of arachidonic acid metabolism in the platelets was initiated by the aspirin-sensitive enzyme, fatty acid cyclooxygenase, and resulted in the formation of thromboxane $B_2$ and a monohydroxy acid.[4] These two compounds were also formed from prostaglandin endoperoxides. The relatively complicated non-prostanoate structure of thromboxane $B_2$ suggested that its formation from prostaglandin endoperoxides occurred by more than one reaction. An intermediate having a fused oxetane-oxane ring structure appeared especially attractive. By a number of chemical studies including trapping experiments with nucleophilic agents, we were able to confirm the presence of an oxetane-oxane structure in the intermediate and to elucidate its complete chemical structure. The compound was called thromboxane $A_2$ and was found to be very unstable in aqueous medium ($t_{1/2} = 30$ sec at 37°).

"At the same time, in collaboration with J. Svensson, who was carrying out his doctoral work at the department of chemistry, we observed a transient formation of very unstable potent pro-aggregating material upon incubation of platelet suspensions with arachidonic acid. This material was identified as thromboxane $A_2$ on the basis of its formation from prostaglandin endoperoxides, instability in aqueous medium, etc.

"Formation and action of thromboxane $A_2$ is the first example of physiological and pathological roles for the prostaglandin-thromboxane system in man. The finding of a new endogenously formed mediator in hemostasis and in the generation of vascular disease has stimulated a large number of biochemical, physiological, and clinical studies. This, I think, is the reason for the frequent citation of our paper."

---

1. **Samuelsson B.** On the incorporation of oxygen in the conversion of 8,11,14-eicosatrienoic acid into prostaglandin $E_1$. *J. Amer. Chem. Soc.* 87:3011-13, 1965.
2. **Hamberg M & Samuelsson B.** On the mechanism of the biosynthesis of prostaglandins $E_1$ and $F_{1\alpha}$. *J. Biol. Chem.* 242:5336-43, 1967.
3. **Hamberg M, Svensson J, Wakabayashi T & Samuelsson B.** Isolation and structure of two prostaglandin endoperoxides that cause platelet aggregation. *Proc. Nat. Acad. Sci. US* 71:345-9, 1974.
   [The *SCI* indicates that this paper has been cited in over 755 publications since 1974.]
4. **Hamberg M & Samuelsson B.** Novel transformations of arachidonic acid in human platelets. *Proc. Nat. Acad. Sci. US* 71:3400-4, 1974.
   [The *SCI* indicates that this paper has been cited in over 790 publications since 1974.]
5. **Moncada S, Gryglewski R, Bunting S & Vane J R.** An enzyme isolated from arteries transforms prostaglandin endoperoxides into an unstable substance that inhibits platelet aggregation. *Nature* 263:663-5, 1976.
   [The *SCI* indicates that this paper has been cited in over 1,260 publications since 1976.]
6. **Samuelsson B.** The leukotrienes: an introduction. (Samuelsson B & Paoletti R, eds.)
   *Leukotrienes and other lipoxygenase products.* New York: Raven Press, 1982. p. 1-17.

## This Week's Citation Classic

CC/NUMBER 17
APRIL 28, 1980

Kates M. Bacterial lipids. *Adv. Lipid Res.* 2:17-90, 1964.
[Division of Biosciences, National Research Council, Ottawa, Canada]

The article is a review of work done up to 1964 on the cellular lipids of bacterial species, the intracellular distribution of these lipids as well as their biosynthetic pathways. Attempts were made to correlate lipid composition with taxonomic classification of bacteria. [The *SCI®* indicates that this paper has been cited over 255 times since 1964.]

Morris Kates
Department of Biochemistry
University of Ottawa
Ottawa, Ontario K1N 6N5
Canada

January 22, 1980

"The idea for this article took several years to develop. Prior to 1958, I had worked mostly on phospholipids of plants and animals. I recall it was in the fall of 1958 that Norman Gibbons, then head of the microbiology section, National Research Council, Ottawa, came into my lab with a test-tube containing a red-coloured greasy blob of material. 'Do you know what this is?' he asked. 'No,' I said. 'Well,' he said, 'this is a sample of extremely halophilic bacteria that *require* 4M NaCl for their growth and survival. They must have some very unusual membrane lipids to be able to function in almost saturated salt! Why don't you have a look at the lipids of these bacteria?'

"So I began to work on the lipids of these halophiles, together with Suren Sehgal, a post-doc with Gibbons, and by the end of the year we knew there was something strange about them since we could detect no fatty acids at all, esterified or free. In fact the lipids were derived from a dialkyl ether of glycerol rather than from a diacyl glycerol as is found for most organisms, in particular a *moderate* halophile *Micrococcus halodenitrificans*[1] that we were also studying for comparison. We published these preliminary findings in 1962[2] and a year later[3] we had established the structure of the glycerol ether as 2,3-diphytanyl-sn-glycerol and the structure of the major phospholipid as the diether analog of phosphatidylglycerophosphate.

"During this period I also became involved with other microbiologists at NRC on a survey of bacterial lipids—Donn Kushner on *Bacillus cereus*[4] (and later with halophiles again), Stan Martin and Gordon Adams[5] on *Serratia marcescens* and Per-Otto Hagen[6] on psychrophilic *Serratia*-like species. I was struck by the great differences in lipid composition among the rather small number of bacterial species that we had examined, and this impression was confirmed on reading the literature of this subject which had already attracted the attention of many first-rate lipidologists in Europe and the US. It occurred to me that perhaps there might be a correlation between lipid composition and bacterial class or family which could be used as an aid in taxonomic classification. I thought it would be helpful for our work and that of other lipidologists to collect all the data available on bacterial lipid composition and cellular distribution and biosynthesis of these lipids.

"The resulting review article appeared about the time that biochemists and microbiologists became aware of the wide diversity of lipids in bacteria and began thinking about their role in membrane function. I think this article was cited frequently because it was the first comprehensive review of the subject and must have attracted the attention of researchers who appreciated the fact that many aspects of the subject were wide-open for investigation. It is indeed gratifying to know that this article has had some effect in stimulating research on bacterial lipids. A more recent review on the lipids of *Escherichia coli* has been published by C.R.H. Raetz."[7]

1. **Kates M, Sehgal S N & Gibbons N E.** Lipid composition of *Micrococcus halodenitrificans* as influenced by salt concentration. *Can. J. Microbiol.* 7:427-35, 1961.
2. **Sehgal S N, Kates M & Gibbons N E.** Lipids of *Halobacterium cutirubrum*. *Can. J. Biochem. Physiol.* 40:69-81, 1962.
3. **Kates M, Sastry P S & Yengoyan L S.** Isolation and characterization of a diether analog of phosphatidylglycerophosphate from *Halobacterium cutirubrum*. *Biochim. Biophys. Acta* 70:705-7, 1963.
4. **Kates M, Kushner D J & James A T.** The lipid composition of *Bacillus cereus* as influenced by the presence of alcohols in the culture medium. *Can. J. Biochem. Physiol.* 40:83-94, 1962.
5. **Kates M, Adams G A & Martin S M.** Lipids of *Serratia marcescens*. *Can. J. Biochem.* 42:461-79, 1964.
6. **Kates M & Hagen P O.** Influence of temperature on fatty acid composition of psychrophilic and mesophilic *Serratia* species. *Can. J. Biochem.* 42:481-8, 1964.
7. **Raetz C R H.** Enzymology, genetics, and regulation of membrane phospholipid synthesis in *Escherichia coli*. *Microbiol. Rev.* 42:614-59, 1978.

# This Week's Citation Classic

Hakomori S. A rapid permethylation of glycolipid, and polysaccharide catalyzed by methylsulfinyl carbanion in dimethyl sulfoxide.
J. Biochem. Tokyo 55:205-8, 1964.
[Dept. Biochem., Inst. Cancer Res., Tohoku Pharmaceutical Sch., Sendai, Japan]

A rapid permethylation of complex carbohydrates such as glycolipids and polysaccharides is described. The method is based on the use of Corey's base, created from dimethylsulfoxide and sodium hydride (methylsulfinyl carbanion). [The *SCI*® indicates that this paper has been cited over 780 times since 1964.]

---

Sen-itiroh Hakomori
Fred Hutchinson Cancer
Research Center
and University of Washington
Seattle, WA 98104

May 16, 1980

"I spent 1956-57 as a Fulbright fellow in Roger Jeanloz's laboratory in Boston and learned the difficulty involved in methylation analysis of complex carbohydrates. At that time at least ten grams of pure complex polysaccharides were needed, and several months were required to obtain 'complete' methylation (which even then was not always completed). Methylation of glycolipid was even more difficult as it forms a micelle. Several workers, including myself, tried methylation of glycosphingolipids, but simply failed. Therefore, I was very surprised to hear Richard Kuhn[1] and his colleagues had found a way, in 1960, to complete methylation of gangliosides within a few days using dimethylformamide and silver oxide or barium oxide.

"In 1963, while I was working at the Tohoku Pharmaceutical College in Sendai, I was trying to de-N-acetylate N-acetylhexosamines in glycoproteins and glycolipids from a purely immunochemical interest. A few reactions were tried in dimethylsulfoxide (DMSO) as it had been used for extraction of glycoproteins from tissues and found an expedient solvent for complex carbohydrates. During a discussion with my colleague, Hiroshi Tomizawa, I learned of a new development of the carbanion created from DMSO by E.J. Corey.[2] I learned more practical properties of the base from Shyo Uda at the chemistry department of Tohoku University. Soon after, I found that the carbanion was strong and that the methylation could take place effectively even for glycolipids. However, I was not quite sure the method would be superior to that of Kuhn. Moreover, I was not very optimistic that the method could be applied to aminosugar containing complex polymers, as the IR spectra of the methylated glycolipid did not show an absorption band at 1550 cm$^{-1}$ (due to -NHCO-) indicating a large extent of N-methylation.

"I thought this would bring us the horrible task of identifying all the N-methylated, partially O-methylated aminosugars. It is now apparent that N-methylation is the most welcome property for GC-MS identification of aminosugar, thanks to the work by Bengt Lindberg[3] in Stockholm and by Klaus Stellner,[4] once in this laboratory.

"The paper was originally submitted to the Journal of Biochemistry in June 1963, but was soon withdrawn because I received much criticism from someone to whom I had sent my preprint. They were reluctant to believe that such a method was feasible and I thought I needed to reinvestigate in detail. It was my friend Tamio Yamakawa who strongly encouraged me to publish the paper, as the method was good enough at least for glycolipids. I resubmitted it in November 1963. The brief paper was published in February 1964. The method has been greatly elaborated in the paper by P.A. Sandford and H.E. Conrad[5] which was published in 1966. Since then the method has been widely accepted. Life as a scientist is totally unpredictable. Frankly, I have mixed feelings as the paper cited is in my subsidiary interest, and my interest developed in quite a different area. I would be happier if some of my other papers, such as glycolipid transformation in neoplasia or discovery of cell surface fibonectin, would have been selected as highly cited articles."

1. Kuhn R, Egge H, Brossmer R, Gauhe R, Klesse W, Lochinger W, Röhm E, Trischmann H & Tschampel D. Über die Gangliosidе des Gehirns. Angew. Chem. 72:805-11, 1960.
2. Corey E J & Chaykovsky M. Methylsulfinylcarbanion. J. Amer. Chem. Soc. 84:866-7, 1962.
3. Björndal H, Hellerqvist C G, Lindberg B & Svensson S. Gas-liquid chromatography and mass spectrometry in methylation analysis of polysaccharides. Angew. Chem. Int. Ed. 9:610-9, 1970.
4. Stellner K, Saito H & Hakomori S. Determination of aminosugar linkages in glycolipids by methylation. Arch. Biochem. Biophys. 155:464-72, 1973.
5. Sandford P A & Conrad H E. The structure of the Aerobacter aerogenes A3(S1) polysaccharide. I. A reexamination using improved procedures for methylation analysis. Biochemistry 5:1508-17, 1966.

# Chapter 3
# Nucleic Acids

The twenty-three Citation Classics described in this chapter have been segregated from those in the chapter on Genetics in volume 1 because they have a distinct chemical orientation and rarely refer to cellular or enzymatic activities associated with the biopolymers. Even Classics that include phrases such as "plant tissues or nucleotide metabolism" in their titles are in actuality discourses on extraction and fractionation of purine and pyrimidine polymers.

Extraction of RNA and DNA from other plant or animal tissue by hot trichloroacetic acid (p. 60) remained popular until the late 1970s, although other solvents were also in use (p. 62). For the direct quantitation by means of phosphorus, sugar, or purine or pyrimidine analysis, the trichloroacetic acid procedure was adequate, but shear forces combined with acid to render the products useless in biological systems. For sequence analysis, sizing, and recombinant technology, the milder detergent–phenol combination is used for RNA and DNA harvesting (p. 64). Hirt's method of detergent extraction of DNA from viruses has been cited 1,135 times.

Dowex separation of the anionic nucleotides was a highly efficient method (p. 65) that, except for time and sensitivity, compared favorably with high-performance liquid chromatography. Perfection of the Dowex chromatographic system was more coolly logical and stepwise in its evolution than the instantaneous and startling success the Randeraths enjoyed with the thin-layer system, although the initial rejection of the latter's manuscript must have tempered their elation. Two-dimensional high-density electrophoresis coupled with a high specific $^{32}$P label in RNA provided an important advance (p. 67) in nucleotide separation and characterization, especially since paper as the second phase had previously proved useless in separating RNA digests.

Quantitation of the DNA by the diphenylamine–Dische reaction as developed by Burton was a widely applied method (over 5,000 citations between 1961 and 1975). Imagine his surprise when his Chicago-born modification of that procedure failed upon his return to England. Salvation of the technique by the addition of sunlight-oxidized acetic acid or acetaldehyde is an interesting limitation of English weather on progress in science. Wyatt's recipe for the hydrolysis of nucleic acids at 16°C at Cambridge continues this theme; he might have perfected the procedure on a day when his laboratory temperature was 11°C!

An important report by Gillespie and Spiegelman on RNA hybridization to immobilized DNA on nitrocellulose paper has had a much greater impact on nucleic acid biology than could have been conceived in 1965 at the time of its discovery or even in 1977 when Gillespie wrote his TWCC. This is the basis for much of the message and gene probing that is a commonplace blotting procedure today. Denhardt describes his Classic on DNA–DNA blotting, a paper cited 1,235 times, as being of little immediate importance until Southern blot analysis for specific DNA sequences became required in recombinant gene technology.

The intercalation of acridine orange between the strands of DNA described by Lerman in 1961 has several interesting features. The forced deviation of the DNA strands of acridine and similar compounds can relate to mutagenesis and to plasmid elimination (curing) of bacteria of these nonchromosomal genes and the highly sensitive detection of DNA or RNA. Ethidium bromide is an acridine dye and is a highly sensitive fluorescent probe for these nucleic acids (p. 77).

| Number 8 | **Citation Classics** | February 21, 1977 |

Schneider W C. Phosphorus compounds in animal tissues. I. Extraction and estimation of desoxypentose nucleic acid and of pentose nucleic acid.
*J. Biol. Chem.* 161: 293-303, 1945.

This new extraction method consists in heating the tissue in 5 ml. of 5 percent trichloroacetic acid for 15 minutes at 90° after removal of phospholipids and acid-soluble phosphorus compounds. The tissue is then cooled and centrifuged. The residue is resuspended in 2.5 ml. of 5 percent trichloroacetic acid and centrifuged. The trichloroacetic acid extracts are combined to form the nucleic acid extract. [The *SCI*® indicates that this paper was cited 1,518 times in the period 1961-1975.]

Walter C. Schneider, Ph.D.
National Institutes of Health
Bethesda, Maryland 20014

November 9, 1976

"It is indeed most gratifying to learn 31 years after publishing my paper that it is one of the most cited scientific papers. It is not entirely surprising to me that this should be the case since methods are the backbone of all scientific research. Consequently, my method for the extraction of nucleic acids from tissues and for their analysis might be expected to be referred to more frequently than average....

"To my mind there were two reasons for the popularity of the method. In the first place it was simple--both nucleic acids could be extracted from tissues with a single hot trichloroacetic acid treatment and then measured colorimetrically. More important, however, was the fact that the method dealt with nucleic acids, molecules that have engrossed the scientific community for more than 30 years and will continue to do so for many more....

"The fact that the paper should be cited so frequently is also a source of great satisfaction to me for another reason. This paper was based upon my doctoral dissertation and was thus my first as a sole author. I was fortunate to have Van R. Potter of the University of Wisconsin guide me into doing this work, which was designed to provide methods for measuring DNA and RNA in fractionations of tissue homogenates by differential centrifugation so that the recovery of subcellular organelles could be measured biochemically. The methods served admirably for this purpose and were instrumental in developing new methods for isolating nuclei, mitochondria, and microsomes, and in studying the subcellular localisation of biochemical functions.

"A few sidelights should also be noted. As I have confessed to friends and colleagues from time to time, the discovery that both nucleic acids could be extracted from tissues with hot trichloroacetic acid was not due to some great inspiration on my part but rather to my misreading one of Zacharias Dische's papers in the *Biochemisches Zeitschrift*. I have noticed a number of times during the intervening years that other scientific breakthroughs could be traced to similar fortunate mistakes--such is the progress of science. Also, imagine my surprise and chagrin upon opening the journal in which my paper appeared to find a paper on the same subject by Gerhard Schmidt and S.J. Thannhauser.[1] Their paper permitted the separation of DNA from RNA, which mine did not, but not the separation of DNA from protein, which mine did. It was immediately obvious to me that the ideal method for measuring nucleic acids would combine the best features of the two methods. I hurried to the laboratory to work out the details and the results were published the following year in the same journal....

"It may be of interest that I continue to use the method essentially unchanged. Although I have followed the literature diligently seeking for improvements that I could adopt, only a few have proved useful. One of these was the neutralisation of the tissue suspension after the initial extractions with cold trichloroacetic acid to avoid loss of protein during the extractions with alcohol. Another was the use of lower concentrations of alkali and shorter extraction times suggested by Fleck and Munro to solubilize RNA. Finally, the more sensitive diphenylamine reaction of Burton and the fluorimetric procedure of Kissane and Robins have proved most useful in more recent years for measuring the small amounts of DNA present in cytoplasmic organelles."

1. Schmidt G & Thannhauser S J. A method for the determination of desoxyribonucleic acid, ribonucleic acid, and phosphoproteins in animal tissues. *J. Biol. Chem.* 161:83-89, 1945.

# Citation Classics

Number 24                        June 13, 1977

Ogur M & Rosen G. The nucleic acids of plant tissues. I. Extraction and estimation of desoxypentose nucleic acid (DNA) and pentose nucleic acid (PNA) from plant tissue. *Archives of Biochemistry* 27:260-76, 1950.

A method is described for the analytical extraction and quantitative estimation of RNA and DNA, enabling the utilization of u.v. absorbance as well as sugar and phosphorus estimation in the validation of the assay. [The *SCI®* indicates that this paper was cited 664 times in the period 1961-1975.]

---

Professor Maurice Ogur
Department of Microbiology
Southern Illinois University
Carbondale, Illinois 62901

February 4, 1977

"A paper dealing with methodology, coming at a time when a field is expanding rapidly, tends to get many citations. This particular paper emerged when interest in the nucleic acids was expanding rapidly and the relationship between the amount of DNA per cell and polidy was being demonstrated. Many biologists were interested in nucleic acid estimation, and the available extraction methods relying either on estimating phosphorus or ribose and deoxyribose, while fairly satisfactory for some cells, ran into major interferences in others.

"I had just completed a Ph.D. in Chemistry at Columbia University in 1948 and was at an important crossroad when an opportunity arose to spend a post doctoral year at the University of Pennsylvania. It was a very stimulating and productive experience for me. I remember my awe at seeing Otto Meyerhof at work, meeting Britton Chance, having many conversations with David Goddard, Ralph Erickson, Ed Cantino, Conway Zirkle, Maurice Sevag, and others, and having my naive ideas worked over by Seymour Cohen's scholarship and critical judgment.

"Ralph Erickson, in the Botany Department, had been working on Lilium for some years and believed that the developing pollen might provide a cell population in sufficient synchrony for the measurement by chemical methods of nucleic acid changes during development. I tackled the problem with great enthusiasm by the available methods, only to be frustrated by the polyuronides and pentosans of plant tissues, which interfered with the orcinol-pentose estimation as a measure of RNA, and by an unknown material, which interfered with the diphenylamine reaction as a measure of DNA. This made me hesitant to rely on phosphorus estimation alone without the check of at least one more of the generic constituents of a nucleotide in molar ratio. I tried the u.v. absorbance of the hot TCA nucleic acid fraction but found that the TCA interfered. It could be removed by steam distillation or solvent extraction, but this seemed too messy. Herman Kalckar had used perchloric acid in the extraction of the acid soluble mononucleotide fraction in a method of estimation utilizing u.v. absorbance. It seemed reasonable therefore that hot perchloric acid might readily replace hot TCA. It also seemed possible that, if the internucleotide link in RNA was more susceptible to alkali than that in DNA, it was worth seeing if it might also be more susceptible to acid as the pre-Feulgen hydrolysis seemed to suggest. Preliminary experiments were encouraging enough to warrant attempting to establish the best conditions of concentration, time and temperature of perchloric extraction to separate RNA and DNA for analytical, though not for preparative, purposes. At this point, the collaboration with Mrs. Gloria Rosen was very helpful. We published a set of conditions which appeared to accomplish this without excessive cross contamination of the fractions in terms of the analytical criteria employed. It also seemed reasonable that our procedure might provide a more carefully standardized pre-Feulgen hydrolysis than the procedure then in use for nuclear cytology.

"Anyone who has developed and published a method must be prepared to receive not only the praise from those who have found it applicable to and useful in their work, but also the blame from those who have encountered unexpected difficulty with it on other materials. I am surprised that the method and various modifications of it appear to have been useful to so many for so long."

# This Week's Citation Classic

CC/NUMBER 24
JUNE 16, 1980

Kay E R M, Simmons N S & Dounce A L. An improved preparation of sodium desoxyribonucleate. *J. Amer. Chem. Soc.* 74:1724-6, 1952.
[Depts. Biochem. and Pathol., Sch. Medicine and Dentistry, Univ. Rochester, Rochester, NY]

An improved method of preparing DNA from various sources is described, making use of the detergent sodium dodecyl sulfate to deproteinize the protein component of DNA-protein complexes. [The *SCI*® indicates that this paper has been cited over 895 times since 1961.]

---

Ernest R. M. Kay
Department of Biochemistry
University of Toronto
Toronto, Ontario M5S 1A8
Canada

April 14, 1980

"Among my recollections of boyhood are the times when my father, who was a science teacher, would bring home a microscope for me to use. I was fascinated by the examination of cells, and intrigued by the nuclei of these cells. My ambition was to learn as much as I could about them, and when the situation presented itself years later I was grateful for the opportunity to study with Alexander Dounce in Rochester. I was his first graduate student.

"Dounce had pioneered in the field of nuclear isolation techniques and studies of the biochemistry of these cell organelles. When I joined his laboratory he was involved in the study of the nature of the binding of DNA in nuclei, and particularly interested in the phenomenon of nuclear gelation.[1] He had commenced some studies with Norman Simmons with the use of the detergent sodium dodecyl sulfate in attempts to isolate DNA from cell nuclei. At the suggestion of Dounce, I worked with Simmons for a while to develop a reproducible procedure using this detergent. Simmons left Rochester soon after we had carried out a few preliminary experiments. The task of developing a procedure for isolating DNA then became a part of my PhD thesis work. This led to developing a method which has been used apparently quite extensively.

"I was familiar with the available methods for isolating DNA and the detergent approach did seem to be a good direction to follow. The sodium dodecyl sulfate had been used earlier by Pirie[2] to solubilize tobacco mosaic virus. In the preliminary experiments, Simmons and I found that the detergent could solubilize DNA from calf thymus chromatin. These experiments had embodied a double salt technique which was rather cumbersome.

"It was decided to attack the problem using the solubility characteristics of DNA-protein complexes in solutions of NaCl alone. The basic procedure was built on earlier work of Mirsky,[3] who had used NaCl solutions to prepare chromosome threads and subsequently extract DNA from them. Using the isolated chromosomes as a starting material, I added detergent solutions and observed the changes in structure of the chromosome threads with the microscope. Experiments were then done using varying concentrations of detergent and suspensions of chromosome threads. These suspensions were centrifuged at high speed and the supernatants analysed for N and P to arrive at the proper detergent concentration for effective solubilization of the DNA. The repurification steps were followed using NaCl solutions and the required detergent concentration. The development of the procedure was a straightforward approach coupling microscopic observations and chemical analysis.

"The procedure appealed to me because of its simplicity, reproducibility, and yield of a pure product, when applied to a variety of tissues. It was at the time of its development a fascinating experience for me to isolate DNA and observe its nature. It is still a source of considerable pleasure for me to help students follow through with the procedure and observe their astonishment at the nature of the product obtained."

1. **Dounce A L.** Cytochemical foundations of enzyme chemistry. (Summer J B & Myrbäck K, eds.) *The enzymes.* New York: Academic Press, 1950. Vol. 1, Part 1. p. 187-266.
2. **Sreenivasaya M & Pirie N W.** The disintegration of tobacco mosaic virus preparations with sodium dodecyl sulfate. *Biochemical J.* 32:1707-10, 1938.
3. **Mirsky A E & Pollister A W.** Nucleoproteins of cell nuclei. *Proc. Nat. Acad. Sci. US* 28:344-52, 1942.

# This Week's Citation Classic

CC/NUMBER 33
AUGUST 17, 1981

Hirt B. Selective extraction of polyoma DNA from infected mouse cell cultures.
*J. Mol. Biol.* **26**:365-9, 1967.
[Swiss Inst. Experimental Cancer Res., Lausanne, Switzerland]

Cells infected by polyoma virus are lysed by the addition of the detergent sodium dodecyl sulfate. The detergent can be precipitated with sodium chloride at 4°. The DNA of the cells, if not sheared, precipitates too. The viral DNA however, which is smaller in size by several orders of magnitude, stays in the supernatant. [The *SCI®* indicates that this paper has been cited over 1,135 times since 1967.]

Bernhard Hirt
Swiss Institute for Experimental
Cancer Research
CH-1066 Épalinges
Switzerland

July 15, 1981

"In the fall of 1964, Roger Weil accepted me in his laboratory, where with two co-workers, he was studying the interaction of polyoma virus with its host cells. Before, I had been working in physics, so everything was new and exciting to me. I was assigned to the task of studying the replication of the viral DNA in cultured cells. From Weil's[1] work I knew that viral DNA is synthesized simultaneously with the DNA of the host, a complicating fact.

"First, I tried to inhibit preferentially the host DNA synthesis by irradiating the cells with x-rays or ultraviolet light prior to infection. But the replication of the viral DNA was inhibited too. I then looked for ways to separate viral DNA from cellular DNA. In their study, Weil, Michel, and Ruschmann[1] had extracted the DNA by a method which in principle had already been used by Avery[2] and co-workers in their historical study: cells were lysed with detergent and the viscosity of the lysate was reduced by pipetting and shaking. The detergent was then precipitated with salt, leaving a mixture of host cellular and viral DNA in the supernatant. Sedimentation velocity analysis in a sucrose gradient showed a fast-sedimenting band of host DNA with a peak of viral DNA on the slow side. For my purpose the resolution of the two DNAs was not sufficient. I wanted the host cellular DNA to sediment faster. I therefore avoided pipetting during the extraction, because shearing forces were known to reduce the molecular weight and the sedimentation coefficient of the DNA. In the following sucrose gradient sedimentation I expected to find a fast-sedimenting peak of cellular DNA, well separated from the slow viral peak. At 11 pm one night in October 1965 I put the samples into the counter and realized after a few minutes that the cellular DNA was not there. I concluded that the experiment had failed, said a word that in English would take four letters and went home to sleep. The rest of the samples had been counted by the time I arrived the next morning and I jumped up when I realized that the modified extraction had produced pure viral DNA. The extraction method was of use immediately in a study on semiconservative replication of polyoma DNA.[3]

"Later I wrote down the procedure and sent it as a letter to the editor of the *Journal of Molecular Biology*. One referee worked hard on it and sent a report three pages long, typed single-spaced, with many suggestions and requests for additional data. Despite some good remarks, the report made me angry by its paternalistic tone (I didn't realize at that time that refereeing papers is not always fun). My manuscript went into a drawer for two and a half months until an American postdoc, David A. Goldstein, came to our lab and helped me rewrite it. I assured the editor that the comments of the referee had been taken into account and the paper was published with no delay.

"The extraction method is simple and effective for isolating small DNA molecules from both bacteria and animal cells. This, together with the fact that many researchers chose to study small DNA viruses, might explain the high number of citations. More recent work in the field is reported by R.A. Weinberg."[4]

1. Weil R, Michel M R & Ruschmann G K. Induction of cellular DNA synthesis by polyoma virus.
   *Proc. Nat. Acad. Sci. US* **53**:1468-75, 1965.
2. Avery O T, MacLeod C M & McCarty M. Studies on the chemical nature of the substance inducing transformation of pneumococcal types. *J. Exp. Med.* **79**:137-58, 1944.
3. Hirt B. Evidence of semiconservative replication of circular polyoma DNA.
   *Proc. Nat. Acad. Sci. US* **55**:997-1004, 1966.
4. Weinberg R A. Integrated genomes of animal viruses. *Annu. Rev. Biochem.* **49**:197-226, 1980.

# This Week's Citation Classic

CC/NUMBER 50
DECEMBER 15, 1980

Scherrer K & Darnell J E. Sedimentation characteristics of rapidly labelled RNA from HeLa cells. *Biochem. Biophys. Res. Commun.* 7:486-90, 1962.
[Department of Biology, Massachusetts Institute of Technology, Cambridge, MA]

The techniques developed for extraction of infectious viral nucleic acid were used to extract cell RNA, resulting in the recognition, for the first time, of high molecular weight nuclear RNA from mammalian cells. The dominant rapidly labeled peaks, '45S' and '35S,' were subsequently found to be ribosomal precursor RNA, and the polydisperse material was characterized as the so-called 'heterogeneous nuclear RNA,' the hnRNA. [The *SCI®* indicates that this paper has been cited over 900 times since 1962.]

James E. Darnell, Jr.
Rockefeller University
New York, NY 10021

October 16, 1980

"My earliest experience with poliovirus RNA synthesis, and my postdoctoral time with Francois Jacob during the year of the prediction and demonstration of messenger RNA in bacteria, had convinced me that the time (1961) was ripe to begin a study of mammalian cell RNA.[1-7] It is obvious from the flood of talented people who soon also came to work on animal cell RNA synthesis that many others agreed. Thus, more or less by accident, our work was an early forerunner in a subsequently popular field. Klaus Scherrer, my first postdoctoral fellow at MIT, and I first began by trying to extract labeled RNA from HeLa cells labeled for 20 to 30 seconds. The available nucleosides were insufficiently radioactive for us to obtain a labeled preparation in this manner, so we lengthened our label times to minutes. With this length of label we were able to reproducibly obtain the high molecular weight RNA. Only recently have we returned to the extremely short label times.[8]

"Two aspects of the 1962 paper probably account for its frequent use as a reference by the large group that has become interested in RNA synthesis. First, the method described allows extraction of the majority of the total cell RNA without (or with minimal) degradation. We were pointed toward this goal by the earliest work of John Colter[9] and E. Wecker[10] who had with similar techniques obtained infectious viral RNA of over $2 \times 10^6$ daltons from animal tissue. The method in outline involved: (1) pulse labeling of cultured cells to observe kinetics of labeling of various RNA components; (2) cell lysis with SDS and RNA extraction with phenol at elevated temperature (60°); (3) zonal sedimentation of RNA to separate discrete species of RNA; (4) radioassay of labeled RNA (newly made) and optical assay of UV absorbance for preexisting RNA. Second, the paper showed that RNA extracted from human cells was different from that of bacteria and this meant there was something to study in the field of animal cell RNA metabolism. Because a major fraction of the RNA from cells labeled for short times was discrete in size but larger than the RNA from cells labeled for long times, the possibility of a precursor to ribosomal RNA was first suggested by this work, although it took another six months or so to suggest that point more strongly. Finally, at the end of an additional 15 years, the probable role of the non-ribosomal polydisperse nuclear RNA, the so-called heterogeneous nuclear RNA, as an mRNA precursor appears likely as well."

1. Brenner S. *Cold Spring Harbor Symp.* 26:101-9, 1961.
2. Gros F, Gilbert W, Hiatt H H, Attardi G, Spahr P F & Watson J D. *Cold Spring Harbor Symp.* 26:111-26, 1961.
3. Novelli G D, Kameyama T & Eisenstadt J M. *Cold Spring Harbor Symp.* 26:133-43, 1961.
4. Nisman B, Cohen R, Kayser A, Fukuhara H, Demailly J, Genin C & Giron D. *Cold Spring Harbor Symp.* 26:145-50, 1961.
5. Hoagland M B. *Cold Spring Harbor Symp.* 26:153-7, 1961.
6. Plesner P. *Cold Spring Harbor Symp.* 26:159-62, 1961.
7. Darnell J E. *Cold Spring Harbor Symp.* 27:149-58, 1962.
8. Derman E, Goldberg S & Darnell J E. *Cell* 9:465-72, 1976.
9. Colter J S, Bird H H & Brown R A. *Nature* 179:859-60, 1957.
10. Wecker E. *Virology* 7:241-3, 1959.

CC/NUMBER 51
DECEMBER 19, 1983

# This Week's Citation Classic™

Hurlbert R B, Schmitz H, Brumm A F & Potter V R. Nucleotide metabolism. II. Chromatographic separation of acid-soluble nucleotides.
J. Biol. Chem. 209:23-39, 1954.
[McArdle Memorial Laboratory, Medical School, Univ. Wisconsin, Madison, WI]

A chromatographic method employing gradient elution of Dowex-1 (formate) columns was developed to systematically resolve nucleoside mono-, di-, and triphosphates. Extracts of tissues were discovered to contain all these derivatives of cytidine, guanosine, and uridine, as well as of adenosine. Related work showed these to be potentially direct metabolic precursors of RNA purines and pyrimidines. [The *SCI*® indicates that this paper has been cited in over 875 publications since 1961.]

Robert B. Hurlbert
Department of Biochemistry
University of Texas
M.D. Anderson Hospital and
Tumor Institute
Houston, TX 77030

September 12, 1983

"Several factors converged at the time of this work to make development of the methodology feasible. First was the commercial availability of the Dowex line of ion-exchange resins, originally developed for nonbiological purposes and then shown by Waldo E. Cohn at the Oak Ridge National Laboratory to be useful for separations of nucleic acid derivatives.[1,2] Second was our purposeful use of the new principle of gradient elution to fan out smoothly *all* the small molecular weight anionic components of tissue extracts, combined with the use of concentrated volatile eluants (of rather low eluting ability but high buffering capacity) to permit complete resolution of co-eluting compounds by rechromatography at a different pH. Third was our suspicion that the immediate precursors (not yet known) of the nucleic acids ought to be lurking in there somewhere and might be related to the nucleotides whose chemistry had recently been described.[3,4] Fourth was the availability of radioisotopes which helped us sort out relationships by following label from $^{14}$C-orotic acid to pyrimidine nucleotide peaks to RNA.[5] It is gratifying that the resolution obtained then was as good as obtained now with modern HPLC, with two differences: what took two days then can be done in two hours with greater sensitivity now.

"I especially want to mention Hanns Schmitz and Anne Brumm (both now deceased, Anne a victim of cancer). We predicted by analogy that triphosphates of cytidine and guanosine ought to exist and should appear just before and after the ATP emerged. I recall vividly Hanns's excitement when a bulge in the ultraviolet 275/260 ratio first revealed CTP to exist in tissues.[6] Also, his almost tearful disappointment after he reported that discovery (last paper on the program at the 1953 American Association for Cancer Research meeting) because the meager audience after the main exodus consisted almost entirely of our lab associates. Anne quietly and with dedication did much of the analytical work on this and related papers;[5,6] both she and Hanns would feel rewarded to learn of this recognition by *Citation Classics*™, that all those overtime hours did help measurably in the massive struggle to gain control over cancer. Van Potter was astute as usual; he let us play around with the procedure at first when it would have been more expedient to stick to business on his original plans as funded, which were quite different. Waldo was somewhat ambiguous when he remarked that he had thought about applying his techniques to tissue extracts but regarded it as 'pearl diving.' It is even harder now to get a grant application funded for such ventures.

"Although then the procedure discovered new nucleotides[7] and helped establish order in the field of nucleic acid biosynthesis, I think its current significance is that it gives a picture, a snapshot, of the metabolic status of the tissue. The simple ribo- and deoxyribonucleotides are not only intermediates in RNA and DNA synthesis, but they and their derivatives are involved as reactants or regulators in almost all metabolic pathways, serving as systemic interconnectors. Thus, their relative and absolute concentrations both reflect and control the physiological function of a cell, its stage in the cell cycle, and its growth potential. Combined with pulse-labeling and treatments by antimetabolites, the method (in modern form) can provide more dynamic insights into control of cell growth. I don't believe it has even yet been fully exploited for this purpose."

1. Cohn W E. The anion exchange separation of ribonucleotides. *J. Amer. Chem. Soc.* 72:1471-8, 1950. (Cited 195 times.)
2. ─────. Some results of the applications of ion-exchange chromatography to nucleic acid chemistry. *J. Cell. Comp. Physiol.* 38(Suppl. 1):21-40, 1951.
3. Baddiley J. Chemistry of nucleosides and nucleotides. (Chargaff E & Davidson J N, eds.) *The nucleic acids: chemistry and biology.* New York: Academic Press, 1955. Vol. I. p. 137-90.
4. Brown D M & Todd A R. Evidence on the nature of the chemical bonds in nucleic acids. (Chargaff E & Davidson J N, eds.) *The nucleic acids: chemistry and biology.* New York: Academic Press, 1955. Vol. I. p. 409-45.
5. Hurlbert R B & Potter V R. Nucleotide metabolism. I. The conversion of orotic acid-6-C$^{14}$ to uridine nucleotides. *J. Biol. Chem.* 209:1-21, 1954.
6. Schmitz H, Hurlbert R B & Potter V R. Nucleotide metabolism. III. Mono-, di-, and triphosphates of cytidine, guanosine, and uridine. *J. Biol. Chem.* 209:41-54, 1954.
7. Mandel P. Free nucleotides in animal tissues. *Progr. Nucleic Acid Res. Mol. Biol.* 3:299-334, 1964.

# This Week's Citation Classic

CC/NUMBER 5
FEBRUARY 2, 1981

Randerath K & Randerath E. Ion-exchange chromatography of nucleotides on poly(ethyleneimine)-cellulose thin layers. J. Chromatography 16:111-25, 1964.
[J. C. Warren Labs., Huntington Memorial Hosp., Harvard Univ., and Biochem. Res. Dept., Harvard Med. Sch. at Massachusetts General Hosp., Boston, MA]

This paper showed that a great number of naturally occurring ring mononucleotides can be separated and identified by poly(ethyleneimine)-cellulose thin-layer chromatography. $R_F$ data for 33 compounds are given, and the factors are discussed which influence the mobility under different elution conditions. The method is compared with other present techniques for separating nucleotides. [The *SCI®* indicates that this paper has been cited over 280 times since 1964.]

Kurt Randerath
Department of Pharmacology
Baylor College of Medicine
Texas Medical Center
Houston, TX 77030

October 17, 1980

"In 1961, while working in F. Cramer's laboratory at the Technische Hochschule in Darmstadt, Germany, I extensively used paper and ion-exchange column chromatography to separate nucleotides. Thin-layer chromatography of lipophilic compounds on silica gel had just then become popular. Hypothesizing that ion-exchange thin layers might give more rapid separations of small amounts of nucleotides than paper or columns, I first attempted to prepare thin layers of ion-exchange materials. ECTEOLA- and DEAE-cellulose layers were soon found to provide rapid separations of nucleotides,[1] but the properties of layers of such chemically substituted ion-exchanges varied somewhat from batch to batch. I therefore explored the possibility of preparing anion-exchange thin layers by incorporating defined cationic chemicals into cellulose thin layers, in analogy to the preparation of reversed-phase thin layers by impregnating silica gel with lipids. Cellulose layers containing lipophilic amines such as n-tetradecylamine were found to give reproducible nucleotides separations but because of the hydrophobicity of such layers the separations were slow. Hydrophilic amines appeared preferable, therefore, but how could one prevent such compounds from migrating with the aqueous mobile phase during chromatography? The idea occurred to me that hydrophilic amines of sufficiently high molecular weights might be physically trapped in thin-layer powders and I immediately started searching for such a material. Only a single compound, polyethyleneimine (PEI, mol. weight 30,000) was commercially available. I remember my disappointment at the low molecular weight of PEI but since the price was only about 4 deutsche marks (then $1), I ordered it anyway. Without knowing that PEI is adsorbed strongly to cellulose (which is the basis for its extensive use in the paper industry), I made a series of cellulose thin layers containing different amounts of PEI.

"I remember my amazement and disbelief when on Saturday, July 28, 1962, I first observed how a mixture of AMP, ADP, and ATP, giving rise to spots smaller than the capital letters in this article, separated completely on a PEI-cellulose thin layer in less than five minutes by development in 1 M NaCl solution. Clearly, this technique had a higher resolving power for nucleotides than any then existing technique.

"During 1963, Erika Randerath and I carried out a systematic study of nucleotide separations on PEI-cellulose thin layers. This work was detailed in this highly cited paper, an earlier version of which had been rejected by another journal. This and our subsequent work on nucleotide separations was done at the Massachusetts General Hospital in Boston where we had moved early in 1963 to join research groups led by H.M. Kalckar and P.C. Zamecnik. A review of this work was published subsequently.[2] It is particularly gratifying that the first full paper written after our immigration to the US has become a *Citation Classic*. I think this may be due to the fact that the paper introduced a powerful, yet simple and inexpensive separation technique, which soon became indispensable to many investigators in nucleotide and nucleic acid research."

1. Randerath K. Thin-layer chromatography of nucleotides. *Angew. Chem. Int. Ed.* 1:435-9, 1962.
2. Randerath K & Randerath E. Thin-layer separation methods for nucleic acid derivatives. *Meth. Enzymol.* 12A:323-47, 1967.

# This Week's Citation Classic

CC/NUMBER 3
JANUARY 19, 1981

Sanger F, Brownlee G G & Barrell B G. A two-dimensional fractionation procedure for radioactive nucleotides. *J. Mol. Biol.* 13:373-98, 1965.
[Medical Research Council Laboratory of Molecular Biology, Cambridge, England]

A two-dimensional procedure for fractionating $^{32}$P-labelled oligonucleotides is described. The first dimension uses high-voltage electrophoresis on cellulose acetate strips and the second electrophoresis on DEAE-cellulose paper. The method is used to fractionate ribonuclease digests of RNA, and micro techniques are described for analysing and sequencing the purified oligonucleotides. [The *SCI®* indicates that this paper has been cited over 870 times since 1965.]

---

F. Sanger
MRC Laboratory of Molecular Biology
University Postgraduate
Medical School
Cambridge CB2 2QH
England

December 16, 1980

"This paper represents my first venture into the field of nucleic acids. My previous work had been concerned with amino acid sequence determination in proteins and I was particularly interested in the development of methods. By 1965, techniques for protein sequencing were well developed and already somewhat standardised, whereas little had been done on the determination of sequences in the other biologically important polymeric molecules—the nucleic acids. This seemed a worthwhile challenge and, together with a bright and enthusiastic PhD student, G.G. Brownlee, and B.G. Barrell, who was a technical assistant at that time, I set to work on the problem.

"Prior to this work the only RNA sequence that had been determined was that of the alanine tRNA, which was done by Holley and his collaborators.[1] The methods used by them were largely developed for protein chemistry and involved rather laborious procedures such as counter-current distribution and ion-exchange chromatography. It seemed to us that in order to be able to sequence the many large nucleic acids present in living matter more rapid and simple methods were needed that could be applied to small amounts of material. In particular we needed a method of fractionating the complex mixture of oligonucleotides obtained by partial digestion of RNA. One important development described in this paper was the use of $^{32}$P-labelled RNA of high specific activity. This made it possible to work on a small scale and to use two-dimensional 'paper' fractionation techniques, which had high resolving power as well as being rapid and simple to carry out. As $^{32}$P can be detected at very low concentrations by radioautography and assayed by counting techniques, $^{32}$P-labelled nucleic acids have been used in most subsequent studies on sequences of RNA and DNA.

"In general, nucleotides do not fractionate well by 'paper' methods and we spent a good deal of time trying out different systems of chromatography and electrophoresis on paper and various modified papers, but usually the products were not well resolved and the radioautographs produced contained only streaks and blotches rather than well-defined bands or spots. The turning point in this work came one morning when Barrell showed me a film he had developed that contained a large number of clear, well-defined spots. This was what we had been looking for and the two-dimensional fractionation we had used formed the basis of the method described in this paper. It was possible to elute the purified nucleotides from the paper and we developed micro methods for analysing and sequencing them. The various methods described in this paper have formed the basis for many subsequent studies on RNA sequences."[2,3]

1. Holley R W, Apgar H, Everett G A, Madison J T, Marquisee M, Merrill S H, Penswick J R & Zamir A. Structure of a ribonucleic acid. *Science* 147:1462-5, 1965.
2. Adams J M, Jeppesen P G N, Sanger F & Barrell B G. Nucleotide sequence from the coat protein cistron of R17 bacteriophage RNA. *Nature* 223:1009-14, 1969.
3. Fiers W, Contreras R, Duerinck F, Haegeman G, Iserentant D, Merregaert J, Min Jou W, Molemans F, Raeymaekers A, Van den Berghe A, Volckaert G & Ysebaert M. Complete nucleotide sequence of bacteriophage MS2 RNA: primary and secondary structure of the replicase gene. *Nature* 260:500-7, 1976.

Number 26     **Citation Classics**     June 27, 1977

Burton, Kenneth. A study of the conditions and mechanism of the diphenylamine reaction for the colorimetric estimation of deoxyribonucleic acid. *Biochemical Journal* 62:315-22, 1956.

The color reaction of Dische (1930) between diphenylamine and deoxyribonucleic acid (DNA) has been studied and modified. The principal modificaations are to add acetaldehyde and to perform the reaction for several hours at 30°C instead of for 3-10 min. at 100°C. Using this modified reaction, the author studied the conditions for the quantitative extraction of DNA from bacteria. [The *SCI*® indicates that this paper was cited 5,037 times in the period 1961-1975.]

Professor Kenneth Burton
Department of Biochemistry
Ridley Building
University of Newcastle Upon Tyne
NE1 7RU England

December 23, 1976

"Of course, I am pleased that one of my papers has been cited so often and am less pleased that most of my other papers have collected very few citations. On reflection, it seems clear that my 1956 paper has not affected the development of biochemistry, despite its frequent citation. If it had not been written, other available methods would have been used instead. It appeared at a time when it was becoming popular to measure DNA. A simple and specific method for measuring, say, arabinose would have been less fashionable but probably of greater value because of the lack of a simple alternative for the measurement of this sugar.

"The title of my paper reflected the fact that the diphenylamine colour reaction of Zaccharias Dische was already a well-used method of estimating DNA and was fairly satisfactory. It was only by chance that I modified it. One evening, I had not enough time to complete the colour development by heating at 100°C and so I left various unheated mixtures of DNA and reagents on the bench overnight. I was astonished next morning to find beautiful intense blue colours with low 'blanks' and straightaway I substituted overnight incubation at 30°C for the development at the higher temperature. (I was not then aware that Patterson had done much the same in 1948. Soon after, I returned from Chicago to England and started work in Oxford. It was an awful moment to find that only very faint colours were formed at 30°C.

"While a student at Cambridge, I had been privileged to hear Gowland Hopkins tell of the discovery of tryptophan and how this had started with studying a temperamental colour reaction. Hopkins and Cole had discovered that the glacial acetic acid needed prior exposure to strong sunlight or--better still--the addition of a small quantity of a suitable aldehyde. The acetic acid I had used in Chicago had indeed been stored in strong sunlight not to be found in Oxford--especially in the winter. Eventually, I selected acetaldehyde as an additional component of the reagent mixture, so obtaining reproducible intense colours on development at 30°. The acetaldehyde probably initiates a chain reaction and many other substances (e.g. hydrogen peroxide) can be used instead.

"The fact that DNA gave a colour that was ostensibly due to deoxyribose on incubation at such a low temperature led me to study whether free deoxyribose was produced and how the backbone chain of the DNA was split. Although I found that free deoxyribose could not be an intermediate, I discovered the diphenylamine-formic acid degradation of DNA which produced pyrimidine tracts quantitatively. This has been of limited value in DNA sequence work (e.g. by Southern in the analysis of satellite DNA) but it has been totally eclipsed by the methods now being so brilliantly developed and exploited by Sanger and his colleagues."

# This Week's Citation Classic

CC/NUMBER 42
OCTOBER 19, 1981

Wyatt G R. The purine and pyrimidine composition of deoxypentose nucleic acids.
*Biochemical J.* 48:584-90, 1951.
[Agricultural Research Council Plant Virus Research Unit,
Molteno Institute, University of Cambridge, Cambridge, England]

A technique for the determination of the purine and pyrimidine composition of DNA is described, using acid hydrolysis, separation of the bases by chromatography on filter paper, detection in UV light, and spectrophotometric estimation. Results for DNA from eight animal and one plant source are reported. [The *SCI*® indicates that this paper has been cited over 865 times since 1961.]

---

G.R. Wyatt
Department of Biology
Queen's University
Kingston, Ontario K7L 3N6
Canada

September 6, 1981

"I am delighted to learn that my 1951 paper has been cited so often, so many years after its publication. Reading this paper now, one feels that it stems from another era, both because of its primitive view of DNA and its perambulating style which would horrify present-day editors.

"I had gone from Canada to Cambridge to pursue a PhD in insect virology in the Agricultural Research Council Virus Unit. My attempts to obtain meaningful results from sick caterpillars proved frustrating, but I fell under the influence of Roy Markham, a young biochemist working on plant viruses (who unfortunately died of cancer in 1979). Markham held the then radical view that understanding of viruses would come through study of their nucleic acids, and he was developing a technique for quantitative analysis of RNA by separation of bases and nucleotides on paper chromatograms, detection with UV light, and estimation in the newly arrived Beckman spectrophotometer. Since the insect viruses in which I was interested contained DNA, Markham suggested that I might try adapting the method for DNA.

"Analysis of DNA by paper chromatography had already been reported from the laboratories of Rollin Hotchkiss[1] and Erwin Chargaff,[2] but the techniques seemed open to improvement. I found that the purine and the pyrimidine bases could be obtained in good yield from a single hydrolysis in formic or perchloric acid. For the chromatographic solvent to overcome the low solubility of guanine, I tested many mixtures of various alcohols with strong acids and selected a combination of isopropanol and HCl that gave compact spots and a well-spaced separation of guanine, adenine, cytosine, and thymine. Animal and plant DNA hydrolysates showed an unexpected fifth spot, which was identified as 5-methylcytosine.[3] I analyzed such DNAs as I could easily obtain and wrote up the method and first results before turning to look at the DNA of some insect and other viruses.

"I believe that this paper has often been cited for methods: conditions for quantitative hydrolysis of DNA, a solvent useful for thin-layer as well as paper chromatography, and a table of extinction coefficients of the DNA bases. In addition, it reported the compositions of several DNAs long before the simpler methods of determination using $T_m$ and buoyant density were developed. This information was useful to Watson and Crick[4] and many others. The discussion in my paper is extraordinarily cautious, however. The remark, 'It is tempting to speculate whether DNA composition may bear some direct relation to genetic structure,' is sufficient to remind us of the DNA revolution that separates 1981 from 1951.

"Those reading my recipe for the isopropanol-HCl solvent may have wondered why 16°C is specified. Working in Britain in 1948-1950, I had come to accept this as a normal room temperature; one day, my notebook records 11°C!

"My sole authorship of this paper reflects the generosity of my advisors; Markham declined to add his name though he gave much help. Data on the base compositions of DNAs from numerous organisms are now available."[5]

1. Hotchkiss R D. The quantitative separation of purines, pyrimidines, and nucleosides by paper chromatography. *J. Biol. Chem.* 175:315-32, 1948.
2. Chargaff E, Vischer E, Doniger R, Green C & Misani F. The composition of the desoxypentose nucleic acids of thymus and spleen. *J. Biol. Chem.* 177:405-16, 1949.
3. Wyatt G R. Recognition and estimation of 5-methylcytosine in nucleic acids. *Biochemical J.* 48:581-4, 1951.
4. Watson J D & Crick F H C. Molecular structure of nucleic acids. *Nature* 171:737-8, 1953.
5. Shapiro H S. Distribution of purines and pyrimidines in deoxyribonucleic acids. (Fasman G D, ed.) *CRC handbook of biochemistry and molecular biology.* Cleveland, OH: CRC Press, 1976. Vol. 2. p. 241-83.

## This Week's Citation Classic

CC/NUMBER 33
AUGUST 16, 1982

Sundaralingam M. Stereochemistry of nucleic acids and their constituents. IV. Allowed and preferred conformations of nucleosides, nucleoside mono-, di-, tri-, tetraphosphates, nucleic acids and polynucleotides. *Biopolymers* 7:821-60, 1969.
[Department of Chemistry, Case Western Reserve University, Cleveland, OH]

Conformational analysis based on single crystal X-ray data of nucleic acid constituents and model compounds, and fiber diffraction data of nucleic acids and polynucleotides, revealed that the nucleoside is constrained to two predominant conformational classes, which generate two favored families of nucleic acid double helices. A comprehensive conformational nomenclature for polynucleotides was developed. [The *SCI*® indicates that this paper has been cited in over 620 publications since 1969.]

Muttaiya Sundaralingam
Department of Biochemistry
College of Agricultural and Life Sciences
University of Wisconsin
Madison, WI 53706

April 23, 1982

"After obtaining my PhD with G.A. Jeffrey in Pittsburgh, I joined Lyle H. Jensen in Seattle in 1962 where I carried out a high precision X-ray analysis of the nucleotide 3'-CMP. A conformational analysis of this molecule led me to think about the stereochemistry of nucleic acids. At that time, the conformational analysis of polynucleotides was regarded as highly complex because of the rotations around the numerous single bonds in the polynucleotide chain. The Watson-Crick base pairing principle was the foundation of nucleic acid secondary structures, but the stereochemical principles governing (a) nucleotide conformation, (b) internucleotide phosphodiesters, and (c) their relation to chain folding were fragmentary. I started to build a unified picture[1,2] of the structural principles governing nucleic acid conformations based on crystallographic studies of nucleotides and molecular modeling. I recognized the crucial role of the sugar in the determination of the conformation of the nucleotide unit, and, ultimately, the polynucleotide sugar-phosphate backbone.

"After a brief spell with Robert Langridge in Boston, I moved in 1966 to Case Western Reserve University where I continued my systematic investigation of the structures of the nucleic acid constituents. Although my detailed paper was submitted to *Biopolymers* in 1968, I had already presented the basic concepts and much of the data in 1966 at the International Symposium on DNA, Jena, German Democratic Republic, and the University of Madras, India, and in 1968, at the ACA summer meeting, Buffalo.

"I found that the restricted rotations about the single bonds (conformational wheels) and the correlations among them (two-dimensional pairwise conformational correlation maps between adjacent bond rotations) greatly reduced the number of conformations preferred for the nucleotide building blocks and the configurations accessible to polynucleotides. In the absence of 3',5'-dinucleoside monophosphate structures, I constructed the internucleotide phosphodiester map ($\omega',\omega$) from model phosphodiester compounds. This map gave the favored regions for both right- and left-handed helices as well as for loop-forming phosphodiesters. I found that the helical conformation was an intrinsic property of the sugar-phosphate backbone, and that the energetics of the sugar-phosphate backbone are important, in addition to base stacking and base pairing interactions, in determining polynucleotide architecture.

"I think this paper is frequently cited because it provided (a) a comprehensive conformational nomenclature for polynucleotides, (b) a foundation for the conformational analysis of nucleotide and polynucleotide structures, and (c) a basis for future theoretical and physicochemical studies of polynucleotides including dynamics in nucleic acids.

"Since the publication of this paper, scores of crystal structures of nucleic acids, including oligonucleotides and transfer RNA have been determined by numerous able colleagues in my group as well as in other laboratories. It is satisfying to see that these studies have amplified and placed on a firmer basis the structural principles of the nucleic acids developed in this paper."[3,4]

1. **Sundaralingam M.** Conformations of the furanose ring in nucleic acids and other carbohydrate derivatives in the solid state. *J. Amer. Chem. Soc.* **87**:599-606, 1965.
2. **Sundaralingam M & Jensen L H.** Stereochemistry of nucleic acid constituents. II. A comparative study. *J. Mol. Biol.* **13**:930-43, 1965.
3. **Sundaralingam M.** Nucleic acid principles and transfer RNA. (Srinivasan R, ed.) *Biomolecular structure, conformation, function & evolution.* Oxford: Pergamon Press, 1980. Vol. 1. p. 259-82.
4. **Sundaralingam M & Westhof E.** The nature of the mobility of the sugar and its effects on the dynamics and functions of RNA and DNA. (Sarma R H, ed.) *Biomolecular stereodynamics.* New York: Adenine Press, 1981. p. 301-26.

# This Week's Citation Classic

CC/NUMBER 3
JANUARY 17, 1983

Wetmur J G & Davidson N. Kinetics of renaturation of DNA.
*J. Mol. Biol.* 31:349-70, 1968.
[Gates, Crellin and Church Laboratories of Chemistry, California Institute of Technology, Pasadena, CA]

This paper describes a comprehensive investigation of the various phenomena which may affect the second order kinetics of renaturation of complementary DNA strands. These phenomena include properties of the DNA, such as strand length and sequence complexity, and properties of the solvent, such as temperature, ionic strength, and viscosity. [The *SCI®* indicates that this paper has been cited in over 855 publications since 1968.]

---

James G. Wetmur
Department of Microbiology
Mount Sinai School of Medicine
of the City University of New York
New York, NY 10029

November 29, 1982

"The work described in this paper was carried out in the laboratory of Norman Davidson when I was a graduate student at the California Institute of Technology. Norman's research had changed from 'pure' physical chemistry to biophysical chemistry of nucleic acids. Yet each of his students took quantum and statistical mechanics and built some sort of machine shortly after his or her arrival in the laboratory. We thought of DNA renaturation as another, albeit important, chemical reaction to which physical principles could be applied in order to elucidate the reaction mechanism. I set out to investigate any variable which might affect DNA renaturation rates. Incidentally, even the T-jump apparatus which I had built turned out to be useful for studying kinetics of some of the faster renaturation reactions.

"We confirmed that DNA renaturation was nearly a second order reaction. The basic mathematical description of such a reaction, as well as the reciprocal plot for determining the rate constant, has been known for about a century.[1] We used these equations to describe the nucleation reaction between complementary sequences in two DNA molecules as well as to describe the total reaction. The equation relating the nucleation rate constant, $k_N'$, and the observed rate constant, $k_2$, was found to be

$$k_2 = k_N' L^{0.5}/N$$

where L and N are the DNA length and sequence complexity respectively.

"An increase in sequence complexity translates into a decrease in the concentration of any particular nucleation site. Britten and Kohne[2] also explored this variable over a wide range of complexities and invented the convenient Cot plot for simultaneous display of the reactions. $k_N'$ depends on the temperature, ionic strength, and solvent (microscopic) viscosity but not on the DNA itself. The only remaining variable of significance is the length of the single strands. The square root dependence came as a surprise. If the molecules zipper up to their ends after each nucleation, why $L^{0.5}$ instead of L? After many discussions, I was persuaded by Norman that an excluded volume effect might limit the availability of nucleation sites. This hypothesis fits well into the framework of all that is known concerning the rate of reassociation of nucleic acids.[3] Further studies with molecules of various shapes and sizes[4] have agreed with the excluded volume hypothesis. A practical offshoot of the studies of microscopic viscosity, which affects $k_2$, and macroscopic viscosity, which does not, is a method for accelerating DNA renaturation.[5]

"I think that this paper is highly cited because it describes the practical and theoretical details of an important process in one place. The paper was the result of three years of work. We avoided the temptation to publish the results piecemeal, a luxury too seldom afforded these days. The work itself was to a great extent a product of the environment in Norman's laboratory in the mid-1960s. I could not have done the theoretical work without the physical chemistry training. I could not have done the experimental work without the willing cooperation and assistance of both my friends in the laboratory and Norman's colleagues throughout Caltech."

---

1. **van 't Hoff J H.** *Studies in chemical dynamics.* London: Williams & Norgate, 1896. p. 12.
2. **Britten R J & Kohne D E.** Repeated sequence in DNA. *Science* 161:529-40, 1968.
3. **Wetmur J G.** Hybridization and renaturation kinetics of nucleic acids. *Annu. Rev. Biophys. Bioeng.* 5:337-61, 1976.
4. **Kinberg-Calhoun J & Wetmur J G.** Circular, but not circularly permuted, DNA reacts slower than linear DNA with complementary linear DNA. *Biochemistry—USA* 20:2645-50, 1981.
5. **Wieder R & Wetmur J G.** One hundred-fold acceleration of DNA renaturation in solution. *Biopolymers* 20:1537-47, 1981.

# Citation Classics

Number 11 — March 14, 1977

**Gillespie D & Spiegelman S.** A quantitative assay for DNA-RNA hybrids with DNA immobilized on a membrane. *Journal of Molecular Biology* 12:829-42, 1965.

A method is described for performing RNA-DNA hybridization with DNA immobilized on a nitrocellulose membrane. The method is simple and convenient and eliminates the competing DNA reaction, allowing reliable quantitation of the RNA-DNA hybridization reaction. [The *SCI®* indicates that this paper was cited 1,227 times in the period 1961-1975.]

---

Dr. David Gillespie
National Cancer Institute
National Institutes of Health
Building 37, Room 6B04
Bethesda, Maryland 20014

December 3, 1976

"The technique of RNA-DNA hybridization using DNA immobilized on a nitrocellulose membrane was developed through insight, hard labor, and a stroke of luck. Most of the insight was provided by Sol Spiegelman, most of the labor by Sally Gillespie, and most of the luck by my errors. Sol immediately recognized the application of an article by Roy Britten describing the immobilization on glass of poly (U) networks formed by irradiation with ultraviolet light and pressured me to form similar networks of denatured DNA on nitrocellulose membranes. After several experiments, all outrageously successful, I inadvertently eliminated the irradiated step and lo! the magic of DNA immobilization on nitrocellulose began. I use the word 'magic' advisedly for even today we do not understand the chemical basis for the DNA immobilization.

"I feel the reason our paper is so often cited is that the protocol we worked out has survived as the simplest, most convenient and most versatile form of the technique. For this, both Sol and I owe my wife, Sally, a large measure of gratitude. She did much of the detailed work that led to the success of the technique and she was never satisfied with an aspect of the method that simply 'worked.' To her the best form of technique was always apparent and striven for; to me this insight never came before repeated botching, and when we reached dead ends Sol was always there solving our problems with one or two words....

"It is one thing to recognize reasons for the success of a classic retrospectively, but it is quite another to envision them during the course of the project or even before it begins. I must admit that as a graduate student I didn't recognize the potential of what I was doing at the time and, in fact, I am still amazed and somewhat bewildered by the longevity of our paper. Sol saw it, however, right from the beginning. He must have said--about once a month in order that I remain sufficiently buoyed to continue--'Gillespie, I'm going to make you famous.' I looked upon the method primarily as a neat trick that would allow me to discover some 'really important' facts of biological interest. These facts, of course, remain to this day in the Library of Congress with my thesis....

"As I look back upon that period while attempting to decide why a 'classic' becomes one, especially in the area of methodology, I keep returning to the notion of developing an unimprovable method. But this notion is so obvious that it would seem to follow that every person developing a technique with the potential of reasonably wide use would end up with such a classic. This leads me to think that the distinction between a classic and a quickly outmoded method lies in the ability of the investigators to see the uses to which the method will be put and evaluate particular parameters accordingly and, as importantly, to take heed of the little irregularities that lead to significant improvements. I mentioned the (lack of) irradiation and magnesium as bits of luck and wisdom earlier, but there were many other smaller points that could have relegated us to the status of a 'good 1965 paper.' For example, we noticed once that DNA filters we had kept for a couple of days in a drawer gave more hybridization to RNA than those DNA filters that were freshly made. The difference was small enough to be ignored, but we didn't ignore it and it led to 'baking' the DNA filters, driving all the water off and causing the DNA to remain more stably bound to the filter during hybridization. Had we not picked up this and several other little irregularities surely someone in 1966 or 1967 would have, and their version would have been the one cited from then on."

# This Week's Citation Classic

Denhardt D T. A membrane-filter technique for the detection of complementary DNA. *Biochem. Biophys. Res. Commun.* 23:641-6, 1966.
[Biological Laboratories, Harvard University, Cambridge, MA]

This paper describes a technique for detecting specific DNA sequences in solution by annealing them to nitrocellulose filters carrying complementary DNA sequences. Prior to the hybridization the filters are incubated with a solution of Ficoll, polyvinylpyrrolidone, and bovine serum albumin. [The *SCI*® indicates that this paper has been cited in over 1,235 publications since 1966.]

David T. Denhardt
Cancer Research Laboratory
University of Western Ontario
London, Ontario N6A 5B7
Canada

August 20, 1982

"I went to Harvard University in late-1964 fresh from my PhD at the California Institute of Technology with the goal of developing an *in vitro* DNA replicating system. For the preceding four years I had worked in Bob Sinsheimer's laboratory on aspects of ΦX174 replication *in vivo* and I wanted to develop a more biochemical approach. The model of ΦX replication that had evolved from our studies on the intact cell had led me to believe that DNA replication was occurring on the cell membrane. Now we needed a method to detect DNA replication in cell extracts. The incorporation of label from radioactive triphosphates into DNA would be ideal if we could develop a simple quantitative procedure to distinguish ΦX DNA from *E. coli* DNA.

"About that time, a publication by Gillespie and Spiegelman[1] appeared. They had extended earlier studies of Nygaard and Hall[2] by first binding single-stranded (SS) DNA to nitrocellulose filters and then using them to quantify complementary RNA sequences in solution. I realized I could use a similar procedure if I could block the nonspecific sticking of the SS DNA to the nitrocellulose without interfering with the annealing reaction. At high ionic strengths denatured DNA and poly[rA] adhere to nitrocellulose (which is also acetylated and fairly hydrophobic) because of the open, unstacked character of the hydrophobic bases; the bases in RNA are more stacked and less available for hydrophobic interactions.[3]

"To prevent the nonspecific binding of the denatured DNA I cast about for suitable compounds. Among the many I tried were Ficoll (a polymer of sucrose), polyvinylpyrrolidone (PVP) (I thought it might resemble an array of bases), and bovine serum albumin (BSA). BSA alone had a profound effect on the nonspecific sticking and together with Ficoll and PVP reduced the background to less than one percent. Before using this procedure to detect ΦX DNA synthesis *in vitro* I thought it wise to demonstrate that it could be used to follow ΦX DNA synthesis *in vivo*. It worked well and the results were published together with the technique in *Biochemical and Biophysical Research Communications*; it was my third independent publication. I was so overwhelmed with reprint requests that the only way I could afford to honor them all was to reduce the six pages to one photographically and send out one-page Xerox copies—perhaps the first 'miniprint' reprint.

"Despite the reprint requests, I saw very few applications of the technique until recombinant DNA technology came into use. Examples of recent applications of DNA-DNA hybridization include the analysis of Southern Blots and the detection of specific cloned sequences in plaques or colonies.[4] Some improvement in the signal-to-noise ratio has been obtained by increasing the concentrations of the several components and including dodecyl sulfate and nonhomologous DNA or poly[rA] in the hybridization reaction. Dextran sulfate also helps to reduce the background and to accelerate the rate of hybridization.[5] This publication has been widely cited because it describes a simple and inexpensive, yet effective, procedure to detect specific DNA sequences."

1. Gillespie D & Spiegelman S. A quantitative assay for DNA-RNA hybrids with DNA immobilized on a membrane. *J. Mol. Biol.* 12:829-42, 1965.
2. Nygaard A P & Hall B D. A method for the detection of RNA-DNA complexes. *Biochem. Biophys. Res. Commun.* 12:98-104, 1963.
3. Cashion P, Sathe G, Javed A & Kuster J. Hydrophobic affinity chromatography of nucleic acids and proteins. *Nucl. Acid. Res.* 8:1167-85, 1980.
4. Wu R, ed. Recombinant DNA. (Whole issue.) *Methods Enzymol.* 68, 1979. 555 p.
5. Wahl G M, Stern M & Stark G R. Efficient transfer of large DNA fragments from agarose gels to diazobenzyloxymethyl-paper and rapid hybridization by using dextran sulfate. *Proc. Nat. Acad. Sci. US* 76:3683-7, 1979.

# This Week's Citation Classic

CC/NUMBER 1
JANUARY 2, 1984

Edmonds M, Vaughan M H, Jr. & Nakazato H. Polyadenylic acid sequences in the heterogeneous nuclear RNA and rapidly-labeled polyribosomal RNA of HeLa cells: possible evidence for a precursor relationship.
*Proc. Nat. Acad. Sci. US* **68**:1336-40, 1971.
[Department of Biochemistry, University of Pittsburgh, PA]

This paper showed that polyA sequences about 200 nucleotides long were covalently bound to messenger RNA (mRNA) and to much larger RNAs in the HeLa cell nucleus. The observation that each mRNA contained a single polyA sequence and the similarity in polyA sequences in the two classes of RNA led us to propose a model in which mRNA molecules are processed from the much longer polyadenylated RNA molecules in the nucleus. [The *SCI*® indicates that this paper has been cited in over 505 publications since 1971.]

---

Mary Edmonds
Department of Biological Sciences
University of Pittsburgh
Pittsburgh, PA 15260

July 27, 1983

"Our discovery of polyA sequences in messenger RNA (mRNA) and heterogeneous nuclear RNA (hnRNA) was the culmination of work begun in the late 1950s at the Institute of Research of Montefiore Hospital in Pittsburgh where Richard Abrams and I were examining extracts of animal cells for RNA synthesis. Within a short time we had found such an activity in extracts from mouse ascites tumor cells.[1] Much to our surprise, however, the enzyme made only polyA and not RNA.[2] Somewhat later we found that the polyA polymerase purified from such extracts actually contained polyA molecules that could be found in boiled extracts of the enzyme.[3] These extracts would stimulate new polyA synthesis when added back to an RNA depleted polyA polymerase.

"These last observations greatly influenced my subsequent thinking about polyA for they suggested that cells must actually contain polyA, and also that such polyA might be relatively easy to isolate. At about that time, Peter Gilham described a method for covalently attaching oligo dT sequences to cellulose that struck us as the ideal way to do it.[4] We quickly prepared oligo dT cellulose by his method and found it would remove polyA from our boiled enzyme extracts. The polyA eluted from this cellulose could then restore polyA synthesis of an RNA depleted polyA polymerase.[5]

"The simplicity and specificity of this method suggested the obvious approach for looking for polyA sequences in animal cells. However, it was not until 1966, after I had moved to the University of Pittsburgh, that I began a systematic search for polyA in $^{32}$P-labeled RNAs from mouse ascites tumor cells. I soon found that small quantities of polyA could be recovered from ribonuclease digested RNA from both nuclei and cytoplasm after passage over oligo dT cellulose.[6]

"In 1967 Maurice Vaughan, who had recently joined our department, helped me carry out the first gel electrophoretic analyses of such polyA. To our surprise the polyAs from both nuclear and cytoplasmic RNA were quite homogeneous and rather large (about 200 nucleotides). Having recently come from J.E. Darnell's laboratory in New York, Vaughan was familiar with the very large RNA molecules that turn over rapidly in the nucleus (hnRNA) and suggested we examine different size classes of this RNA for polyA sequences. Indeed, similar polyA sequences turned up in all sizes of hnRNA including the very largest. Although my experiences with polyA polymerases had led me to think of polyA as a distinct set of homopolymers, the association of small polyA sequences with huge RNA molecules suggested a covalent linkage to RNA instead. Since we had already found that small added polyAs tended to co-sediment with large hnRNAs in sucrose gradients, it was clear that such artifacts had to be ruled out to establish a covalent attachment to RNA.

"In 1970 Hiroshi Nakazato, who had just arrived in my laboratory from Japan, undertook the hybridization competition experiments described in the paper that effectively ruled out intermolecular hybridization as the source of the polyA sequences in large hnRNA. With this evidence for a covalent linkage firmly established, we submitted our paper on polyA sequences in mRNA and hnRNA and proposed on the basis of the similarity of their polyA sequences that hnRNAs were the precursors of mRNA.

"The frequent citation of this paper may come from its revelation of an unsuspected structural feature of mRNA also shared by many hnRNA molecules. This, of course, provided strong support for the long suspected precursor role for hnRNA, but probably more important was the realization that the polyA sequence could be used to purify mRNA. In fact, Nakazato quickly showed how the mRNA of HeLa cells could be purified on oligo dT cellulose with the techniques now in widespread use."[7]

---

1. Edmonds M & Abrams R. Incorporation of ATP into polynucleotide in extracts of Ehrlich ascites cells. *Biochim. Biophys. Acta* **26**:226-7, 1957.
2. ——————. Polynucleotide biosynthesis: formation of a sequence of AMP units from ATP by an enzyme from thymus nuclei. *J. Biol. Chem.* **235**:1142-9, 1960.
3. ——————. Nature of a polynucleotide required for polyribonucleotide formation from adenosine triphosphate with an enzyme from thymus nuclei. *J. Biol. Chem.* **237**:2636-42, 1962.
4. Gilham P T. Complex formation in oligonucleotides and its application to the separation of polynucleotides. *J. Amer. Chem. Soc.* **84**:1311, 1962.
5. Edmonds M & Abrams R. Isolation of a naturally occurring polyadenylate from calf thymus nuclei. *J. Biol. Chem.* **238**:PC1186-7, 1963.
6. Edmonds M & Caramela M G. The isolation and characterization of AMP-rich polynucleotides synthesized by Ehrlich ascites cells. *J. Biol. Chem.* **244**:1314-24, 1969.
7. Nakazato H & Edmonds M. The isolation and purification of rapidly labeled polysome-bound RNA on polythymidylate cellulose. *J. Biol. Chem.* **247**:3365-7, 1972.

## This Week's Citation Classic

CC/NUMBER 5
FEBRUARY 1, 1982

Hewish D R & Burgoyne L A. Chromatin sub-structure. The digestion of chromatin DNA at regularly spaced sites by a nuclear deoxyribonuclease.
*Biochem. Biophys. Res. Commun.* 52:504-10, 1973.
[Sch. Biological Sciences, Flinders Univ. South Australia, Bedford Park, South Australia]

When rat liver cell nuclei were incubated such that an endogenous deoxyribonuclease digested the nuclear DNA, the DNA products were found to have a regular series of discrete sizes. The deoxyribonuclease digested purified DNA in a random manner, and it was concluded that proteins within the nuclei restricted the access of the nuclease to regularly spaced sites on the nuclear DNA. [The *SCI*® indicates that this paper has been cited over 655 times since 1973.]

Dean R. Hewish
Division of Protein Chemistry
CSIRO
Parkville, Victoria 3052
Australia

October 14, 1981

"In 1973, when we published our evidence for a repeating substructure within the cell nucleus, researchers in the field of chromatin structure were eager for any new technique which would give insight into the structure of the DNA-protein complex. Several laboratories were apparently nearing the point at which they could have carried out our experiments. It is perhaps ironic, however, that although our research was not directed toward an investigation of chromatin substructure, we were uniquely well placed to carry out such a study.

"For some years, members of our laboratory had been studying the properties of isolated rat liver cell nuclei, in order to obtain an *in vitro* system for investigating DNA replication. One type of nuclear preparation was found to be unusual, in that the nuclei contained high levels of a deoxyribonuclease which rapidly digested the nuclear DNA when magnesium and traces of calcium ions were present.[1] Obviously, the nuclei were of little value for the study of DNA synthesis, as they underwent auto-destruction under the very conditions necessary for DNA synthesis. It was noticed that, although the DNA was extensively degraded, very small fragments were not produced. We reasoned that the deoxyribonuclease must have some function in the intact cell, and I spent some time studying its properties, with few results of interest. As I was nearing the end of my PhD project, the time came to review the work and to decide whether the preparation was worth further investigation. In the course of one discussion, we decided to analyse the fragments of DNA produced by the enzyme in intact nuclei. To be honest, our recollections of our exact motives are hazy, but we were influenced by a previous publication of Williamson,[2] who had described a regular series of size classes among DNA fragments in the cytoplasm of cultured cells. We had wondered for some time whether the breakdown of the nuclear DNA in our system was in any way related to the production of those cytoplasmic fragments.

"Accordingly, I allowed a batch of nuclei to auto-digest for varying times, purified the DNA, and separated the fragments, by electrophoresis, on the basis of size. The results were dramatic. A series of discrete fragments was produced, indicating that the enzyme was limited in its action to certain sites on the chromatin and we therefore had the first evidence for a regular substructure within the nuclear chromatin. We proposed that the nuclear proteins were responsible for this regular structure, a supposition which has been shown to be correct.

"The reason for the large number of citations is clear. The technique which we introduced for probing the structure of chromatin was both very simple and effective. Other workers have since used different deoxyribonucleases to distinguish both lower and higher orders of chromatin structure and the results have correlated with electron microscopic and X-ray diffraction studies to produce a consistent and detailed picture of the subunit organization of the eukaryote chromosome. For a recent review, see the *Annual Review of Biochemistry*."[3]

1. Burgoyne L A, Waqar M A & Atkinson M R. Calcium dependent priming of DNA synthesis in isolated rat liver nuclei. *Biochem. Biophys. Res. Commun.* 39:254-9, 1970.
2. Williamson R. Properties of rapidly labelled deoxyribonuclease acid isolated from the cytoplasm of primary cultures of embryonic mouse liver cells. *J. Mol. Biol.* 51:157-68, 1970.
3. McGhee J D & Felsenfeld G. Nucleosome structure. *Annu. Rev. Biochem.* 49:1115-56, 1980.

# This Week's Citation Classic™

CC/NUMBER 51
DECEMBER 17, 1984

Lerman L S. Structural considerations in the interaction of DNA and acridines.
J. Mol. Biol. 3:18-30, 1961.
[Med. Res. Council Unit for Mol. Biol., Univ. Cambridge, England and Dept. Biophys., Univ. Colorado Med. Ctr., Denver, CO]

Molecular parameters inferred from hydrodynamics and fiber diffraction agree with the supposition that aminoacridines impose a substantial departure from the Watson-Crick helical structure upon binding to DNA. The polycyclic cation is sandwiched between otherwise adjacent base pairs in the partially unwound helix. The results are stereochemically plausible and conflict with other hypotheses [The SCI® indicates that this paper has been cited in over 950 publications since 1961.]

---

Leonard S. Lerman
Genetics Institute
225 Longwood Avenue
Boston, MA 02115

August 14, 1984

"I began to think about the structure of complexes of DNA with polycyclic aromatic molecules while engaged in studies on genetic transformation in pneumococcus. Rough stereochemical calculations confirmed the plausibility of an intercalated structure in which the double helix unwinds slightly to open a space of the appropriate thickness into which a carcinogenic polycyclic hydrocarbon (or a similar mutagen) might fit. The real problem seemed to be to devise an experiment that would provide structural evidence. While optical and dialysis-equilibrium measurements gave meaningful thermodynamic data, they did not permit decisive structural interpretations that might distinguish intercalation from any other binding geometry. Lengthening of the molecule seemed to be easily measurable without elaborate instrumentation. The upper limit of first-order binding[1] implied a length extension of 45 or 50 percent, which could be demonstrable as a change in the intrinsic viscosity of DNA, assuming that the increment in length was not compensated by increased flexibility. The attachment of nonintercalated cations would be expected to give a small decrease in the intrinsic viscosity because of diminished electrostatic repulsions between parts of the molecule. The first experiment showed a large viscosity change in the right direction, opposite to the effect of putative nonintercalators.

"Most of this paper is concerned with the examination of physical properties and parameters other than viscosity. Although no single experiment could stand by itself to demonstrate intercalation, each ruled out alternative structural hypotheses while remaining consistent with the intercalated structure. I prepared X-ray diffraction patterns from proflavine-containing DNA fibers and constructed a satisfactory molecular model during my sojourn with Crick at the MRC laboratory in Cambridge. In subsequent papers,[2-4] I reported the use of two nonparallel transition moments in flow orientation measurements to define the plane of intercalation and demonstrated the unreactivity of intercalated amino groups toward nitrous acid. The work was summarized together with new studies indicating the intercalation of other types of molecules, the thermal stabilization of DNA, the selective stabilization of the double (ribo) helix over the triple, a better analysis of the fluorescence orientation data, and other results.[4] One of my drawings from that paper has been widely reproduced, although usually credited to later authors. In reference 5, I showed the intercalation of a carcinogenic pentacyclic acridine and an extensive set of negative measurements on nonintercalators. A new, clear demonstration of complexes of DNA with benzopyrene was included.

"The analysis of the intercalated structure offered a key to understanding the existence of two types of mutation that were mutually compensatory when in close proximity. On genetic grounds and by analogy to intercalation, these were identified by the Brenner-Crick group as insertions and deletions. As far as I know, my hypothesis[2] that pairwise intercalation can introduce unequal crossing over is the only proposal for a mechanism that specifically depends on the intercalated structure. Other hypotheses that invoke only helix stabilization ignore the nonmutagenicity of nonintercalating helix stabilizers.

"This paper was the first to use the term 'intercalation' with respect to DNA. It was analogous to use of the term in calendar structure (e.g., February 29th is an intercalated day). It also describes the insertion of metal atoms between layers of graphite in a sense similar to the DNA usage. A review on structure has been given by Berman and Young.[6] The frequent citation of this paper may be attributable to the relevance of the proposed structure to the mechanism of action of some antibiotics, of some antiparasitics including antimalarials, of some anticancer agents, and of substances that effect loss or alteration of extrachromosomal DNA, as well as carcinogens and mutagens. It is also cited because the structure represented the first well-defined departure of DNA from simple helicity."

---

1. Peacocke A R & Skerrett J N H. The interaction of aminoacridines with nucleic acids.
   Trans. Faraday Soc. 52:261-79, 1956. (Cited 425 times since 1956.)
2. Lerman L S. The structure of the DNA-acridine complex. Proc. Nat. Acad. Sci. US 49:94-102, 1963.
   (Cited 570 times.)
3. ............. Amino group reactivity in DNA-aminoacridine complexes. J. Mol. Biol. 10:367-80, 1964.
   (Cited 105 times.)
4. ............. Acridine mutagens and DNA structure. J. Cell. Comp. Physiol. 64(Suppl. 1):1-18, 1964.
   (Cited 310 times.)
5. ............. The combination of DNA with polycyclic aromatic hydrocarbons. Proceedings of the Fifth National
   Cancer Conference, 1964. Philadelphia: Lippincott, 1965. p. 39-48.
6. Berman H M & Young P R. The interaction of intercalating drugs with nucleic acids.
   Annu. Rev. Biophys. Bioeng. 10:87-114, 1981.

# This Week's Citation Classic

CC/NUMBER 35
AUGUST 27, 1984

Le Pecq J B & Paoletti C. A fluorescent complex between ethidium bromide and nucleic acids: physical-chemical characterization. *J. Mol. Biol.* 27:87-106, 1967.
[Unité de Physico-Chimie and Unité de Biochimie et Enzymologie, Institut Gustave-Roussy, Villejuif, France]

This paper describes the interaction of a trypanocidal drug, ethidium bromide, with nucleic acids. It is shown that this compound binds specifically to double stranded nucleic acids by intercalating its planar aromatic ring between two adjacent base pairs according to a model first proposed by L.S. Lerman.[1] The binding of ethidium to DNA and RNA is accompanied by a very large increase of its fluorescence quantum yield. It can therefore be used as a fluorescent probe of nucleic acid structure. [The *SCI®* indicates that this paper has been cited in over 690 publications since 1967.]

Jean-Bernard Le Pecq
Unité de Biochimie et Enzymologie
Institut Gustave-Roussy
94800 Villejuif
France

May 25, 1984

"In the early 1960s, a lot of people were interested in studying the interaction of nucleic acids with various ligands such as simple cations, dyes, mutagens, and proteins. Known DNA-binding ligands, such as acridines, were often compounds endowed with pharmacological properties. Therefore, we thought it could be of interest to check whether drugs or biologically active substances having structures related to that of acridines could also be DNA ligands. We then noticed the papers of Dickinson et al.[2] and Newton[3] which reported that the trypanocidal action of phenanthridine drugs could result from interference with nucleic acid metabolism. We then asked Dickinson from Boots Pure Drug Company for samples of these compounds. We received a few milligrams of several derivatives. The most abundant sample was ethidium. I therefore began my study with this compound, which appeared later to be the most interesting one.

"I performed very simple experiments to start with, using differential spectroscopy to get evidence of DNA complexes. I immediately noticed that ethidium was able to complex with DNA because of the change of color of ethidium upon mixing with DNA. One day, it occurred to me that the ethidium-DNA mixture was shining a little bit. I therefore thought that ethidium could become fluorescent after DNA binding. At once, I prepared solutions of ethidium with and without DNA and went down three floors to another laboratory which had a fluorometer. Before doing any measurements, I put the cuvettes under the exciting beam and watched. I saw at once that the ethidium-DNA mixture emitted an intense, beautiful, orange-red fluorescence whereas the fluorescence of ethidium alone could hardly be seen with the naked eye. I immediately realized that this observation could be very useful and started quantitative fluorescence measurements the same day. It later appeared that, indeed, fluorescence was a very sensitive technique for characterizing the DNA binding of ethidium. Because of its fluorescent properties, ethidium became a widely used fluorescent probe of nucleic acids. It is commonly used to stain DNA bands in gel electrophoresis to detect them by fluorescence. Many citations to this work are probably related to the fluorescence properties of ethidium.

"At the time I wrote this paper, I moved to the California Institute of Technology for a postdoctoral stay with Davidson. I met Bauer and Vinograd whose labs were next to mine. I introduced ethidium to them. They were looking for intercalating reagents to study closed circular DNA and they were interested in ethidium properties. From their work,[4] ethidium became a widely used compound in this field, mainly to prepare circular DNA by ultracentrifugation in CsCl density gradient. A review has recently discussed the main uses of ethidium in the field of nucleic acid research."[5,6]

1. Lerman L S. Structural considerations in the interaction of DNA and acridines. *J. Mol. Biol.* 3:18-30, 1961.
   (Cited 935 times.) [*Citation Classic* in press.]
2. Dickinson L, Chantrill B H, Inkley G N & Thompson M J. The antiviral action of phenanthridinium compounds. *Brit. J. Pharmacol.* 8:139-42, 1953. (Cited 20 times since 1955.)
3. Newton B A. The mode of action of phenanthridines: the effect of ethidium bromide on cell division and nucleic acid synthesis. *J. Gen. Microbiol.* 17:718-30, 1957. (Cited 85 times since 1957.)
4. Bauer W & Vinograd J. The interaction of closed circular DNA with intercalative dyes. I. The superhelix density of SV40 DNA in the presence and absence of dye. *J. Mol. Biol.* 33:141-72, 1968. (Cited 435 times.)
5. Morgan A R, Lee J S, Pulleyblank D E, Murray N l & Evans D H. Ethidium fluorescence assays. Part I. Physicochemical studies. *Nucl. Acid. Res.* 7:547-69, 1979.
6. Morgan A R, Evans D H, Lee J S & Pulleyblank D E. Ethidium fluorescence assays. Part II. Enzymatic studies and DNA-protein interactions. *Nucl. Acid. Res.* 7:571-94, 1979.

# This Week's Citation Classic

CC/NUMBER 23
JUNE 8, 1981

Radloff R, Bauer W & Vinograd J. A dye-buoyant-density method for the detection and isolation of closed circular duplex DNA: the closed circular DNA in HeLa cells.
*Proc. Nat. Acad. Sci. US* 57:1514-21, 1967.
[Norman W. Church Lab. for Chemical Biol., California Inst. Technol., Pasadena, CA]

A buoyant-density method for the isolation and detection of closed circular DNA is described. The method is based on the reduced binding of the intercalating dye, ethidium bromide, by closed circular DNA. In an application of this method we have found that HeLa cells contain, in addition to closed circular mitochondrial DNA of mean length, 4.81 microns, a heterogeneous group of smaller DNA molecules which vary in size from 0.2 to 3.5 microns and a paucidisperse group of multiples of the mitochondrial length. [The *SCI*® indicates that this paper has been cited over 845 times since 1967.]

Roger J. Radloff
Department of Microbiology
School of Medicine
University of New Mexico
Albuquerque, NM 87131

May 13, 1981

"During the 1960s, Jerome Vinograd and most of the students and postdoctoral fellows in his laboratory were studying intensively the biological and physical properties of the closed circular DNA found in the papovaviruses and in mitochondria. In 1963, I became a graduate student in that laboratory and William Bauer, now at the State University of New York at Stony Brook, also joined the laboratory as a student shortly thereafter. Initially, I worked on several projects dealing with the structure and properties of the closed circular double-stranded DNA of polyoma virus, a papovavirus, while he began working with the chemistry of the interaction between the dye, ethidium bromide, and DNA isolated from the papovavirus, simian virus 40 (SV-40). His work showed that closed circular DNA forms a band at equilibrium in CsCl buoyant density gradients in the analytical ultracentrifuge at a greater density than does the linear DNA in the presence of ethidium bromide because less dye is bound by the closed circular DNA.

"In the latter part of 1966, we began working together, under the guidance of Vinograd, to see if we could develop a useful preparative procedure for the separation of closed circular from linear DNA in CsCl buoyant density gradients containing ethidium bromide. I initially attempted the experiments with purified closed circular and linear polyoma viral DNA and was successful in separating the two forms. Within several months, the method was extended to the separation of closed circular DNA from polyoma virus infected cells and from noninfected cells. The closed circular DNA from noninfected HeLa cells was found to consist primarily of molecules the size of mitochondrial DNA. In addition, however, two other groups of circular molecules were found at low frequency. One group was a heterogeneous population of molecules which vary in size but are smaller than the mitochondrial DNA, while the other group contained molecules which were multiples of the mitochondrial length. The significance of the closed circular DNA larger and smaller than mitochondrial DNA is still not fully understood.

"I believe this paper has been cited frequently for two reasons. One is that it was published near the beginning of a series of a great many experiments with the closed circular DNAs found in viruses, mitochondria, and bacteria. A second reason is that it contains a simple and reliable method that provides good separation for closed circular DNA molecules from linear DNA molecules.

"Many aspects of this research remain in my memory. What stands out most clearly is that nearly every experiment tried was successful. That was an exciting experience and one which I believe occurs infrequently for most scientists.

"Vinograd died in 1976. He was instrumental in the development of a number of well-known techniques in centrifugation and in the elucidation of the structure of closed circular DNA in addition to his other research accomplishments. All who knew or worked with him admired and respected his intuition, his abilities, and his patience. A review of this field was recently published by Bauer."[1]

1. **Bauer W R.** Structure and reactions of closed duplex DNA. *Annu. Rev. Biophys. Bioeng.* 7:287-313, 1978.

| Number 12 | **Citation Classics** | March 20, 1978 |

Huang R C & Bonner J. Histone, a suppressor of chromosomal RNA synthesis. *Proc. Nat. Acad. Sci. US* **48**:1216-22, 1962.

Chromatin isolated from pea embryos possesses the ability to carry out the DNA-dependent synthesis of RNA from the four riboside triphosphates. The present paper concerns the roles in such synthesis of the several components of chromatin. [The *SCI*® indicates that this paper was cited 679 times in the period 1962-1976.]

Professor Ru-Chih C. Huang
Department of Biology
Johns Hopkins University
Baltimore, Maryland 21218

April 25, 1977

"There is an old saying: 'When you know nothing about Zen, a mountain is a mountain; when you know something about Zen, a mountain is no more a mountain; when you know more things about Zen, a mountain again is a mountain.' If one substitutes the word gene for Zen, the saying tells exactly of our knowledge about chromatin.

"The work selected as one of the most cited papers was done when little was known about the molecular approach to gene expression. The term chromatin as an interphase state of chromosome was beginning to be accepted as a biochemical working usage. The isolation and purification of chromatin was still approached with more art than science, until the pea embryo was chosen. From then on the procedure became more standardized and a 'pea popper' was installed in the basement of Kerckhoff Laboratory at Caltech to handle the mass quantity of preparation. The well defined genetic system of peas was actually not fully explored but the precision in the differentiated and synchronous state of germinating pea embryos helped to establish the biochemical procedures.

"The desire to study the process of RNA synthesis was novel in the early 1960s. While chromatin as the genetic machinery can now be isolated intact, Michael Chamberlin, then a graduate student of Paul Berg, was able to obtain purified enzymes which would catalyze the polymerization reaction of RNA using DNA as the template.[1] Recognizing the significance of the combination, I went to Stanford, learned the procedure and upon returning to Caltech, applied it to chromatin studies. The results were not unexpected, namely that chromatin is rather inefficient in the support of DNA-dependent RNA synthesis, whether in a native or reconstituted form with either prokaryotic or eukaryotic RNA polymers. Because of the repressive function of the histones on RNA template activity, the named suppressor was used, which served its illustrative purpose although sometimes confused with the connotation of the prokaryotic system.

"The science of chromatin has gone from nothing → something → more things. The speculation of Stedman and Stedman in 1950[2] prompted the work cited, which apparently has generated a plethora of interest and studies on chromatin. Many controversial issues arose as further details were attended to. These have included the function of various chromosomal components, the fidelity of the RNA polymerases, and tissue specificity of the chromatins. However, the basic scheme that histone, being the major class of protein in the chromosomes, functions by blocking the transcription, remains unchallenged. It was a bold assumption and the experiments in 1962 provided important documentation.

"Today, not only the structure of chromatin has been revealed in microscopic details but also the function has been examined at a unique gene level. The devotion of the 1977 Cold Spring Harbor Symposia to this topic attests this progress.

"It may be added that during the midst of excitement of the work in 1962, I gave birth to a daughter, Suzanne. After seven days in the maternity ward at Pasadena, I was back at the bench and have stayed there ever since, even now as a full professor at Johns Hopkins. James Bonner too has continued to do work personally as well as inspiring others with brilliant ideas and thoughts."

### REFERENCES
1. Chamberlin M & Berg P. Deoxyribonucleic acid-directed synthesis of ribonucleic acid by an enzyme from escherichia coli. *Proc. Nat. Acad. Sci. US* **48**:81, 1962.
2. Stedman E & Stedman E. Cell specificity of histones. *Nature* **166**:780-1, 1950.

Number 37 **Citation Classics** September 12, 1977

Born, Gustav V R. Aggregation of blood platelets by adenosine diphosphate and its reversal. *Nature* 194:927-9, 1962.

A photometric method is introduced with which platelet aggregation and its inhibition by ADP and related compounds respectively are investigated quantitatively. [The *SCI®* indicates that this paper was cited 681 times in the period 1961-1975.]

---

Professor Gustav V.R. Born
Department of Pharmacology
Cambridge University
Hills Road, Cambridge, CB2 2QD
England

July 28, 1977

"To be told that the paper on platelet aggregation I wrote fifteen years ago is 'one of the most cited articles ever published' is astonishing and interesting. I can think of three possible reasons. First,. the paper helped to make platelets interesting research objects by suggesting functions for two, then recent, observations which had been made independently but seemed to be related. I found platelets to contain extraordinarily high concentrations of ATP, some of which broke down during clotting. Hellem and colleagues discovered the highly specific induction of platelet aggregation by ADP which is, of course, the first breakdown product of ATP. The paper suggests that these processes are connected through the formation of ADP from ATP in cells damaged by or involved in vascular injury. My current comment would be that that has turned out to be an oversimplification as far as the platelets themselves are concerned, which release ADP mostly from a different pool than that in which ATP breakdown occurs. On the other hand, in spite of much new knowledge about the role of prostaglandins and other endogenous agents in platelet aggregation, much evidence supports the original proposition that ADP formed from ATP in other cells, including red cells,[1] initiates platelet aggregation in the circulation, including that through artificial organs.[2]

"Secondly, the paper introduced the photometric method which later acquired the horrible name of 'aggregometry,' whereby it became possible to quantify and analyse platelet reactions *in vitro*. Not long before I had been following ribonuclease activity 'turbidimetrically' and I simply made the banal adaptations appropriate for measuring changes in light transmission associated with the aggregation of platelets in plasma. For some time the measurements remained wholly empirical; then the relation between the optical and the cellular events was established and quite recently the optical observations have been explained on the basis of light-scattering theory.[3]

"Already before publication the photometric method was rapidly taken up by visitors to whom it had been demonstrated; and after the paper appeared 'aggregometry' soon became widely used, presumably because of its simplicity and reproduceability. It has been responsible for major discoveries in platelet function, notably the 'second wave' of aggregation. This is the optical manifestation of the platelet 'release reaction,' much investigated since as an example of exocytosis and because the discovery of its inhibition by acetyl salicylic acid is the origin of the Aspirin trials in coronary thrombosis.

"This brings up the third reason for the popularity of the 1962 paper. It showed that platelet aggregation by ADP can be inhibited and reversed by the closely related substances AMP or ATP. The last paragraph of the paper reads: 'If it can be shown that ADP takes part in the aggregation of platelets in blood vessels it is conceivable that AMP or some other substance could be used to inhibit or to reverse platelet aggregation in thrombosis.' Cross and I soon found that AMP was much less inhibitory than adenosine which, in turn, is much less effective than several inhibitors unrelated to ADP which have been discovered since, such as some prostaglandins. Elucidation of these inhibitory mechanisms has made rapid progress, not only because platelets are advantageous models for other cell systems but also because it may result in important advances in drug treatment. I am glad to have contributed to this."

1. Born G V R, Bergquist D & Arfors K E. Evidence for inhibition of platelet activation in blood by a drug effect on erythrocytes. *Nature* 259:233-5, 1976.
2. Richardson P D, Galletti P M & Born G V R. Regional administration of drugs to control thrombosis in artificial organs. *Transactions of the American Society of Artificial Internal Organs* 22:22-9, 1976.
3. Latimer P, Born G V R & Michal F. Applications of light-scattering theory to the optical effects associated with the morphology of blood platelets. *Archives Biochemistry & Biophysics* 180:151-9, 1977.

# This Week's Citation Classic

CC/NUMBER 37
SEPTEMBER 15, 1980

Cohen S S, Flaks J G, Barner H D, Loeb M R & Lichtenstein J. The mode of action of 5-fluorouracil and its derivatives. *Proc. Nat. Acad. Sci. US* 44:1004-12, 1958.
[Depts. Biochem. and Pediat., Univ. Pennsylvania Sch. Med., Philadelphia, PA]

The antitumor agent, 5-fluorouracil, and its deoxyribosyl derivative are converted to the deoxyribonucleotide in *E. coli* and provoke thymine deficiency and 'thymineless death.' Fluorodeoxyuridylate, isolated from the bacteria or synthesized enzymatically *in vitro*, is an irreversible inhibitor of the thymidylate synthetase isolated from phage-infected bacteria. [The *SCI®* indicates that this paper has been cited over 495 times since 1961.]

Seymour S. Cohen
Department of
Pharmacological Sciences
State University of New York
Stony Brook, NY 11794

September 2, 1980

"When G.R. Wyatt and I discovered 5-hydroxymethylcytosine (HMC) in 1952,[1] my laboratory began to study its biosynthesis. In 1953, Hazel Barner and I found that a thymine deficiency led to the death of growing bacteria,[2] and we suggested that this might explain the antitumor effects of some antifolates. We had also observed a phage-induced synthesis of thymine and HMC in bacteria auxotrophic for thymine. By 1957, Joel Flaks and I had found that extracts of phage-infected bacteria contained large amounts of two virus-induced enzymes which made the viral pyrimidines as deoxyribonucleotides.[3] Such extracts provided the most active sources of thymidylate synthetase, permitting a study of inhibitors of this enzyme apparently crucial to DNA synthesis and cell survival. In that year, Charles Heidelberger and Robert Duschinsky had discovered that 5-fluorouracil markedly inhibited tumors in mice,[4] and they asked me to apply our bacterial and phage systems to clarifying the mode of action of the analog on thymine synthesis. I readily accepted their invitation, with the results presented in the abstract. I also suggested to Duschinsky that fluorocytosine might be a selective antifungal agent and was pleased to learn in 1959[5] that fluorocytosine was specifically inhibitory to these organisms.

"Heidelberger has extended many detailed studies with fluorouracil in human cancer and helps to lead a cancer center at USC. Duschinsky has retired to almost full-time skiing in Switzerland. As a result of experiments in 1956 with spongothymidine[6] and our first experiences in cancer chemotherapy, I became interested in the potentialities of the D-arabinosyl nucleosides and problems of chemotherapy in general. Flaks is currently a professor of biochemistry at the University of Pennsylvania. Our other co-workers have raised families and have then returned to the laboratory.

"Although it is now some 20 years since the discovery that fluorouracil provokes 'thymineless death,' the nature of these events is less than crystal clear. Most workers today do believe that antitumor therapy with fluorouracil, as well as agents such as inhibitors of dihydrofolate reductase, e.g., amethopterin, does produce a thymine deficiency. The importance of this effect in chemotherapy has led to increasingly detailed studies of the pure synthetase and reductase and their inhibition. The primary sequences of these key enzymes determined by an infecting parasite and host should be quite different. A thorough comparison of the parasite- and host-determined enzymes may then provide a rational approach to the development of a chemotherapy necessary to selectively inhibit or kill an infecting organism."

1. Wyatt G R & Cohen S S. The bases of the nucleic acids of some bacterial and animal viruses: the occurrence of 5-hydroxymethylcytosine. *Biochemical J.* 55:774-82, 1953.
2. Barner H D & Cohen S S. The induction of thymine synthesis by T2 infection of a thymine requiring mutant of *Escherichia coli*. *J. Bacteriology* 68:80-8, 1954.
3. Flaks J G & Cohen S S. The enzymic synthesis of 5-hydroxymethyldeoxycytidylic acid. *Biochim. Biophys. Acta* 25:667-8, 1957.
4. Heidelberger C, Chaudhuri N K, Danneberg P, Mooren D, Griesbach L, Duschinsky R, Schnitzer R J, Pleven E & Scheiner J. Fluorinated pyrimidines, a new class of tumour-inhibitory compounds. *Nature* 179:663-6, 1957.
5. Duschinsky R. Personal communication.
6. Cohen S S & Barner H D. Studies on the induction of thymine deficiency and on the effects of thymine and thymidine analogues in *Escherichia coli*. *J. Bacteriology* 71:588-97, 1956.

## This Week's Citation Classic
CC/NUMBER 41
OCTOBER 12, 1981

Cleaver J E. Repair replication of mammalian cell DNA: effects of compounds that inhibit DNA synthesis or dark repair. *Radiat. Res.* 37:334-48, 1969.
[Lab. Radiobiology, Univ. California Med. Ctr., San Francisco, CA]

In this paper DNA repair in ultraviolet damaged human cells was measured by the incorporation of labeled thymidine or deoxycytidine during the $G_1$, $G_2$, and mitotic stages of the cycle. Three groups of compounds were investigated as to whether they inhibited repair: (1) compounds that inhibit the synthesis of precursors of DNA, (2) compounds that bind to DNA, and (3) compounds acting nonspecifically. [The *SCI*® indicates that this paper has been cited over 320 times since 1969.]

James E. Cleaver
Laboratory of Radiobiology
School of Medicine
University of California
San Francisco, CA 94143

April 30, 1981

"It is a most pleasant surprise that this paper has become a most-cited paper in our field of DNA repair in mammalian cells. The key to its frequent citation is that one of the compounds investigated, hydroxyurea, has turned out to be an invaluable adjunct to routine assays of DNA repair. This drug provides a way of specifically inhibiting semiconservative DNA replication without at the same time interfering with DNA repair. Subsequently, it has been found that the blanket assumption that hydroxyurea has no effect on repair is probably wrong. Both hydroxyurea, and cytosine arabinoside, also used in this paper, are found to interfere with late stages of DNA repair and this very inhibition is now used as another assay for the frequency of repair events.[1]

"This early study proved to be one of the first papers in what has now become a vast, extensively worked area of DNA repair in eukaryotic cells that has importance in radiobiology and chemical carcinogenesis. Practical importance has even come out of this early study because measurements of repair are being used for identification of environmental agents that are potentially mutagenic and carcinogenic. DNA repair assayed by thymidine labeling in the presence of hydroxyurea, as first described in this paper, is one of the test systems under current evaluation in the Environmental Protection Agency's Gene-Tox Program.

"This particular study was part of a series of investigations we did leading to the discovery of human diseases which were ultraviolet sensitive with high levels of carcinogenesis because of defects in a DNA repair system. That was first reported in a paper that also became a *Citation Classic*.[2] The parallel growth of investigations of repair using inhibitors and genetic defects has been invaluable in laying out a role for damage and repair as early events in carcinogenesis from environmental agents.[3,4]

"This work has been recognized by my receiving both the Radiation Research Society's Award for Research in 1973 and the Lila Gruber Memorial Award for Cancer Research from the American Academy of Dermatology in 1976. The receipt of these gave me great pleasure, as also has the continuous support we have received from the Atomic Energy Commission (now the Department of Energy) for the research carried out from 1966 to the present."

1. **Hiss E A & Preston R J.** The effect of cytosine arabinoside on the frequency of single strand breaks in DNA of mammalian cells following irradiation or chemical treatment.
   *Biochim. Biophys. Acta* 478:1-8, 1977.
2. **Cleaver J E.** Defective repair replication of DNA in xeroderma pigmentosum.
   *Nature* 218:652-6, 1968.
   [Citation Classic. *Current Contents/Life Sciences* 24(30):22, 27 July 1981.]
3. **Cleaver J E & Painter R B.** Absence of specificity in inhibition of DNA repair replication by DNA-binding agents, co-carcinogens, and steroids in human cells. *Cancer Res.* 35:1773-8, 1975.
4. **Cleaver J E.** DNA damage, repair systems and human hypersensitive diseases.
   *J. Environ. Pathol. Toxicol.* 3:53-68, 1980.

Chapter 4

# Proteins and Amino Acids

Prior to the current popularity of DNA research, the study of proteins— as enzymes, as structural proteins, as blood components—dominated biochemistry. Two factors in greatest need for success in this study and in most other areas of biochemistry were convenient separation and purification methods, and reliable, inexpensive, and rapid quantitative procedures. The importance of these two is sharply emphasized by a recognition that nearly half of the Citation Classics in this grouping of fifty-two articles on proteins relate to those two subjects.

The first six Classics describe quantitative methods for proteins. The rapid and simple ultraviolet absorption technique developed by Waddell was obviously limited when other absorbing agents were in the protein-containing sample. This objection was overcome by the Lowry, Rosebrough, Farr, and Randall adaptation of the Folin–Wu method for sugars to the determination of proteins. The Lowry method, cited over 100,000 times between 1961 and 1982, is the King of the Classics. No other Classic approaches it in citation frequency. But even the Lowry method was improved upon by Miller, although forceful mixing and gentle heating are hardly major changes in the substance of the method. Prior use of the biuret method in either the macro- or micro-form as described by Gornall, Bardawill, and David (6,430 citations) proved very satisfactory for most protein determinations. These methods were not suitable for quantitating proteins fixed in polyacrylamide gels, and this fault was overcome by Coomassie blue staining (p. 91). This dye is now also adapted to the determination of proteins in solution.

Other brilliant contributions by biochemists at the Swedish University of Uppsala do not overshadow the molecular sieving discovery cited in the next two Classics written by Porath and Flodin. The separation and sizing of proteins based on molecular weight (and conformation) via gel filtration is so simple a procedure that science students in junior high school can do it successfully. It is perhaps this very simplicity, like Fleming's description of the bacterium-destroying mold *Penicillium,* that resulted in its failure to draw forth discussion after the initial description. Although the article published in 1959 was not immediately popular, by 1963 the acceptance of gel filtration for molecular weight determination of proteins was universal.

One disadvantage of Sephadex and other gels for molecular weight determinations is the size of the sample needed. Polyacrylamide gel electrophoresis (PAGE) of proteins in small columns (disc electrophoresis) was rapid, and suitable determinations could be made with microgram quantities of protein (p. 95). Moreover, the addition of sodium dodecyl sulfate (SDS) and heating to the original technique of Ornstein and Davis allowed an accurate molecular weight determination for proteins which had a different conformation in their original state (p. 96) . The "PAGE scientists," like others, had to overcome the initial rejection of their articles and frustration from lack of research funds. The combination of electrophoretic separation with molecular sizing was initially no more acceptable to journal editors (p. 99) than was PAGE sizing (p. 98): both were at first rejected but after publication became Classics, with citation scores of 2,870 and 1,440 respectively.

Another modification of PAGE to permit the detection of $^3$H-labeled molecules by fluorography became an extremely popular method (3,660 citations), since it was both rapid and sensitive. The Classic by Laskey, a companion to this Classic by Bonner and Laskey, should also be mentioned here. Limited hydrolysis of proteins followed by their SDS-PAGE separation proved to be a rapid method for determining whether or not proteins of identical mobility were genuinely identical (Cleveland's Classic with 1,830 citations).

Additional separation procedures with high citation scores are the separation of dansyl-derivatives of amino acids (1,505 citations), isoelectric focusing of proteins (1,130 citations), and affinity chromatography (1,580 citations). To perform his first separations of dansyl amino acids on polyamide films, Wang literally took the shirt off his professor's back. The ampholytes needed for isoelectric focusing were created by an equally imaginative but more coldly scientific approach (p. 104). The application of affinity chromatography to protein purification has been expanded by the coupling of highly specific monoclonal antibodies to agarose or other immobilized supports. Antibodies are also useful in the purification of low molecular weight materials.

Other papers on proteins include that described by Deutsch, who, when purifying plasminogen, rediscovered the coprecipitability of calcium and phosphate ions, and that described by Clegg, who unabashedly recalls his resignation/firing ceremony from one post-doctoral position before he began the characterizations of hemoglobins that led to his Classic. The last paragraph of Clegg's "This Week's Citation Classic" (TWCC) contains an important message for granting committees. These two Classics are followed by twelve review articles covering the spectra of hemoproteins, the properties of milk proteins, the chemistry of collagen, glycoproteins, lipoproteins, protein denaturation, and insulin (pp. 116–124). The three review articles on lipoproteins (pp. 125–127) created the analytical base necessary for clinical evaluations of hypo- or hyperlipemic conditions. Fleischman's TWCC on the four-peptide structure of γ-globulin would be equally appropriate in the chapter on Immunology in volume 1. R. R. Porter was a co-author of the paper discussed, and this was a portion of his work responsible for his Nobel award in 1972.

This chapter concludes with commentaries on a few Classics related to the quantitation of amino acids, either free or when peptide bound into proteins. Sulfur-containing amino acids could be quantitated by $p$-mercuribenzoate titration; tryptophan and tyrosine, by their powerful absorption at 220 nm, or by chemical means (pp. 129–133). The concluding papers are on automated amino acid analysis and inhibition of protein synthesis by puromycin, an analog of aminoacyl adenosine.

# This Week's Citation Classic

CC/NUMBER 42
OCTOBER 19, 1981

Waddell W J. A simple ultraviolet spectrophotometric method for the determination of protein. *J. Lab. Clin. Med.* **48**:311-14, 1956.
[Dept. Pharmacology, Univ. North Carolina School of Medicine, Chapel Hill, NC]

The method utilizes the ultraviolet absorbancy of the peptide bond instead of that of aromatic amino acids. This permitted greater sensitivity, accuracy, and specificity than was possible with methods previously described. [The *SCI®* indicates that this paper has been cited over 700 times since 1961.]

---

William J. Waddell
Department of Pharmacology and Toxicology
School of Medicine
Health Sciences Center
University of Louisville
Louisville, KY 40292

August 25, 1981

"Younger biologists might well wonder why the National Institute of Neurological Diseases and Blindness supported the apparently isolated development of a quantitative method for protein in solution. The National Institutes of Health (NIH) grant which supported the work in the mid-1950s actually was for the study of the chemical and physical disposition of antiepileptic agents. However, at that time investigators were freer to use their judgment to follow valuable leads during the progress of the work even if these leads were not at all a part of the original application.

"We decided to investigate the effect of acid-base alterations in dogs on the disposition of the anticonvulsants which are weak acids. Infusion of sodium bicarbonate intravenously promoted redistribution of phenobarbital from the brain to blood and hastened the renal excretion of this drug.[1]

"The infusion of sodium bicarbonate intravenously in dogs also produced hemolysis.[2] The question that arose was whether the membrane of these cells was contracting or whether the cell was increasing in volume. It was thought that this could be measured from the dilution of a protein solution in which the red cells were suspended. The protein methods available did not satisfy the investigator. This protein method, which is now a *Citation Classic*, was devised to answer the red cell question and then published as a general method since it had advantages of greater simplicity, rapidity, sensitivity, accuracy, and specificity than other methods. It is for this reason that the method is widely used.

"In addition, this research project clarified the principle by which weak acids and bases distribute among compartments of different pH values; this resulted in a method for determination of intracellular pH values.[3] A metabolic product (DMO) of one of the anticonvulsants appeared to be almost ideal for intracellular pH measurements and is now widely used for that purpose.

"The current wisdom of granting agencies in requiring tightly planned, detailed, and highly focused applications which are carefully followed by the investigator might answer some questions. In my opinion, however, it is unfortunate that funding is a thing of the past for investigations that are broad but that acquire original, unanticipated, and useful knowledge."

1. Waddell W J & Butler T C. The distribution and excretion of phenobarbital. *J. Clin. Invest.* **36**:1217-26, 1957.
2. Waddell W J. Lysis of dog erythrocytes in mildly alkaline isotonic media. *Amer. J. Physiol.* **186**:339-42, 1956.
3. Waddell W J & Butler T C. Calculation of intracellular pH from the distribution of 5,5-dimethyl-2,4-oxazolidinedione (DMO). Application to skeletal muscle of the dog. *J. Clin. Invest.* **38**:720-9, 1959.

# Citation Classics

Number 1                                                              January 3, 1977

Lowry O H, Rosebrough N J, Farr A L & Randall R J. Protein measurement with the Folin phenol reagent. *J. Biol. Chem.* **193**:265, 1951.

...The authors assert that the use of the Folin phenol reagent for the measurement of proteins "has not found great favor for general biochemical purposes." This study is concerned with modifying the Folin phenol reagent procedure by treating protein solutions "with copper in alkali." By recording the color change after the copper treatment, and measuring the quantity of protein present with a Beckman spectrophotometer, the authors determined that "measurement of protein with copper and the Folin reagent" is more sensitive and simpler than other procedures.

*Professor Oliver H. Lowry:*

..."It is flattering to be 'most cited author,' but I am afraid it does not signify great scientific accomplishment. The truth is that I have written a fair number of methods papers, or at least papers with new methods included. Although method development is usually a pretty pedestrian affair, others doing more creative work have to use methods and feel constrained to give credit for same[1]... Nevertheless, although I really know it is not a great paper (I am much better pleased with a lot of others from our lab), I secretly get a kick out of the response....

"Perhaps you would be interested in a little about the history of the method. Back in 1922, Wu, who worked with Folin, applied the reagent to proteins, without $CU^{2+}$, so it was based on the tyrosine and tryptophane contents of the protein. This procedure was used sporadically for some time and had a reputation for erratic results, probably because traces of contaminating $CU^{2+}$ would increase the readings. Herriot, in a 1935 footnote to another paper, mentioned that $CU^{2+}$ enhanced readings with protein, and in 1941 published a short communication describing the $CU^{2+}$ enhancing effect for 7 proteins, and giving convincing evidence that the enhancement was attributable to reaction with peptide bonds.

"Before I came to St. Louis we had need of a micro method for protein, studied the reaction some more, and came up with a revised procedure which we felt was an improvement, particularly in regard to application to a variety of situations. Actually, however, we had made few fundamental changes from the method of Herriot, and had never really intended to publish it.

"When I came to St. Louis in 1947 Earl Sutherland, who was here then, adopted our procedure, but for several years complained that he had to cite it as 'personal communication,' and, he inquired, why didn't we write it up? So we finally went to work and did the necessary things: studied the reaction more thoroughly, tested it a lot of ways, described its virtues and disadvantages, compared the results with a Kjeldahl procedure, investigated what interfered, etc.

"This was a lot of work and the three co-authors helped in various ways. The greatest help was from Miss Nira Rosebrough (now Mrs. Nira R. Roberts) who became one of the best technicians I have ever had. She left us in 1957, worked for awhile with Dr. Rosen in Buffalo, and then quit science to raise a family. Dr. A. Lewis Farr (M.D.) was a post-doctoral student who had an outstanding record in medical school but decided after a year or so to go into private practice in his hometown in Greenville, Mississippi. Mrs. Randall was a technician who stayed at most a year, then left with her husband, and I don't believe has been in science since, but I have lost track of her.

"I...am puzzled why the paper is so often cited, and cited as such. I would like to think it is partly because we studied it pretty thoroughly and it is still applicable in most cases without modification, whereas the original Kjeldahl method, for example, has had innumerable major modifications and microfications, and people cite the particular modification they use. Another reason why our method isn't simply referred to by name is that it's quite a mouthful to say – 'Lowry, Rosebrough, Farr, and Randall.' The method apparently filled a need in the beginning--and a lot of people measure proteins in their work. Once it became established by people like Sutherland and Kornberg, other people may have thought it was the method to use, or at least checked the procedure they were using against it."[2]

---

1. **Lowry O H.** Personal communication to D.J.D. Price, November 11, 1969.
2. **Lowry O H.** Personal communication to E. Garfield, August 5, 1976.

# This Week's Citation Classic

CC/NUMBER 9
MARCH 3, 1980

Miller G L. Protein determination for large numbers of samples.
*Analyt. Chem.* 31:964, 1959.
[Pioneering Res. Div., Quartermaster Res. and Engineering Ctr., Natick, MA]

Protein analysis of large numbers of samples by the Lowry method was facilitated by addition of the color reagents with enough force to ensure complete mixing and by brief heating of the mixtures to ensure full development of color intensity. [The *SCI®* indicates that this paper has been cited over 830 times since 1961.]

Gail L. Miller
1333 Prospect Hill Road
Villanova, PA 19085

January 4, 1980

"The resolution of fungal cellulases by starch-block electrophoresis was found to be improved with the use of longer blocks and narrower sections; this led to the production of large numbers of fractions that required analysis for protein content. Because of the time-dependent development of color in the Lowry protein method, analysis of numerous samples necessitated the addition of the color reagents and the reading of the color intensities on a controlled time schedule for each sample. To avoid these restrictions, a modified procedure was developed, whereby the different steps could be performed more at one's convenience. The analysis was facilitated by addition of the color reagents with enough force to ensure complete mixing and by brief heating of the mixtures to ensure full development of color intensity. A few weeks of easy experimentation were all that were required to solve this problem, and wide use of the modified procedure by investigators with similar needs was anticipated. The article describing the work was rejected successively by two journals, then accepted by *Analytical Chemistry* after editorial compression of the text to only a few sentences and elimination of kinetics curves that demonstrated the incomplete color development in the original Lowry procedure and the complete color development in the modified procedure.

"The more charitable of my friends were pleased by the recognition, pointed out by *Current Contents®*, that this article received;[1] less charitable friends noted that methods articles are likely to be cited more frequently than other types. The number of requests for reprints was small compared to the number of citations. This was probably due to the shortness of the article, since it could easily be copied by hand, if not memorized.

"Recognition in any field doubtless has elements of chance in it, and perhaps one should not be too particular as to the form in which it comes. It seemed to me to be ironic, however, that this particular contribution should receive recognition instead of others that I had made. For example, my work on tumor-specific cytotoxic heterologous antiserum against human cancer cells required over two years of difficult, obstacle-beset experimentation, and I viewed the final accomplishment as a major contribution.[2] The published work attracted a thousand requests for reprints but no recognition in terms of grant support for further work. The inexorable consequence of the lack of grant support was a loss of job, 18 months' unemployment, and eventual change of career. Now a technical writer in the Carcinogenesis Testing Program at the National Institutes of Health, I am glad to be able to contribute in some way to the cancer effort, but I should have much preferred the challenge of laboratory experiments."

1. **Garfield E.** Highly cited articles. 39. Biochemistry papers published in the 1950s.
   *Current Contents* (25):5-12, 20 June 1977.
2. **Miller G L & Wilson J E.** Specificity of cytotoxic heterologous antiserum to cultured human cancer cells. *J. Immunology* 101:1078-82, 1968.

# This Week's Citation Classic

NUMBER 13
MARCH 26, 1979

Gornall A G, Bardawill C J & David M M. Determination of serum proteins by means of the biuret reaction. *J. Biol. Chem.* 177:751-66, 1949.

The properties of stabilized biuret reagents were investigated in order to define the optimal concentrations of copper, alkali, tartrate and potassium iodide. An 'ideal' reagent and a simple procedure are described for the determination of total protein, albumin and globulins in serum or plasma. [The *SCI*® indicates that this paper has been cited over 6,430 times since 1961.]

Allan G. Gornall
Department of Clinical Biochemistry
Banting Institute
University of Toronto
Toronto, Canada M5G 1L5

April 12, 1978

"During the second World War it was my privilege to serve as a Clinical Chemist in the Canadian Naval Hospital in Halifax. In the period from 1942-1946 laboratory analyses were performed on about 200 cases of hepatitis. Serum proteins were approximated by specific gravity in copper sulfate solutions.

By 1947, I was back in Toronto as Assistant Professor of Pathological Chemistry and C. J. Bardawill, a recent graduate in medicine, was my first graduate student. His assigned research project was to study biochemical abnormalities in experimental liver injury. We needed a simple, rapid, and accurate method for serum proteins and the biuret reaction seemed to offer the best prospects. Bardawill gained useful analytical experience by comparing several biuret reagents and we settled on a modified version of Weichselbaum's 'dilute' reagent. The paucity of data on the behavior of these reagents (and some unfounded reservations on my part concerning the analytical prowess of my colleague) prompted me to investigate in more detail the factors affecting the behaviour of biuret reagents stabilized with tartrate. These experiments were performed between January and June of 1948 and the paper was sent for publication in July.

"Also, as a result of my war-time Hospital experience, I had persuaded the Dean of Medicine to support the establishment of a Clinical Investigation Laboratory in our Department under my supervision. Its purpose was to foster the investigation of pathological conditions in patients, by a collaboration between members of the Clinical Departments and members of the Department of Pathological Chemistry. Maxima David was the technologist in this laboratory. She had a clever idea that instead of setting up a separate serum dilution for total protein, we could use an aliquot of the sodium sulfate dilution immediately after mixing. This proved correct, the method was thus improved and she earned her right to co-authorship. Bardawill obtained his Master's degree and went on to become a fine internist; his death at an early age was a sad loss for Canadian medicine.

"The success of our biuret method for serum proteins resulted in large measure from the timing of its publication. Here was a clearly established procedure in a major journal that any laboratory interested in serum proteins could set up and use. Hospital laboratory services were developing all around the world. Many clinical studies must have included measurements of serum proteins, thus accounting for the numerous citations. We ordered a modest 300 reprints and ran out of them within weeks. My hope is that the method has made a significant contribution to the care and welfare of patients."

# This Week's Citation Classic

Itzhaki R F & Gill D M. A micro-biuret method for estimating proteins.
*Anal. Biochem.* 9:401-10, 1964.
[Department of Radiotherapeutics, University of Cambridge, England]

This method of estimating proteins depends on their interaction with alkaline copper sulphate. It is rapid, fairly sensitive, and reasonably independent of type of protein. It is unaffected by the presence of high concentrations of DNA. [The *SCI*® indicates that this paper has been cited in over 830 publications since 1964.]

---

Ruth F. Itzhaki
Department of Virology
University Hospital of South Manchester
Withington Hospital
West Didsbury, Manchester M20 8LR
England

July 29, 1982

"Mike Gill and I developed the micro-biuret method in the department of radiotherapeutics, University of Cambridge. I was struggling then to characterise chromatin. The few publications on that subject were of little help, appearing to bear no relationship to one another. Also, I was engaged in a constant battle with some refractory equipment, homemade, for an esoteric technique known as electric birefringence. It was a 'do-it-yourself' laboratory with a tradition of *laissez-faire*. Even if one was a newly fledged PhD, one managed on one's own; there was no question of being guided by a senior worker nor of any technical assistance. But on the whole a technician would have been an encumbrance as I was feeling my way slowly and painfully in a subject which at that time interested no one in Cambridge apart from the head of my department. The general attitude was understandable; chromatin was demonstrably messy in its properties — unlike whiter-than-white DNA — and the fact that in the living cell the latter was yoked to proteins and RNA was immaterial.

"Initially, a technician would have been useful. I was trying to analyse the effects of radiation on chromatin — though I soon realised that it was necessary to characterise the chromatin first and irradiate it after. My early efforts involved giving the chromatin a dose of X rays and dashing to my laboratory to look at the birefringence properties, apparatus permitting, before they went too far into decline because of postirradiation effects. Another pair of hands would have been invaluable, if only to beat the birefringence equipment into submission.

"The main problems with chromatin were that there was neither a standard method of preparation nor a defined product. Even the gross composition was uncertain. It was obviously necessary to have a quick method for estimating protein in the presence of large amounts of DNA. Gill, who was working with nuclei, had a similar requirement and so he and I devised the micro-biuret method. It was less sensitive than the Folin-Lowry[1] but was much quicker and simpler.

"I suppose that its ease of use accounts for its popularity. But I would have preferred my subsequent studies on chromatin structure — the first to use DNAases[2] and polylysine[3] as probes — and on distribution of carcinogen-bound sites in chromatin,[4] to be better known. However, it is rewarding that both approaches triggered off a number of studies by others, even if circumstances precluded my continuing them myself.

"I cannot help feeling nostalgic for Cambridge. The city was — and is — so beautiful. Also, one was able to do the work one thought necessary; there was no pressure to move off the topic from any higher authorities sublimely indifferent to, or ignorant of, its nature. Lastly, there was more scope for individualism, as opposed to teamwork, and for simple experimentation. Techniques are becoming increasingly complex. Engaged now in Southern blots, *in situ* hybridisation, and embarking on recombinant DNA work, I am highly dependent on a continuity of scientific assistance — not easy to maintain in days of financial constraints. The consolation is, of course, that one can now ask far more searching questions."

---

1. **Lowry O H, Rosebrough N J, Farr A L & Randall R J.** Protein measurement with the Folin phenol reagent. *J. Biol. Chem.* **193**:265, 1951. [Citation Classic. *Current Contents* (1):7, 3 January 1977.]
2. **Itzhaki R F.** The arrangement of proteins on the deoxyribonucleic acid in chromatin. *Biochemical J.* **125**:221-4, 1971.
   [The *SCI* indicates that this paper has been cited in over 60 publications since 1971.]
3. ─────── . Structure of deoxyribonucleoprotein as revealed by its binding to polylysine. *Biochem. Biophys. Res. Commun.* **41**:25-32, 1970.
   [The *SCI* indicates that this paper has been cited in over 50 publications since 1970.]
4. **Cooper H K, Margison G P, O'Connor P J & Itzhaki R F.** Heterogeneous distribution of DNA alkylation products in rat liver chromatin after *in vivo* administration of N, N-DI[$^{14}$C]methylnitrosamine. *Chem.-Biol. Inter.* **11**:483-92, 1975.
   [The *SCI* indicates that this paper has been cited in over 35 publications since 1975.]

## This Week's Citation Classic

CC/NUMBER 40
OCTOBER 3, 1983

Chrambach A, Reisfeld R A, Wyckoff M & Zaccari J. A procedure for rapid and sensitive staining of protein fractionated by polyacrylamide gel electrophoresis. *Anal. Biochem.* 20:150-4, 1967.
[Endocrinology Branch, Natl. Cancer Inst., and Lab. Immunology, Natl. Inst. Allergy and Infectious Diseases, Natl. Insts. Health, Bethesda, MD]

A staining procedure for protein zones on polyacrylamide gel is described which a) provides a clear background without destaining; b) requires only 1.5 hours; c) maintains the protein band throughout in the reliable fixative, 12.5 percent trichloroacetic acid (TCA); and d) is sensitive to 2 μg of protein per zone. [The *SCI*® indicates that this paper has been explicitly cited in over 1,035 publications since 1967.]

A. Chrambach
Department of Health & Human Services
National Institutes of Health
Bethesda, MD 20205

July 8, 1983

"When I joined the National Institutes of Health in 1966, Ralph Reisfeld and I were the sole full-time devotees of polyacrylamide gel electrophoresis at that institution. June Zaccari and Mary Wyckoff were his and my assistants, respectively. Working with the glycoprotein, human chorionic gonadotropin (hCG), I found it impossible to fix and stain it with the standard procedure at that time, using one percent Amidoblack in 7.5 percent acetic acid as the stain-fixative. Ralph had a similar interest in improving fixation and staining of IgG chains, parathyroid hormone, pokeweed mitogen, and phytohemagglutinin. We decided to undertake a joint systematic search for a stain effective in all of those cases. A matrix of stain-fixative combinations was set up, including the stains Coomassie Brilliant Blue R-250, Amidoblack, Poinceau Red, Bromphenolblue, and Fluorescein, and the fixatives picric, perchloric, acetic, sulfosalicylic, tannic, phosphotungstic, and trichloroacetic acid (TCA) as well as mercuric chloride.

"TCA proved the only fixative capable of fixing hCG bands at least for a few hours and was compatible with one of the dyes—Coomassie Blue R-250. Testing TCA at various concentrations, we made the chance observation that Coomassie Blue solutions in ten to 12.5 percent TCA stained without an appreciable background, while both lower and higher concentrations of TCA gave a background stain. This is presumably due to the fact that in this narrow range, the dye solution is saturated and partitions into the relatively hydrophobic protein zone. This view is supported by the fact that Coomassie Blue R-250 is partially insoluble in ten percent TCA, and precipitates after about one day from solution, temporarily brought about by rapid mixing of a solution of the dye in 20 or 25 percent TCA with an equal volume of water (a crucial recipe omitted through oversight from the report but provided later in footnote 4 of reference 1). The report is marred by as yet another accidental omission, i.e., failure to warn against the shrinkage of gels after two weeks of storage. Prolonged storage of the gels stained by the reported procedure is important because protein zones on gels stored in ten percent TCA containing a small amount of the dye take up progressively more Coomassie Blue R-250. Thus, storage can be used to increase detection sensitivity.

"If in spite of these oversights the report proved popular, this is undoubtedly due to the fact that it presented the first no-background, rapid staining procedure in polyacrylamide gel electrophoresis. Since that time, the procedure has been developed in two ways: 1) Substitution of Coomassie Blue R-250 by G-250 allows for preparing a stable 0.25 percent stock solution of the dye.[2] That procedure has the disadvantage, however, that storage is in the poor fixative, acetic acid, and that a band intensification with storage time is not achieved. 2) In conjunction with a rapid diffusion of SDS from gels, it allows for the staining of SDS-proteins within a few hours without a background."[3]

1. Rodbard D & Chrambach A. Estimation of molecular radius, free mobility, and valence using polyacrylamide gel electrophoresis. *Anal. Biochem.* 40:95-134, 1971. (Cited 430 times.)
2. Diezel W, Kopperschlaeger G & Hofmann E. An improved procedure for protein staining in polyacrylamide gels with a new type of Coomassie Brilliant Blue. *Anal. Biochem.* 48:617-20, 1972. (Cited 235 times.)
3. An der Lan B C, Sullivan J V & Chrambach A. Effective fixation and rapid staining of proteins in SDS-PAGE. (Stathakos D, ed.) *Electrophoresis '82.* Berlin: Walter de Gruyter, 1983. p. 225-33.

# This Week's Citation Classic
CC/NUMBER 19
MAY 11, 1981

Porath J & Flodin P. Gel filtration: a method for desalting and group separation.
*Nature* 183:1657-9, 1959.
[Institute of Biochemistry, University of Uppsala, and Research Laboratories, Pharmacia, Uppsala, Sweden]

The introduction of cross-linked dextran (Sephadex) provided for the first time a practical and inexpensive chromatographic method for the separation of solutes according to their molecular dimensions. [The *SCI*® indicates that this paper has been cited over 640 times since 1961.]

Jerker Porath
Institute of Biochemistry
Biomedical Center
University of Uppsala
S-751 23 Uppsala
Sweden

April 15, 1981

"One day in the early 1950s we observed that certain low-molecular weight substances in a mixture were separated from each other upon elution from an electrophoresis column packed with starch grains—despite the fact that, by a mistake, no electric current had been applied. My assistant and I studied the phenomenon systematically and found that the test substances chiefly grouped together in two main classes: a rapidly migrating, high-molecular and a more slow-moving, low-molecular class. Polypeptides were found to occupy an intermediate position. I had no doubt about the mechanism: differential exclusion according to molecular size. In addition I observed that several dyes migrated—often with strong retardation—as well-defined zones. Unfortunately the starch had defects as a sieving medium that we could not eliminate. However, I thought I had made a genuine discovery of an unknown phenomenon. That was not the case, but I believe I was the first to recognize its potentialities. I hoped for a better gel-medium.

"Some years later Per Flodin and I discussed the possibility of synthesizing an ideal convection medium for zone electrophoresis. Flodin remembered that there was a bottle of cross-linked dextran on one of the laboratory shelves at Pharmacia pharmaceutical company. Perhaps that gel-forming substance might be suitable? Eventually I got the bottle in my hand. A column was packed and the adsorption properties we tested as we used to do routinely. We never ran an electrophoresis experiment (Stellan Hjertén did somewhat later). The test substances travelled in a way that strongly reminded me of the experiments on starch: perhaps the ideal molecular sieve had been discovered?

"When I announced the method for the first time at a conference in Bruges (1959) there was no response. Within the first years the publication in *Nature*, which is now considered as a *Citation Classic*, evoked little interest.

"We proposed that gels should be tailor-made for different molecular size ranges. In the beginning, I often encountered critics telling me that it was ridiculous to believe in the future of a method that had such a limited separation interval ($R_F = 0.5-1.0$).

"The 'gel filtration' method directly gave the impetus to 'gel permeation chromatography' of non-polar substances on polystyrene gels. This very simple technique for fractionation of biomaterials and synthetic polymers can be applied to sample quantities from microliters to hundreds of liters and from a fraction of a microgram to hundreds of grams or more. It has also turned out to be a useful tool for estimation of molecular size and for the study of association-dissociation phenomena. More recent work in the field has been reported by H. Determann[1] and T. Kremmer and L. Boross."[2]

1. **Determann H.** *Gel chromatography.* Berlin: Springer-Verlag, 1968. 195 p.
2. **Kremmer T & Boross L.** *Gel chromatography.* Chichester, England: Wiley, 1979. 300 p.

# This Week's Citation Classic™

CC/NUMBER 52
DECEMBER 26, 1983

Flodin P & Killander J. Fractionation of human-serum proteins by gel filtration.
*Biochim. Biophys. Acta* 63:403-10, 1962.
[Research Laboratory, AB Pharmacia, and Department of Clinical Chemistry, University of Uppsala, Sweden]

In this paper, the preparative separation of human serum proteins according to molecular size is reported. A new type of dextran gel, Sephadex G-200, was used and the separated fractions analyzed by paper electrophoresis, immunoelectrophoresis, gel diffusion, serological titration, and density gradient ultracentrifugation. [The *SCI*® indicates that this paper has been cited in over 625 publications since 1962.]

Per Flodin
Department of Polymer Technology
Chalmers University of Technology
S-412 96 Göteborg
Sweden

September 12, 1983

"The first Sephadex types developed separated only low molecular weight proteins and not those of a size corresponding to the blood serum proteins. It was a disappointment since our interests were pretty much focused on preparative separation of biological macromolecules and primarily of proteins. The reason was that the first gels were made in blocks which were subsequently ground to a size suitable for column packing. The irregular form caused high resistance to flow and compression of the softer gels with low cross-linking density. This limited the use of the gels to a relatively low molecular weight region.

"In 1960, a method to make gels in spherical bead form was developed in my laboratory.[1,2] The beads made it possible to push the limit of separation upward to include, e.g., the blood proteins. At that time, Johan Killander and I decided to look into the possibility of fractionating and identifying serum protein components with the new G-200 gel type. His background in clinical chemistry matched my separation experience and very soon interesting results began to appear.

"Early in 1962, I accepted a job in another company. Before leaving Uppsala, I wanted to obtain a PhD and was therefore in a hurry to write a thesis. It was finished during the spring and the present paper was part of it. Parenthetically, it may be worth mentioning that the summary of the thesis was a monograph,[2] 12,000 copies of which were sent on request to interested scientists during the following year.

"The method we described provided a way to make preparative fractionations of serum proteins with simple equipment and essentially without denaturation. The latter was shown by the immunological tests we applied. Since a vast number of scientists, for obvious reasons, study blood serum proteins, we expected a considerable interest. To judge from the number of reprints I still have in my possession, other papers in the gel filtration series were more popular. However, the number of citations seems to confirm our earlier belief.

"The paper has no doubt contributed to the awards I have received for the gel filtration method and the Sephadex gels. In 1963, I received, together with Jerker Porath, the Arrhenius medal of the Swedish Chemical Society; in 1968, the gold medal of the Swedish Academy of Engineering Sciences; and in 1979, the Tswett medal in chromatography.

"A recent comprehensive review of the subject can be found in *Gel Chromatography*."[3]

1. **Flodin P.** *Process for preparing hydrophilic copolymerization and product obtained thereby.*
   US patent 3,208,994. 28 September 1965.
2. ............ *Dextran gels and their applications in gel filtration.* Uppsala, Sweden: Pharmacia AB, 1962. 85 p.
3. **Kremmer T & Boross L.** *Gel chromatography: theory, methodology, applications.* London: Wiley, 1979. 298 p.

# This Week's Citation Classic

CC/NUMBER 12
MARCH 23, 1981

Whitaker J R. Determination of molecular weights of proteins by gel filtration on Sephadex. *Anal. Chem.* 35:1950-3, 1963.
[Dept. Food Science and Technology, Univ. California, Davis, CA]

An excellent linear correlation between the logarithm of molecular weight of a protein and the ratio of its elution volume, V, to the void volume, $V_o$, of the column was found for chromatography of proteins on Sephadex G-100 and G-75, cross-linked dextrans. [The *SCI*® indicates that this paper has been cited over 1,100 times since 1963.]

John R. Whitaker
Department of Food Science
and Technology
University of California
Davis, CA 95616

March 6, 1981

"The research was the result of a need to determine the molecular weights of ficin, papain, and bromelain on which I was working and the relative inaccessibility of an analytical ultracentrifuge. Stimulus for the research was provided by the work of P. Andrews published in *Nature* in 1962 on use of agar columns for molecular weight determination of some proteins.[1] Molecular weight determination on agar columns proved unsuccessful largely because of the ion-exchange properties of the acidic components of agar and the nonuniformity of the material. Early in my investigations, particular attention was given to the effect of proper equilibration of the column with buffer, suppression of the smaller number of ionic (carboxyl) groups on Sephadex, and calibration of the column with essentially spherical proteins in order to avoid shape effects. I was ecstatic with the initial excellent results with standard proteins in terms of reproducibility, agreement with known molecular weights, and ease of determination. My enthusiasm was severely dampened when I discussed the results with colleagues more versed in protein chemistry than I. They assured me that molecular weight determinations by gel filtration were not to be taken seriously and that gel filtration could never replace the ultracentrifuge for molecular weight determinations.

"For some six months I let the investigation lapse as I pondered the merits of bringing the data to a publishable stage. Toward the end of the summer of 1962, having decided my colleagues could be wrong, I went on to complete the work and to submit it for publication. Fortunately, the editor of *Analytical Chemistry* considered it worthy of publication.

"The paper recognized the method would not give correct molecular weights for certain types of proteins. By chance, I had included ovomucoid, lysozyme, and hemoglobin in the research. Results with these proteins indicated that proteins containing appreciable carbohydrate would give apparent molecular weights higher than the true value and that proteins which readily dissociate into subunits or absorb to the gel would give apparent molecular weights lower than the true value. Interestingly, the proteins ficin, papain, and bromelain for which I developed the method do not behave normally in gel filtration. For reasons which still elude me, these proteins give apparent molecular weights some 20 percent too low on either Sephadex or Bio-gel columns.

"The method caught on rapidly for a number of reasons. Gel filtration proved to be a rapid, highly reproducible, and inexpensive method of determining molecular weights with as much confidence as by ultracentrifugation and with the additional advantage that molecular weights can be determined on crude preparations provided the protein of interest has measurable biological activity. Its success has been due to the contributions of many other scientists—P. Andrews, G.K. Ackers, etc.—who early recognized the potential of the method and who later provided a theoretical basis for understanding its performance and use not only in determining molecular weights but also shapes of molecules.[1,2]

"J. Porath, P. Flodin, and co-workers at Pharmacia in Sweden developed cross-linked dextrans to serve as inert stabilizing media for electrophoresis. Little did they anticipate they would prove to be such powerful analytical tools for separation and molecular weight determinations.[3-5] A recent review of this field has been published by Ackers in *Proteins*."[6]

1. Andrews P. Estimation of molecular weights of proteins by gel filtration. *Nature* 196:36-9, 1962.
2. Steere R L & Ackers G K. Restricted-diffusion chromatography through calibrated columns of granulated agar gel; a simple method for particle-size determination. *Nature* 196:475, 1962.
3. Flodin P. *Dextran gels and their applications in gel filtration.* Uppsala, Sweden: Pharmacia, 1962. 85 p.
4. Flodin P & Porath J. Molecular sieve processes. (Heftmann E, ed.) *Chromatography.*
   New York: Reinhold, 1961. p. 328-43.
5. Granath K A & Flodin P. Fractionation of dextran by gel-filtration method. *Makromol. Chem.* 48:160-71, 1961.
6. Ackers G K. Molecular sieve methods of analysis. (Neurath H & Hill R L, eds.) *Proteins.*
   New York: Academic Press, 1975. Vol. 1. p. 1-94.

## This Week's Citation Classic

CC/NUMBER 6
FEBRUARY 9, 1981

Reisfeld R A, Lewis U J & Williams D E. Disk electrophoresis of basic proteins and peptides on polyacrylamide gels. *Nature* 195:281-3, 1962.
[Merck Sharp and Dohme Research Labs., Rahway, NJ]

Polyacrylamide gel electrophoresis has been modified to make possible the separation of basic proteins and peptides. This technique permits excellent resolution with samples as small as 50 μg and within as little as 20 minutes. Crystalline trypsin was shown to contain a minor component with the mobility of chymotrypsin. Basic proteins and peptides also shown to contain more than one electrophoretic component include ribonuclease, protamine sulfate, globin, and lysine vasopressin. [The *SCI*® indicates that this paper has been cited over 1,765 times since 1962.]

R.A. Reisfeld
Scripps Clinic and Research Foundation
10666 North Torrey Pines Road
La Jolla, CA 92037

January 14, 1981

"One of the more exciting biochemical developments in 1960 was the appearance of acrylamide gel electrophoresis or 'disc electrophoresis,' a term coined by Leonard Ornstein and B.J. Davis at Mt. Sinai Medical Center, New York, who were instrumental in developing this procedure.[1,2] I met Ornstein by chance in 1961 and became immediately fascinated by the sensitivity, simplicity, and elegance of disc electrophoresis for the analysis of proteins and peptides. I was indeed so highly impressed by the discriminatory powers of this procedure that I convinced my colleagues, U.J. Lewis and Don E. Williams, who at that time worked with me at Merck, Rahway, New Jersey, to visit Ornstein and Davis at Mt. Sinai Medical Center and to let them show us how to perform disc electrophoresis. My colleagues and I immediately started to use this method, which aided us considerably in efforts to purify and to characterize pituitary hormones. It is difficult to describe the excitement, expectation, and sometimes frustration in those early years, when we watched the 'destaining' of the gels and the appearance of one or multiple components on the polyacrylamide gels.

"As an illustration that a powerful method can sometimes be performed by simple and inexpensive means, it is worthwhile to mention that, following Ornstein's and Davis' example, we prepared our original electrophoresis apparatus from components bought literally at the 'five and dime store,' utilizing the carbon rods from old flashlight batteries as electrodes. In fact, Ornstein and Davis in the early phase of their work had to run from the basement of the hospital to the roof in order to get enough sunlight (not always easy to find in New York City) to aid the catalysis of polymerization of the 'stacking gels.'

"We were stimulated to develop our method by demands of colleagues who wanted to analyze basic proteins which could not be resolved in the Ornstein-Davis system which had a 'running pH' of 9.6.

"When I left Merck in 1963 to join the laboratory of immunology at the National Institutes of Health (NIH), I was indeed fortunate to get there at a very exciting time in the field of immunology, i.e., when there was an intensive effort made by some excellent and imaginative investigators to resolve the complex structure of the antibody molecule. I was fortunate to be among the first to introduce disc electrophoresis to the field of immunology and to be able to utilize this powerful technique to gain some knowledge about the immunoglobulins. I also was privileged to be able to teach disc electrophoresis techniques to many bright and imaginative young scientists at NIH who then used it to excellent advantage. The disc electrophoresis technique has over the years become an even more versatile, useful, and powerful research tool, especially through the work of J. Maizel at Albert Einstein College of Medicine, who introduced the use of ionic detergents to estimate molecular weights,[3] and also through the sustained and imaginative efforts of my former colleague, Andreas Chrambach at NIH, who together with his associates, T.M. Jovin and D. Rodbard, vastly expanded and improved the applicability of the method, and developed its underlying theoretical principles to put it on a sound mathematical basis."[4,5]

1. **Ornstein L.** Disc electrophoresis—I—background and theory. *Ann. NY Acad. Sci.* 121:321-49, 1964.
2. **Davis B J.** Disc electrophoresis—II—method and application to human serum proteins. *Ann. NY Acad. Sci.* 121:404-27, 1964.
3. **Maizel J V.** Mechanical fractionation of acrylamide gel electropherograms: radioactive adenovirus proteins. *Science* 151:988-90, 1966.
4. **Rodbard D & Chrambach A.** Estimation of molecular radius, free mobility, and valence using acrylamide gel electrophoresis. *Anal. Biochem.* 40:95-134, 1971.
5. **Jovin T M, Dante M L & Chrambach A.** *Multiphasic buffer system output.* Springfield, VA: National Technical Information Service, 1970. PB 196085 to 196092.

# This Week's Citation Classic

CC/NUMBER 48
NOVEMBER 26, 1979

Hedrick J L & Smith A J. Size and charge isomer separation and estimation of molecular weights of proteins by disc gel electrophoresis.
*Arch. Biochem. Biophys.* 126:155-64, 1968.
[Department of Biochemistry and Biophysics, University of California, Davis, CA]

A simple method was devised relating the electrophoretic mobility of a protein determined by disc gel electrophoresis to its size and charge characteristics. The method is applicable to a single protein or to mixtures of proteins, provided a specific detection test is available. Knowing the relative size and charge of proteins is not only useful for their differential characterization but also as a predictive aid in their purification. [The SCI® indicates that this paper has been cited over 800 times since 1968.]

Jerry L. Hedrick
Department of Biochemistry
and Biophysics
University of California
Davis, CA 95616

October 4, 1978

"This paper is a specific application of the disc gel electrophoretic technique brilliantly developed by L. Ornstein and B.J. Davis in the late 1950s.[1,2] I was first introduced to the technique by Bob Metzenberg in 1961 while I was a graduate student at Wisconsin. I was very excited by its potential. When I arrived at the University of Washington, Seattle, in 1962 as a postdoctoral fellow, I rapidly applied the method to the study of the two forms of an enzyme. At that time, the isomeric forms of the enzyme were thought to differ predominantly in terms of size and minimally in terms of charge. However, the results I obtained were not consistent with the then accepted paradigm. Even though the validity of my observations could be accepted, the interpretation of them could not and, accordingly, I shelved this result till my first academic position at Davis.

"In 1966, I invited Al Smith to join me in Davis after I obtained my first federal grant. As I was a new independent investigator coincident with the beginning of the reduction in federal research funds, we had minimal resources with which to work. Our electrophoretic equipment was made of bits and pieces from discarded equipment and plasticware purchased at the local grocery store—carbon electrodes salvaged from dead flashlight batteries and baby blue 'Popeye' cereal bowls with a 'magic eye' for use as electrode reservoirs.

"With this crude but functional equipment, our approach to the problem was purely an empirical one, the usual case in method development. A very simple method of estimating the relative size and charge of a protein was eventually found by determining its electrophoretic mobility as a function of gel concentration. A log-linear plot of the data gave the sought-after straight line relation. The slope of the line was related to the size of the protein and the intercept, the charge.

"The excitement I felt about our discovery was heightened as this was my first independent creation as an assistant professor. The thrill was abruptly dampened when our paper was rejected by a leading biochemistry journal as being inappropriate. We submitted it to another journal where it was rapidly reviewed and accepted. We subsequently applied the method to the enzyme isomer problem and showed that the paradigm existing in 1962 was incorrect and extended the method to the case of proteins binding noncharged ligands.[3] This extension, in contrast to the original paper, has gone virtually unnoticed.

"I believe the paper has been popular because of its wide applicability and simplicity. The method itself is simple as are the interpretations of the results. Unfortunately, many of the recent putative theoretical attempts to mathematically relate mobility and the size and charge of a protein have neither simplified nor explained the fundamental principles of gel electrophoresis, but rather obfuscated them. The axiom relating simplicity and acceptability seems verified by the popularity of this paper."

1. **Ornstein L.** Disc electrophoresis. I. Background and theory. *Ann. NY Acad. Sci.* 121:321-49, 1964.
2. **Davis B J.** Disc electrophoresis. II. Method and application to human serum proteins. *Ann. NY Acad. Sci.* 121:404-27, 1964.
3. **Hedrick J L, Smith A J & Bruening G E.** Characterization of the aggregated states of glycogen phosphorylase by gel electrophoresis. *Biochemistry* 8:4012-9, 1969.

# This Week's Citation Classic

CC/NUMBER 10
MARCH 9, 1981

Dunker A K & Rueckert R R. Observations on molecular weight determinations on polyacrylamide gel. *J. Biol. Chem.* 244:5074-80, 1969.
[Biophysics Lab. and Dept. Biochemistry, Univ. Wisconsin, Madison, WI]

Electrophoresis in SDS-polyacrylamide gels provides a simple, inexpensive method for separating proteins, determining their relative proportions and estimating their molecular weights. This method has revolutionized the study of viruses, chromosomes, ribosomes, and membranes. [The *SCI*® indicates that this paper has been cited over 680 times since 1969.]

A. Keith Dunker
Biochemistry/Biophysics Program
Washington State University
Pullman, WA 99164

January 22, 1981

"By 1965 my thesis advisor, Roland Rueckert,[1] and, independently, Jacob Maizel[2] had shown the picornavirus coat to contain several proteins. Rueckert and Maizel both used polyacrylamide gel electrophoresis in the presence of 8 M urea. My problem was to determine the relative proportion and the molecular weight of each coat protein.

"I attempted to improve the separation of the virus proteins in the gels containing 8 M urea by trying every pH at which polyacrylamide remains stable, without success. Inadvertently, I even tried pH values in which polyacrylamide is only metastable.

"Then came the remarkable paper by Shapiro, Vinuela, and Maizel[3] showing that, in gels containing sodium dodecyl sulfate (SDS), proteins migrate during electrophoresis according to their molecular weights. Thus, in one experiment we could dissolve and separate the picornavirus capsid proteins, determine their sizes and, by integrating the protein peaks, estimate their relative proportions. What would have been a monumental task (and probably several PhD theses) had suddenly become one relatively simple experiment. However, before SDS gels would be useful to us, the reliability and the limitations of the method needed to be established, especially for virus capsid proteins.

"We studied about 20 proteins including several virus capsid proteins and also chemically modified proteins to determine the effects of size, charge, and shape on the mobility of SDS-protein complexes. To our utter amazement, large changes in the charge and in the internal disulfide bonding registered as small, second order perturbations on the molecular weight estimates. Virus capsid proteins gave the correct molecular weight values. Extensions of this early work have been published.[4]

"Once we were able to set limits on the reliability of SDS gels, we used this method to study the picornavirus capsid proteins.[5] 'Observations on molecular weight determinations on polyacrylamide gel' was published belatedly as an afterthought. Despite our tardiness, our paper was received and published only about a month after the blockbuster by Weber and Osborn.[6] Occasionally I wonder what the outcome would have been of publishing our SDS-gel work before, rather than after, our studies on the picornavirus capsid."

1. **Rueckert R R.** Studies on the structural protein of ME virus. *Fed. Proc.* 23:160, 1964.
2. **Maizel J V.** Evidence for multiple components in the structural protein of type 1 polio virus. *Biochem. Biophys. Res. Commun.* 13:483-9, 1963.
3. **Shapiro A L, Vinuela E & Maizel J V.** Molecular weight estimation of polypeptide chains by electrophoresis in SDS-polyacrylamide gels. *Biochem. Biophys. Res. Commun.* 28:815-20, 1967.
4. **Dunker A K & Kenyon A J.** Mobility of sodium dodecyl sulphate-protein complexes. *Biochemical J.* 153:191-7, 1976.
5. **Rueckert R R, Dunker A K & Stoltzfus C M.** The structure of mouse-elberfeld virus: a model. *Proc. Nat. Acad. Sci. US* 62:912-19, 1969.
6. **Weber K & Osborn M.** The reliability of molecular weight determinations by dodecyl sulfate-polyacrylamide gel electrophoresis. *J. Biol. Chem.* 244:4406-12, 1969.

# This Week's Citation Classic

CC/ NUMBER 33
AUGUST 17, 1981

Panyim S & Chalkley R. High resolution acrylamide gel electrophoresis of histones.
*Arch. Biochem. Biophys.* 130:337-46, 1969.
[Dept. Biochemistry, Univ. Iowa, Iowa City, IA]

A high resolution gel electrophoresis of histones is described, capable of distinguishing five major groups of calf thymus histones and histone fractions whose mobilities differ by as little as one percent. The applicability of this technique for a comparison of histones from a wide variety of species is discussed. [The *SCI*® indicates that this paper has been cited over 1,440 times since 1969.]

Sakol Panyim
Department of Biochemistry
Faculty of Science
Mahidol University
Rama VI Road
Bangkok 4
Thailand

July 9, 1981

"When I was a graduate student in the laboratory of Roger Chalkley at the University of Iowa, the structure and function of histones was one of the exciting areas of research. The conjecture by Stedman and Stedman[1] that histones were gene repressors led many investigators to search for tissue and species specific histones. I was persuaded by Chalkley to look for specific histones in various tissues and animals along the evolutionary scale. I was then trying to find a simple method which could resolve histones into a minimum of five types as shown by the chemical fractionation of Phillips and Johns.[2] I started by trying various existing methods including ionic exchange (IRC-50) chromatography, electrophoresis, and chemical fractionation. We came to a conclusion that polyacrylamide gel electrophoresis would best suit our need. However, there were conflicting reports about the number of histone bands on polyacrylamide gel electrophoresis. In the course of the study we realized that one of the reasons for the conflicting reports was proteolysis of histones, and fortunately we found that sodium bisulfite was a very effective inhibitor against the proteolysis.

"With intact histones in our hands, I tried all available polyacrylamide gel systems and found that the best procedures were that of Bonner et al.[3] and that of Johns.[4] However, the Bonner et al. procedure failed to separate H2B from H2A while Johns's method was incapable of resolving H2B from H3. By careful analysis of these two techniques, it became apparent to us that they mainly differed in urea concentration. We thought that by changing the urea from 6.25 M (Bonner et al.[3]) to 0 M (Johns[4]) there should be the urea concentration capable of resolving H2B from H2A from H3 and thus, capable of resolving histones into five main types (H1, H3, H2B, H2A, and H4). To our delight I found that at 2.5 M urea histones were resolved into five bands, some of which still showed microheterogeneity. Our delight turned to frustration when the manuscript was rejected. Fortunately, the editor of *Archives of Biochemistry and Biophysics* promptly considered it worthy of publication. For a recent review of this field the reader can refer to I. Isenberg.[5]

"Trying to rationalize why this paper became a Citation Classic, I believe that this paper has been highly cited because it was published at the peak of a need for a simple and reliable method capable of separating histones into five main types. There were a very large number of reprint requests as soon as the paper appeared. I wonder if the paper had been published in 1971, when I finished my PhD, would it still be highly cited?"

1. Stedman E & Stedman E. Cell specificity of histones. *Nature* 166:780-1, 1950.
2. Phillips D M P & Johns E W. A fractionation of the histones of group F2a from calf thymus. *Biochemical J.* 94:127-30, 1965.
3. Bonner J, Chalkley G R, Dahmus M, Fambrough D, Fujimura F, Huang R C, Huberman J, Jensen R, Marushige K, Ohlenbusch H, Olivera B & Widholm J. Isolation and characterization of chromosomal nucleoproteins. *Method. Enzymol.* 12B:32-7, 1968.
4. Johns E W. The electrophoresis of histones in polyacrylamide gel and their quantitative determination. *Biochemical J.* 104:78-82, 1967.
5. Isenberg I. Histones. *Annu. Rev. Biochem.* 48:159-91, 1979.

# This Week's Citation Classic

CC/NUMBER 51
DECEMBER 20, 1982

O'Farrell P H. High resolution two-dimensional electrophoresis of proteins.
*J. Biol. Chem.* **250**:4007-21, 1975.
[Dept. Molecular, Cellular and Developmental Biology, Univ. Colorado, Boulder, CO]

This paper described a method for high resolution analytical separation of proteins by two-dimensional gel electrophoresis. Proteins are separated in the first dimension according to their isoelectric points, and in the second dimension according to their molecular weights. [The *SCI®* indicates that this paper has been cited in over 2,870 publications since 1975.]

Patrick H. O'Farrell
Department of Biochemistry and Biophysics
School of Medicine
University of California
San Francisco, CA 94143

November 4, 1982

"As a graduate student in Boulder, Colorado, in 1972, I had isolated a series of mutations affecting development of the colonial algae, *Volvox*, and had visions of defining the normal pattern of gene expression during development and the effects that mutations had on this regulatory program. This had proved to be a powerful approach in analyses defining the mechanisms regulating gene expression in bacteriophage. Through interactions with my adviser, Jacques Pène, and another faculty member, Larry Gold, I became familiar with these bacteriophage systems. It was apparent that the reason for the success of the experimental approach was that sodium dodecylsulfate gel electrophoresis[1] provided a method capable of separating, identifying, and quantitating the majority of the bacteriophage proteins. To adapt this approach to the study of development in a eukaryote, such as *Volvox* (which has roughly 100 times the genetic complexity of a bacteriophage), I needed a separation system with much higher resolution. Although an adequate separation procedure did not exist, the use of two-dimensional methods for increased resolution was a well-established approach for chromatography and electrophoresis. Consequently, I set out to combine the two most powerful electrophoresis methods available.

"I had some early indications that this work was going to have considerable impact. Based on little more than a preliminary autoradiogram, the National Science Foundation awarded us a small grant for the development of the technique, and a number of investigators began to inquire whether the method was ready to be applied. Pène left Boulder at this time; I stayed on to complete the work and my degree under the sponsorship of David Hirsh.

"While I was preparing my thesis, Gold and Hirsh organized a course that helped to introduce the method and contributed to the recognition of the paper. Scientists from institutions representing various areas of the country were invited to Boulder, where I presented a four-day course on the techniques and their applications. The two-dimensional electrophoresis method eventually played an important role in the research programs of several of the course participants, such as Fotis Kafatos and Peter Geiduschek.

"After the course, I defended my thesis, submitted the manuscript to the *Journal of Biological Chemistry*, and moved to the department of biochemistry and biophysics, University of California, San Francisco, to pursue postdoctoral studies with Gordon Tomkins. I was more than a little surprised when a few months later I received two unfavorable reviews and a letter rejecting my manuscript; it had been reviewed by half a dozen scientists prior to submission. The reviews from the journal concluded that the manuscript appeared to be 'highly speculative in places and to be extrapolated in terms of usefulness far beyond what the author has any reason to expect.' With the cooperation and help of members of the journal's editorial board, the initial rejection decision was reversed.

"This paper has been frequently cited because the method has found a wide range of applications. Although I never returned to examine patterns of gene expression in *Volvox*, many workers have now applied the method to examine the changing patterns of proteins made during development. Additionally, the method has played a major role in diverse developments extending from the identification of the bacterial recA protein (thereby assisting in the elucidation of recombinational mechanisms and the induction of these functions by DNA damage) to the detection of the phosphorylation targets of viral transforming proteins. Although other techniques are now in the forefront of advances in molecular biology, refinements of electrophoretic separation methods[2] have continued to extend their usefulness."

1. Laemmli U K. Cleavage of structural proteins during the assembly of the head of bacteriophage T4. *Nature* **227**:680-5, 1970.
2. O'Farrell P Z, Goodman H M & O'Farrell P H. High resolution two-dimensional electrophoresis of basic as well as acidic proteins. *Cell* **12**:1133-42, 1977.
   [The *SCI* indicates that this paper has been cited in over 480 publications since 1977.]

99

## This Week's Citation Classic

Bonner W M & Laskey R A. A film detection method for tritium-labelled proteins and nucleic acids in polyacrylamide gels. *Eur. J. Biochem.* **46**:83-8, 1974.
[Medical Research Council Lab. Molecular Biol., Univ. Postgrad. Med. Sch., Cambridge, England]

CC/NUMBER 1
JANUARY 3, 1983

A sensitive fluorographic method is described which enables one to detect $^3H$ in polyacrylamide gels using X-ray film. The method is also useful for $^{14}C$ detection, being ten times as sensitive as autoradiography. [The *SCI*® indicates that this paper has been cited in over 3,660 publications since 1974.]

William M. Bonner
National Cancer Institute
National Institutes of Health
Bethesda, MD 20205

November 9, 1982

"Ron Laskey and I met in John Gurdon's Developmental Biology Laboratory in Cambridge. Like many researchers in the early-1970s, we had found that biological samples could be compared much more easily and accurately on polyacrylamide gels formed in slabs rather than tubes. Furthermore, with radioactive samples, the slab gel could be dried onto paper and autoradiographed, thereby eliminating the tedious and sometimes fickle procedure of slicing tube gels into 1 mm thick slices for scintillation counting. Slab gel autoradiography was good for $^{32}PO_4$, adequate for $^{14}C$ and $^{35}S$, but hopeless for $^3H$, since its beta particle was too weak to exit from the gel.

"We had read Randerath's paper[1] on fluorography of $^3H$ on thin layer plates and therefore suspected that if a scintillant such as PPO could be placed inside the polyacrylamide matrix, $^3H$ could be detected on film. The problem was that PPO is totally insoluble in water, and we felt that the chances of finding a solvent compatible with both PPO and polyacrylamide were close to zero. However, we decided to try anyway, and pulled some solvents off the shelf, one of which was DMSO. We soon found that it could replace the water in the polyacrylamide matrix and in addition could solubilize PPO quite well. We soaked a radioactive gel in DMSO-PPO and put it at -70°C with a piece of film.

"Both Laskey and I were primarily occupied with other projects and had treated these experiments more as a pastime than as a serious endeavor. However, when we developed that film and saw the image, the realization struck us that we had in essence solved the problem. We changed our priorities, and the paper was submitted two weeks before I left the Medical Research Council. Laskey later found that the film response in fluorography was nonlinear, and devised a procedure to make it linear.[2] Several years later, I devised a sensitive fluorography procedure for such solid supports as thin layer plates, papers, and membranes.[3]

"This publication has been highly cited because it removed the last major deficiency of polyacrylamide slab gels—the inability to detect $^3H$ on film. Fluorography is now used with polyacrylamide gel electrophoresis in most fields of molecular biology; indeed, several commercial products for fluorography are available. Recently, suitable water soluble scintillators have been found or developed, which promise to make fluorography even easier."

1. Randerath K. An evaluation of film detection methods for weak β-emitters, particularly tritium. *Anal. Biochem.* **34**:188-205, 1970.
2. Laskey R A & Mills A D. Quantitative film detection of $^3H$ and $^{14}C$ in polyacrylamide gels by fluorography. *Eur. J. Biochem.* **56**:335-41, 1975.
3. Bonner W M & Stedman J D. Efficient fluorography of $^3H$ and $^{14}C$ on thin layers. *Anal. Biochem.* **89**:247-56, 1978.

# This Week's Citation Classic™
CC/NUMBER 41
OCTOBER 8, 1984

Cleveland D W, Fischer S G, Kirschner M W & Laemmli U K. Peptide mapping by limited proteolysis in sodium dodecyl sulfate and analysis by gel electrophoresis.
*J. Biol. Chem.* 252:1102-6, 1977.
[Department of Biochemical Sciences, Princeton University, NJ]

This paper describes a rapid method for identification and characterization of proteins. The technique, which is especially suitable for analysis of proteins that have been isolated following electrophoresis in detergent-containing polyacrylamide gels, exploits partial enzymatic proteolysis and analysis of the cleavage products by gel electrophoresis. [The *SCI®* indicates that this paper has been cited in over 1,830 publications since 1977.]

---

Don W. Cleveland
Department of Biological Chemistry
School of Medicine
Johns Hopkins University
Baltimore, MD 21205

June 18, 1984

"As a result of its speed, resolving power, adaptability, and ease of use, polyacrylamide gel electrophoresis in the presence of the detergent sodium dodecyl sulfate is the most widely utilized method for the determination of both the purity and molecular mass of polypeptides in protein samples. This simple, but powerful, technique, first popularized by Weber and Osborn[1] and improved by Laemmli,[2] remains the centerpost of available methods for polypeptide analysis and characterization more than 16 years after its introduction.

"With the increased resolution of polyacrylamide gels, however, came a recurrent, companion problem: did polypeptides that shared indistinguishable mobility on such gels but were isolated using different methods or source materials represent biochemically related proteins or not? It was precisely this question that led to the initial attempt one Saturday morning to produce from a purified polypeptide a characteristic 'peptide map' of proteolytic fragments (now known as a Cleveland map). This was to be accomplished by intentional addition of a protease that would digest the substrate polypeptide, thereby leaving a series of cleavage products that could then be resolved using the previously mentioned, ubiquitous method of gel electrophoresis. Because of the difficulty in resolving very small peptide fragments, however, digestion conditions that produced large, stable fragments were required. This was achieved by the fortuitous but unexpected discovery that most commonly used proteases, which under normal conditions digest proteins into small fragments, yield large digestion products when the digestions are done in the presence of sodium dodecyl sulfate. From this observation, a very useful analytic technique (not to mention a 'classic' paper) was born.

"At that point, only the problem of publication remained. A manuscript was submitted to the *Journal of Biological Chemistry*, but to our amazement this was *rejected WITHOUT review* by an associate editor who inferred erroneously that a more comprehensive paper was to be sent elsewhere. Happily, when we inquired, we encountered the unusual situation of having the same editor instruct us that a resubmitted manuscript *without revision* would be accepted forthwith."

---

1. Weber K & Osborn M. The reliability of molecular weight determinations by dodecyl sulfate polyacrylamide gel electrophoresis. *J. Biol. Chem.* 244:4406-12, 1969. (Cited 16,735 times.)
2. Laemmli U K. Cleavage of structural proteins during the assembly of the head of bacteriophage T4. *Nature* 227:680-5, 1970. (Cited 23,635 times.)

# Citation Classics

Number 45      November 7, 1977

Katz A M, Dreyer W J & Anfinsen C B, Jr. Peptide separation by two-dimensional chromatography and electrophoresis. *Journal of Biological Chemistry* 234:2897-900, 1959.

The authors describe a modified method for the separation of peptides on filter paper that utilizes chromatography followed by high voltage electrophoresis cooled by a non-explosive organic solvent. [The *SCI*® indicates that this paper was cited 619 times in the period 1961-1975.]

---

Dr. Arnold Martin Katz
University of Connecticut
Medical Center
Farmington, Connecticut 06232

February 28, 1977

.."I am somewhat amused to find this paper listed amongst the 'most cited articles ever published' as this contribution is primarily methodological in scope, and the method wasn't even original to us! The value of this detailed description of a modification of the peptide 'fingerprinting' method used earlier by V.M. Ingram probably lies primarily in our use of 'varsol' in plastic, instead of toluene in glass, for cooling during the high voltage electrophoresis of 18¼ by 22½ inch filter paper chromatograms. This modification was necessitated by the configuration of our NIH lab, where the hood (in which we did the electrophoresis) was adjacent to the only exit. The thought of a spark—generated by the 2000 volt, 250 milliamp power supply—igniting 10 gallons of toluene in a huge and fragile glass jar served as a most effective stimulus to this paper. Repeated demonstrations to visiting scientists (and girlfriends) of the explosive response of a few drops of toluene to a lighted match provided a vivid reminder of the hazards of the Ingram method. A number of us spent much time looking for other organic solvents, especially those which wouldn't permeate or dissolve plastics, which could be used in these peptide separations. As I recall, the use of varsol (a light petroleum fraction with a flash point well above room temperature) stemmed from several visits to the National Bureau of Standards and one to a local gas station.

"The remainder of this method was relatively standard; the use of n-butanol-acetic acid-water for paper chromatography was well established in Dr. Anfinsen's laboratory, and high voltage paper electrophoresis had been employed with considerable elegance by many, including Sanger's group. The combination of these principles, which separated peptides in two dimensions by methods based on different properties of each peptide, had, as indicated earlier, been used by Ingram to identify specific amino acid substitutions in abnormal human hemoglobins. By performing the chromatography first, unlike Ingram who did the electrophoresis first, we were able to improve the resolution of the 'fingerprints' by taking advantage of the percolation of the buffer—applied to dried paper chromatograms—towards the origin, which tended to 'sharpen' the bands.

"At the onset, I indicated with some chagrin that this methodological paper, and not one of my more recent conceptual articles, has become a 'best seller'. Yet history's judgement may not be wholly inappropriate. This paper was, in fact, the only full-length paper to have come from my two-year tenure as Research Associate in Dr. Anfinsen's laboratory at the NIH. I was, at that time, most jealous of my friends who were able to grind out large numbers of papers laden with data while I struggled for almost two years with methodology. In retrospect, however, I have come to appreciate the opportunity I had to learn many different techniques that proved invaluable in my later work. It is for this reason that the paper I wrote with Bill Dreyer and Chris Anfinsen not only calls up vivid memories of my scientific youth, but also serves as a reminder of the value of methodology in scientific observations."

# This Week's Citation Classic

CC/NUMBER 35
AUGUST 27, 1984

Woods K R & Wang K T. Separation of dansyl-amino acids by polyamide layer chromatography. *Biochim. Biophys. Acta* 133:369-70, 1967.
[New York Blood Center, New York, NY and Dept. Chemistry, National Taiwan University, Taipei, Taiwan]

The paper describes polyamide separation of dansyl derivatives. The terse, simple instructions were apparently easy to follow reproducibly in many laboratories, establishing the method as a standard analytical technique. [The *SCI*® indicates that this paper has been cited in over 1,505 publications since 1967.]

---

K.R. Woods
New York Blood Center
310 East 67th Street
New York, NY 10021

May 24, 1984

"I have not met Kung-Tsung Wang to this day. He became my pen-pal collaborator for a short while and made a significant contribution to our repertoire of methods for unraveling protein structure.

"During 1966, I was privileged to serve the Rockefeller Foundation on a brief scientific mission to India and took the opportunity en route to visit Brian Hartley's laboratory in Cambridge where intensely fluorescent dansyl derivatives were being separated and identified on filter paper by high-voltage electrophoresis.[1] The wattage requirement was sufficient for an electrocution; thus, the apparatus and its lethal power supply had to be secured in a special limited-access room. The British are so heroic! I lacked the courage to duplicate that apparatus in my laboratory.

"After returning from India intending to find a safer way to separate dansyl-amino acids, I noticed an article by Wang et al. describing use of nylon-coated Mylar films for separating dinitrophenyl amino acids.[2] I was curious and wrote to Wang suggesting that his substrate should have just the right properties for partition chromatography of dansyl-amino acids. He promptly replied with an offer not to be refused. Explaining that his department could not afford dansyl-reference standards or reagents to synthesize them, he proposed an exchange. I went to the balance table; tipped a few crystals of each of my reference standards onto weighing papers; folded, labeled, and sealed them with Scotch tape; and sent them to Wang in an ordinary airmail envelope. Soon I received a return letter with a 2.5 × 15 cm. polyamide-Mylar strip enclosed. Many separation experiments were performed using that strip, rinsing away the dansyl compounds with ammonia or acetone between each trial. We quickly confirmed that solvent systems later to be described in our publication enabled consistently comparable separations both in Taipei and New York.

"And now for the rest of the story. About four years ago, I received a telephone call from the late Stanford Moore, an infrequent though hardly unique occasion. 'Woods?' he paused awaiting confirmation that indeed it was I, 'I have just returned from Taiwan. Have you ever met K.T. Wang?' 'No,' I replied, 'I have not.' 'I looked him up,' said Moore, 'and he's quite an undaunted scientist.' Moore went on to tell me he had learned that when Wang was a graduate student and came up with the idea of making polyamide laminates for chromatographic separations, it was Wang's professor who contributed the only readily available polyamide—his best white nylon shirt! Wang dissolved the shirt in formic acid (presumably having removed the buttons) and went to work coating window glass to make his first polyamide layers.[3]

"Now if it should happen that readers of this account have enjoyed the precision and simplicity of using polyamide layers for separating dansyl-amino acids, you will share my admiration for Wang's resourcefulness and appreciate the generosity of his professor who literally, for the sake of Wang's experiments, gave him the shirt off his back!

"Why do I think this publication has been so frequently cited? It is brief, fewer words than this commentary, and the technique is elegant for its simplicity, providing a rapid and sensitive means for separating dansylated products of proteins and peptides and identifying their N-terminal amino acids.

"An example of its persistence as a standard technique appeared as recently as Váradi and Patthy's report from Budapest delineating a tripeptide segment of fibrinogen responsible for plasminogen binding.[4] The technique provided convenient help in sorting out essential enzymic fragments derived from fibrinogen."

---

1. **Gray W R & Hartley B S.** A fluorescent end-group reagent for proteins and peptides. *Biochemical J.* 89:59P, 1963. (Cited 640 times.)
2. **Wang K T, Huang J M K & Wang I S Y.** Polyamide layer chromatography of dinitrophenyl amino acids. *J. Chromatography* 22:362-8, 1966.
3. **Wang K T, Wang I S Y & Lin A L.** Polyamide thin layer chromatography. *J. Chin. Chem. Soc. Ser. II* 8:241-50, 1961.
4. **Váradi A & Patthy L.** $\beta(Leu_{121}\text{-}Lys_{122})$ segment of fibrinogen is in a region essential for plasminogen binding by fibrin fragment E. *Biochemistry—USA* 23:2108-12, 1984.

# This Week's Citation Classic

CC/NUMBER 40
OCTOBER 6, 1980

Vesterberg O & Svensson H. Isoelectric fractionation, analysis, and characterization of ampholytes in natural pH gradients. IV. Further studies on the resolving power in connection with separation of myoglobins. *Acta Chem. Scand.* **20**:820-34, 1966.
[Biochem. Dept., Nobel Med. Inst., Stockholm, and Dept. Phys. Chem., Chalmers Inst. Technol., Gothenburg, Sweden]

With a new system of buffer substances it was possible to create pH gradients suitable for isoelectric focusing of proteins. The method offered outstanding properties at separation for preparative and analytical purposes. The isoelectric point of proteins could be determined easily and was reproducible. [The *SCI®* indicates that this paper has been cited over 1,130 times since 1966.]

Olof Vesterberg
Department of Occupational Health
Division of Chemistry
National Board of Occupational
Safety and Health
S-171 84 Solna
Sweden

August 19, 1980

"The techniques for analytical and preparative separation of proteins have taken great strides during the last two decades. Seventeen years ago, the principle of isoelectric focusing was known by just a few scientists. At that time, I was lucky to start working with Harry Svensson at the Karolinska Institute, Stockholm. He had just worked out the theoretical basis of isoelectric focusing.[1] However, suitable buffer substances (carrier ampholytes) were lacking. Specialists in organic synthesis were contacted, but without success. One day it became obvious to me that it should be possible to make a large number of buffer substances with a suitable distribution of acidic and basic dissociation constants (pK values). During my studies I learned that pK values of polyvalent protolytes could be influenced by various substituents in the molecules, including the protolytic groups themselves. Thus, by coupling carboxylic acids to polyvalent amines, it should be possible to synthesize ampholytes useful in isoelectric focusing. After having considered various ways to make such buffer substances, I was able to prepare the first synthetic carrier ampholytes. They were shown to give the predicted pH course on electrolysis and also to effect protein separation by isoelectric focusing. After these successful initial results, I worked very hard. Myoglobin, cytochrome c, and lactoperoxidase were now intensively examined by isoelectric focusing. Their isoelectric points were directly determined, and their heterogeneity was confirmed.

"Very interesting calculations were made concerning the resolving power of the method and also about the relation between differences in charge and in isoelectric points of very similar multiple molecular forms.[1] It was concluded that a charge difference of a few tenths of an electronic unit, or a difference in isoelectric point of about 0.01 pH unit, was sufficient to allow separation by isoelectric focusing. This implied a theretofore unattained resolving power. Isoelectric points of proteins could be determined in a very simple direct way (much easier than with other methods) and with a high degree of repeatability.[2] Thus a very valuable physico-chemical characteristic of proteins became easily available.

"More than 1,000 requests for reprints were received after publication of this article. I believe that the above text and mentioned advantages can explain this unusual interest. LKB Produkter's (Sweden) action to make apparatus and chemicals commercially available shortly after the article appeared also contributed to the success of the new method. The article became, so to speak, the definitive breakthrough of isoelectric focusing since it in all details substantiated theoretical predictions and gave clear evidence of an unsurpassed resolving power.

"In cooperation with others I have up till now published almost 60 papers in which isoelectric focusing has been used, and a review was recently published.[3] Internationally, there are presently about 5,000 papers dealing with the method.

"Recently, I have been developing a new sensitive method for quantification of proteins by immunoelectrophoresis in tubes.[4] Due to a very wide applicability it is also expected to attract the interest of many."

1. Svensson H. Isoelectric fractionation, analysis, and characterization of ampholytes in natural pH gradients. I. The differential equation of solute concentrations at a steady state and its solution for simple cases. *Acta Chem. Scand.* **15**:325-41, 1961.
2. Vesterberg O. Isoelectric fractionation, analysis, and characterization of ampholytes in natural pH gradients. V. Separation of myoglobins and studies on their electro-chemical differences. *Acta Chem. Scand.* **21**:206-16, 1967.
3. ──────. Isoelectric focusing. A review of analytical techniques and applications. *Int. Lab.* **1978**:61-8, 1978.
4. ──────. Quantification of proteins with a new sensitive method—zone immunoelectrophoresis assay. *Hoppe-Seylers Z. Physiol. Chem.* **361**:617-24, 1980.

# This Week's Citation Classic

CC/NUMBER 22
JUNE 2, 1980

Cuatrecasas P. Protein purification by affinity chromatography: derivatizations of agarose and polyacrylamide beads. *J. Biol. Chem.* 245:3059-65, 1970.
[Lab. Chem. Biol., Nat. Inst. Arthritis and Metabolic Diseases, NIH, Bethesda, MD]

The preparation of agarose and polyacrylamide bead derivatives for the purification of a variety of proteins and enzymes is described. The methodologies described permit the attachment of ligands directly or through extended hydrocarbon chains to immobilized supports. [The *SCI*® indicates that this paper has been cited over 1,580 times since 1970.]

Pedro Cuatrecasas
Wellcome Research Laboratories
Burroughs Wellcome Company
Research Triangle Park, NC 27709

May 14, 1980

"One of the more gratifying aspects of scientific work is the knowledge that one's own contributions have helped and influenced other scientists and thus furthered the overall progress of science. The manner by which this paper may have had an impact and the reason for its frequent citation are difficult to assess, but the methods and suggested applications may have spurred interest and encouragement. This was one of the first papers in this field, which has since seen literally thousands of publications.

"The term 'affinity chromatography' was first christened in 1968 when my colleagues, M. Wilchek and C.B. Anfinsen, and I used biospecific adsorption to purify several enzymes.[1] The basic concepts described in that and the present paper were, as in virtually all scientific discoveries, heavily based on previous knowledge. The fundamental ideas were simple and rational and surely had been in the minds of others. As is frequently the case, the time was probably ripe for more formally promulgating the principles and procedures in the simplest of terms in order to help establish the generality of the method.

"The impetus for this work arose from studies of the active site of micrococcal nuclease by affinity labeling with specific inhibitors. The reasoning was simply that if an inhibitor could be directed irreversibly to the active site, then why couldn't the enzyme be made to bind by its *active site* to an inhibitor irreversibly bound to a solid polymer or support. It is on this historical note that the idea and term 'affinity chromatography' were conceived.[1]

"Although excellent examples of the basic concepts existed prior to this work, these had apparently been considered by others as isolated and unique examples. The paper under consideration attempted to formalize the approach and methodology in order to focus attention on its general applicability and feasibility. By detailing the concepts and techniques, and by describing specific chemical manipulations of possible general utility, it was hoped that others would also perceive that the purification of biologically active macromolecules by biospecific adsorption was a potentially valuable tool that could be approached *systematically* in various fields of biochemistry and biology. The fact that affinity chromatography is now common nomenclature in biochemistry, without need for citation, attests to the validity and inherent obviousness of the concepts championed in our early publications.[2]

"The paper tried to describe the basic procedural principles in practical terms, and it detailed numerous simple chemical strategies for derivatizing ligands and solid supports for use as specific adsorbents. It described the importance of interposing spacers between the matrix backbone and the ligand. The general feasibility of deliberately designing insoluble supports of virtually any ligand was stressed. Illustrative examples were presented to provoke interest; in addition to enzymes, the applicability to hormone receptors, cyclic nucleotide-binding proteins, SH-group-containing proteins and intact cells was described.

"Since this publication, much progress and improvements have been made both in technology and in specific applications; the methods are now routine for purifying receptors, binding proteins, and for cell separations."[2]

1. Cuatrecasas P, Wilchek M & Anfinsen C B. Selective enzyme purification by affinity chromatography. *Proc. Nat. Acad. Sci. US* 61:636-43, 1968.
2. Jakoby W B & Wilchek M, eds. Affinity techniques. Enzyme purification: part B. *Methods Enzymol.* 34:3-810, 1974.

105

# Citation Classics

**Number 2** — January 9, 1978

Boucher R, Veyrat R, de Champlain J & Genest J. New procedures for measurement of human plasma angiotensin and renin activity levels. *Can. Med. A. J.* 90:194-201, 1964.

The authors describe an improved procedure for angiotensin isolation and determination and a new method for the measurement of plasma renin activity. [The *SCI*® indicates that this paper was cited a total of 667 times in the period 1964-1976.]

---

Dr. Roger Boucher, Director
Laboratory of Biochemistry
on Hypertension
Clinical Research Institute of Montreal
Montreal, Quebec H2W 1R7, Canada

February 14, 1977

"It is difficult to foresee the importance that a publication will have. I guess no one does publish with the premeditated idea his article will one day be one of the most cited articles published! This is extremely flattering and a great honor. It is also a reflection of the important interest in the field of experimental and clinical hypertension and the great need for an accurate and precise method for measuring plasma renin.

"My work received impetus when Dr. Jacques Genest, realizing that the disturbances found in aldosterone regulation could not explain by themselves the basic mechanisms of essential hypertension, was studying the relationships to those disturbances of the renal pressor system. When he demonstrated in 1959-60 that angiotensin was a major factor controlling aldosterone, he asked me to devise a method for measurement of plasma angiotensin and of renin. We had many discussions about the physiological and clinical applications of such a method. The key element in the method was the finding that the Dowex resin [$50W-X2(NH_4^+)$] not only protected the formed angiotensin from the proteolytic enzymes, but also did not modify the kinetics of the renin-substrate reaction and permitted the purification of the angiotensin. This work was presented at the International Symposium on "Angiotensin-Aldosterone-Sodium and Hypertension" held at the Chantecler Hotel in Ste-Adèle, Canada, in October 1963. It was organized by our multidisciplinary hypertension research group of the Clinical Research Institute of Montreal. Dr. Robert Veyrat, a Swiss research fellow from Geneva, who unfortunately died prematurely in March 1973, and Dr. Jacques de Champlain, a fellow of the Medical Research Council of Canada and now Professor of Physiology at the University of Montreal, contributed in many ways.

"As it turned out, the method for measurement of plasma angiotensin was not used extensively because of the large amounts of blood required to perform the assay. But I guess that the method for plasma renin activity determination, which was subsequently used by almost every other worker in the field and modified by a number in minor ways, resulted in crucial contributions to our knowledge of the physiopathology of the renin-angiotensin system, and its importance in the diagnosis of renovascular hypertension and primary aldosteronism and in the regulation of aldosterone secretion and of sodium.

"We have always shared the view that good methods are a major factor in scientific progress."

# This Week's Citation Classic

CC/NUMBER 26
JUNE 28, 1982

Greenfield N & Fasman G D. Computed circular dichroism spectra for the evaluation of protein conformation. *Biochemistry* 8:4108-16, 1969.
[Graduate Department of Biochemistry, Brandeis University, Waltham, MA]

The circular dichroism (CD) spectra of several proteins, whose secondary structures were known from X-ray diffraction studies, were fitted by a linear combination of the CD spectra of poly-L-lysine in the α helical, β pleated sheet, and random coil forms. The agreement between the structures predicted by CD and those found by X-ray crystallography was good. [The *SCI®* indicates that this paper has been cited over 850 times since 1969.]

Norma Greenfield
Department of Obstetrics and Gynecology
College of Physicians & Surgeons
Columbia University
New York, NY 10032

March 3, 1982

"This paper is an extension of work that was done with Betty Davidson when I had just started graduate research in Gerald Fasman's lab. My thesis project was to study the spectral transitions of model amides to help understand the optical activity of proteins. Davidson, a postdoctoral fellow in the lab, was studying factors affecting the conformational transitions of poly-L-lysine, as monitored by optical rotatory dispersion (ORD). It was felt that this basic polypeptide might ultimately serve as a model for the histones. She expected to observe a helix-coil transition when the peptide was heated and was surprised when instead it aggregated to give a spectrum which she had not seen before. This new spectrum was identified as that of the anti-parallel pleated sheet, or β form.

"We realized with some excitement that we inadvertently had obtained the first good model reference ORD spectra for the β pleated sheet conformation of proteins in solution. We therefore tried to see if we could predict the secondary structure of a protein by fitting its ORD spectrum with a linear combination of the spectra of the α helix, β pleated sheet, and random coil. I had just finished taking a two-week course in how to program in FORTRAN so the job of curve fitting was given to me.[1] The results using ORD unfortunately were not in good agreement with those determined from X-ray crystallography.

"When we obtained a CD attachment for our spectropolarimeter we repeated our study because CD gives better resolution of spectral transitions than ORD. This time our fits were good; in retrospect they were actually better than we reported. When we did our studies there were only preliminary reports on the X-ray crystallography of chymotrypsin. Our predicted values turned out to agree very well with what was later found.

"I think that the paper has been cited as often as it has for several reasons. 1. It contained useful methods to calculate protein conformation whether or not one had a computer. 2. It gave precise reference data for the CD spectra of poly-L-lysine in the three reference conformations with clear illustrations. 3. When it was written very little was known about the secondary structure of globular proteins since only six proteins had been examined by X-ray crystallography. CD proved to be an excellent method to show that β structure was as ubiquitous in globular proteins as the well-characterized α helix.

"Today, calculations of protein conformation from CD spectra usually employ reference data obtained from the analysis of the CD spectra of proteins whose structure has been determined by X-ray crystallography.[2-5] These methods give somewhat improved correlation between CD spectroscopic and X-ray crystallographic estimations of secondary structure.

"It is amazing to me that I have on my desk an inexpensive computer which is more powerful and easier to use than the one at Brandeis University which took up a large room and cost over 100 times as much; and the statistical and mathematical techniques that I once found so complicated I now apply routinely in areas ranging from enzyme kinetics to difference spectral measurements of cholesterol binding to cytochrome P-450. It was so difficult. Now it is so simple."

1. **Greenfield N, Davidson B & Fasman G D.** The use of computed optical rotatory dispersion curves for the evaluation of protein conformation. *Biochemistry* 6:1630-7, 1967.
2. **Saxena V P & Wetlaufer D P.** A new basis for interpreting the circular dichroism spectra of proteins. *Proc. Nat. Acad. Sci. US* 68:969-72, 1971.
3. **Chen Y H, Yang J T & Chau K H.** Determination of the helix and form of proteins in aqueous solution by circular dichroism. *Biochemistry* 13:3350-9, 1974.
4. **Provencher S W & Glockner J.** Estimation of globular protein secondary structure from circular dichroism. *Biochemistry* 20:33-7, 1981.
5. **Hennessey J P, Jr. & Johnson W C, Jr.** Information content in the circular dichroism of proteins. *Biochemistry* 20:1085-94, 1981.

**Number 22**     **Citation Classics**     May 29, 1978

**Nemethy G & Scheraga H A.** Structure of water and hydrophobic bonding in proteins. 1. A model for the thermodynamic properties of liquid water. *J. Chem. Phys.* **36**: 3382-400, 1962.

The authors describe the structure of liquid water in terms of an equilibrium between small hydrogen-bonded clusters and non-hydrogen-bonded molecules, and derive the distribution of molecules among five hydrogen-bonding states from an approximate partition function. (The *SCI®* indicates that this paper was cited 603 times in the period 1962-1976.)

George Nemethy
Department of Chemistry
Cornell University
Ithaca, NY 14853

December 13, 1977

"It is somewhat surprising that a statistical-thermodynamic paper would be in the list of the 'most cited articles.' However, this seems to be an example of being in the right place at the right time: the model of water structure which I developed as my Ph.D. thesis with Professor Scheraga in 1962 appeared at a time when its subject matter commanded much attention, and when the paper could serve as a stimulating starting point for studies in various fields.

"It had been recognized for many decades that the unusual properties of water as a liquid and as a solvent could be ascribed to the presence of intermolecular association, due to hydrogen bonding between the water molecules. In addition, in the 1950's, the recognition gained ground that interactions in biological macromolecules are affected profoundly by the structure of water and by its changes. Such interactions in turn decisively influence the spatial structure and therefore the biological activity of these molecules. In fact, the aim of our studies with Dr. Scheraga was not only to derive a reasonable model for the structure of water but to apply it to a quantitative explanation of the phenomenon of hydrophobic interaction, i.e., the interaction of nonpolar solute groups in an aqueous medium.

"Almost all of the older models recognized that water can be described in terms of a mixture of a hydrogen-bonded, bulky aggregation of water molecules (in a sense 'ice-like' in structure, although this term has led to much abuse and misunderstanding) with molecules not possessing hydrogen bonds and packed more densely. In most models proposed prior to about 1960, such a mixture was described in terms of a simple two-state thermodynamic equlibrium.

"The model which we proposed with Dr. Scheraga in 1962 was more elaborate. It had three main features. (1) Five, rather than just two states of water molecules were used to describe hydrogen-bonding equilibria, and an attempt was made to assign self-consistent physical properties to these states. (2) An approximate statistical-mechanical formulation was used. (3) The model and its mathematical formulation could be extended in a simple fashion to provide a physical explanation and a quantitative description of the unusual thermodynamic behavior of nonpolar solutes in water. This extension of our model was based on fundamental ideas proposed and developed by some earlier workers, notably by H.S. Frank and by W. Kauzmann.

"At the time of the appearance of the paper, interest in both water structures and hydrophobic interactions was widespread. The combination of the intuitive simplicity of the model with its power of quantitative prediction made it appealing to many workers in the field of aqueous solutions and of polymers. In spite of many shortcomings, the paper seemed to provide a useful framework at a time when discussion of the role of water in macromolecules had become fashionable."

# This Week's Citation Classic

Gregory R A & Tracy H J. The constitution and properties of two gastrins extracted from hog antral mucosa. Part I. The isolation of two gastrins from hog antral mucosa. Part II. The properties of two gastrins isolated from hog antral mucosa. *Gut* 5:103-17, 1964.
[Physiological Lab., Univ. Liverpool, England]

Almost 60 years after its discovery, the hormone gastrin was isolated as a pair of heptadecapeptide amides of identical amino acid constitution. They differed only in that the single tyrosine residue present was sulphated in one member of the pair. The pure hormone was shown to have a wide range of actions on motor and secretory functions in the digestive system. [The *SCI®* indicates that this paper has been cited in over 670 publications since 1964.]

R.A. Gregory
Physiological Laboratory
University of Liverpool
Liverpool L69 3BX
England

September 9, 1983

"In July 1959, Hilda Tracy and I reluctantly undertook to try to make a gastrin extract which would stimulate gastric secretion in a conscious dog; we needed it to clarify the mechanism by which fat in the duodenum inhibited gastric secretion. I was unenthusiastic; exactly 20 years before, I had gone from University College London to learn about gastrointestinal physiology from Andrew Ivy at Northwestern University, Chicago. My first task was to make a gastrin extract described in a then recent paper[1] and test it on a conscious dog. The result was a spectacular disaster; my extract would have done credit to a soup kitchen and was an excellent emetic, but it did not stimulate gastric secretion.

"By the time gastrin and I met again, I had become chairman of physiology at the University of Liverpool. The laboratory in which Hilda and I eventually isolated gastrin had—unknown to us—been much frequented by a young demonstrator, John Sydney Edkins, before he moved to London, where in 1905 he made the experiments on anaesthetised cats which convinced him that there existed a gastric hormone; he named it 'gastrin.'[2] After his brief moment of fame, there set in what Bill Paton of Oxford once described as 'the tide of undiscovery'; it eventually seemed clear that gastrin was probably the ubiquitous histamine, and Edkins's hypothesis fell into a limbo of doubt and disrepute.

"In 1938, Simon Komarov of Montreal made a histamine-free extract which was active in anaesthetised cats; but it had little or no action in conscious dogs and its effect was resistant to atropine, which inhibited gastric secretion to feeding in conscious animals and man. There was general doubt that 'Komarov's gastrin'[1] represented the hormone. It was in fact Komarov's extract that I had tried to make in Ivy's laboratory; Komarov later excused my incompetence by telling me the circumstances which accounted for his description of the method being difficult to follow—another story!

"Hilda and I eventually found a new method of extraction which gave a histamine-free potent product effective in the conscious dog. We were not contemplating isolation, and sent a short paper describing the method to a well-known journal. It was rejected with a curt note that the work might be of interest if taken further. Hilda was nettled: 'We will!' she said—and the outcome was the paper which is the subject of this commentary. On the way, we showed that the Zollinger-Ellison syndrome results from the inappropriate elaboration by the pancreatic tumour or its metastases of a gastrin-like stimulant[3] which we later isolated and found identical with the human antral hormone.

"The gastrin story has been recounted in many lectures[4,5] and reviews,[6,7] in which I have endeavoured to acknowledge our great debt to distinguished colleagues, including George Kenner (deceased) of Liverpool, Ieuan Harris (deceased) of Cambridge, Kenner's pupils Bob Sheppard (now in Cambridge) and Kan Agarwal (now in Chicago), and Morton Grossman (deceased) of CURE, Los Angeles.

"Perhaps this paper has been widely quoted because gastrin was the first gastrointestinal hormone to be sequenced and synthesised; its isolation ended more than a half century of uncertainty and initiated a great number of studies on its numerous actions, most of which were identified as properties of the hormone for the first time

1. **Komarov S A.** Gastrin. *Proc. Soc. Exp. Biol. Med.* 38:514-16, 1938.
2. **Edkins J S.** On the chemical mechanism of gastric secretion. *Proc. Roy. Soc. London B* 76:376, 1905.
3. **Gregory R A, Tracy H J, French J M & Sircus W.** Extraction of a gastrin-like substance from a pancreatic tumour in a case of Zollinger-Ellison syndrome. *Lancet* 1:1045-8, 1960. (Cited 280 times.)
4. **Gregory R A.** The gastrointestinal hormones: a review of recent advances. The Bayliss-Starling Lecture 1973. *J. Physiology* 241:1-32, 1974.
5. ──────. Some aspects of the gastrointestinal hormones—past, present and future. *Jpn. J. Gastroenterol.* 77:1-23, 1980.
6. ──────. A review of some developments in the chemistry of the gastrins. *Bioorg. Chem.* 8:497-511, 1979.
7. ──────. Heterogeneity of gut and brain regulatory peptides. *Brit. Med. Bull.* 38:271-8, 1982.

## This Week's Citation Classic

NUMBER 11
MARCH 12, 1979

Johns E W. Studies on histones 7. Preparative methods for histone fractions from calf thyumus. *Biochem. J.* 92:55-9, 1964.

The author describes two preparative methods for the isolation of the four main groups of histones, F1, F2A, F2B and F3, from calf thymus. [The *SCI®* indicates that this paper has been cited over 625 times since 1964.]

E.M. Bradbury
Portsmouth Polytechnic
Biophysics Laboratory
Portsmouth, England

April 14, 1977

"The methods developed for the isolation of histones by E.W. Johns and his colleagues, particularly D.M.P. Phillips, have provided the major impetus to most of the chemical and physico-chemical studies on histones. The reason for this success lay in the development of simple clean methods for the isolation and purification of large quantities of the individual histones. In this paper two methods were given for the large scale isolation of four histone groups in one preparation. These were called F1, F2A, F2B, and F3. A subsequent paper[1] showed how F2A could be separated into two components to give F2A1 and F2A2. Thus complete separation of the five histone classes, now called H1, H2A, H2B, H3 and H4, was achieved. Approximately 400 mg of each of these fractions could be isolated from 100 g of calf thymus, and the histones could be further purified by conventional techniques. Other methods used at the time employed column separation, and these were unable to separate the arginine-rich histones H3 and H4 and gave only a few mgs of the other histones.

"Serendipity always plays a part in major scientific achievements and some of the procedures established were the results of happy accidents. Method 1 came about because histones were being tested for RNase activity. Samples of histones were incubated with RNA, and perchloric acid (PCA) added to precipitate all high molecular weight materials. As a control, PCA was also added to solutions of the individual histones. All precipitated except the very lysine-rich histone H1. PCA was then used for the specific isolation of H1. Method II was discovered because ethanol and acid were being used to inhibit proteases and to make other proteins less soluble. However, only about half of the histone was extracted. This proved to be a specific method for the separation of histones H2A, H3 and H4 from H1 and H2B. Another feature of this paper is that it contained the first reference to the now very interesting HMG chromosomal proteins as a contaminant of H1.

"It is quite clear that these methods formed the basis for the successful histone sequencing and for many of the detailed chemical and physical studies of the properties of histones. In a decade from now I wouldn't be surprised to be writing about the HMG group of histones."

### REFERENCE

1. Johns E W. A method for selective extraction of histone fractions F2(A)1 and F2(A)2 from calf thymus deoxyribonucleoprotein at pH 7.
   *Biochem. J.* 105:611-4, 1967. [The *SCI®* indicates that this paper has been cited over 140 times since 1967.]

| Number 49 | **Citation Classics** | December 5, 1977 |

Ratnoff O D & Menzie C. A new method for the determination of fibrinogen in small samples of plasma. *Journal of Laboratory and Clinical Medicine* 37:316-20, 1951.

The authors describe a simplified method for the determination of fibrinogen in plasma which permits the accurate and rapid determination of the amount of fibrinogen in samples of normal plasma as small as 0.1 ml, one tenth the amount used in methods previously described. [The *SCI*® indicates that this paper was cited 668 times in the period 1961-1975.]

Dr. Oscar D. Ratnoff
Department of Medicine
Case Western Reserve University
Cleveland, Ohio 44106

February 2, 1977

"The flattery implicit in having a paper among the top 500 is tempered by the realization that this is indeed a minor work, providing no new biological insights. When I returned to Johns Hopkins University School of Medicine after World War II, I began a series of studies of the fibrinolytic properties of plasma. Inevitably, I was drawn to an investigation of Jobling's[1] report that the crisis in pneumonia was associated with an increase in the proteolytic activity of serum. Analysis of the events that took place during the course of pneumonia required a rapid but accurate technique for measuring the concentration of fibrinogen in plasma, since this protein is a natural substrate of plasmin, a fibrinolytic enzyme that can be generated from its precursor, plasminogen. Earlier methods were either inaccurate, required large amounts of plasma, or were time-consuming.

"With the able collaboration of Mr. Calvin Menzie, who was Dr. C. Lockard Conley's technician, I devised a technique for capturing the fibrin that formed when diluted plasma was clotted by thrombin. We wound the fibers upon crushed glass, allowing us to separate the fibrin from other proteins with a minimum of effort. We then determined the concentration of fibrin by the Folin-Ciocalteau technique, standardized against both a gravimetric and a chemical analysis for protein. Accurate results required only 0.5 ml of plasma, much less than other methods available at the time. I probably didn't explain this well in my text, for a prominent journal rejected the manuscript with the comment that the results I obtained with my micromethod were the same as with macromethods, and thus represented no advance. This seems to lend a touch of irony under the present circumstances. In any case, while still in Baltimore, I used the method in several projects, including one long forgotten study in which Clay and I[2] demonstrated that, contrary to previous dogma, hypofibrinogenemia need not follow total hepatectomy in dogs if the technique were sufficiently careful.

"We have abandoned our original method, preferring the modification of Ogston, Ogston and Bennett[3] in which the fibrin formed in diluted plasma is caught on a glass cloth filter and is thus more readily washed. Additionally, we form the clot in the presence of 0.005M epsilon-aminocaproic acid, to reduce possible losses from fibrinolysis. More recently, one of my young colleagues, Dr. M. Mortazavi, has found a commercially available, semiautomated adaptation of the method of Ellis and Stransky[4] to be satisfactory, and perhaps more accurate in hypofibrinogenemic states, but this method has not been tested with plasmas containing qualitatively abnormal fibrinogens."

REFERENCES
1. Jobling J W, Petersen W & Eggstein A A. The serum ferments and antiferment during pneumonia. *Journal of Experimental Medicine* 22:568-89, 1915.
2. Clay R C & Ratnoff O D. Modified one-stage hepatectomy in the dog: with some notes on the effect of hepatectomy on the coagulability and proteolytic activity of the blood. *Bulletin of the Johns Hopkins Hospital* 88:457-72, 1951.
3. Ogston D, Ogston C M & Bennett N B. Arterio-venous differences in the components of the fibrinolytic enzyme system. *Thrombosis et Diathesis Haemorrhagica* 16:32-7, 1966.
4. Ellis B C & Stransky A. A quick and accurate method for the determination of fibrinogen in plasma. *Journal of Laboratory and Clinical Medicine* 58:477-88, 1961.

# This Week's Citation Classic

CC/NUMBER 45
NOVEMBER 7, 1983

Deutsch D G & Mertz E T. Plasminogen: purification from human plasma by affinity chromatography. *Science* 170:1095-6, 1970.
[Department of Biochemistry, Purdue University, Lafayette, IN]

Plasminogen was prepared from human plasma by affinity chromatography on L-lysine coupled to agarose. The single step procedure produced plasminogen in approximately 90 percent yield and the product exhibited multimolecular forms on acrylamide gel electrophoresis. [The *SCI®* indicates that this paper has been cited in over 765 publications since 1970.]

---

Dale G. Deutsch
Department of Pathology
School of Medicine
and
University Hospital
State University of New York
Stony Brook, NY 11794

August 2, 1983

"This work on plasminogen was performed at Purdue University where I became a graduate student after two years in the Peace Corps. I worked in Edwin Mertz's laboratory where my project was to purify and characterize plasminogen. I succeeded four or five other graduate students who purified plasminogen from any species of animal whose blood they could obtain. The multiple step procedures were very laborious and after about a year of drying out and repacking monstrous columns, I became discouraged and drifted out of the laboratory. My lack of good results made me embarrassed to approach Mertz (who concluded I was goofing off) and he suggested I get a master's degree. To make matters worse, my course work wasn't going well and the chance to help take over the administration building, as part of an antiwar demonstration, came as a welcome relief during this period.

"The idea for the preparation of the affinity adsorbent came about as a result of a series of events. While traveling from Purdue to my home in New York one vacation, I visited my childhood friend, Alan Kaufman, who had started his own medical practice. In our scientific and clinical discussions, he reminded me that ε-aminocaproic acid was used in severe hemorrhage cases (it binds plasminogen and prevents its activation to plasmin[1]) to control bleeding. Some weeks later I thought of hooking up lysine to a column by its α-amino group (which yields an ε-aminocaproic acid functional group) to pull plasminogen out of blood. I mentioned this to my housemate and fellow biochemistry graduate student, George Doellghast. He, in turn, knew of a new procedure which allowed coupling of small ligands by their amino groups to agarose[2] using the CNBr reaction originally described for insolubilizing proteins.[3]

"In the summer of 1969, I worked out the initial method using just 10 ml of human plasma and a small affinity column. When I told Mertz about the new method he immediately recognized its utility and from then on I had his total support and encouragement until I graduated with my PhD. We published an abstract in *Federation Proceedings* and experienced a few setbacks while trying to get it in final form for publication in *Science*. For a while my recovery went from 90 percent to five percent which I discovered was due to switching from phosphate to Tris buffer for washing the column. I also spent weeks trying to measure the purity of the product by a spectrophotometric active site titration method. With the help of John Knox, a fellow graduate student who worked on Mertz's high lysine corn project, we determined that I had phosphate in my enzyme preparation and calcium in my assay buffer and I was measuring the kinetics of calcium phosphate precipitation. A committee at Purdue refused to patent the procedure because they thought it had no commercial value.

"This paper seems to be widely quoted because it is a technique paper which provides a simple and reliable method to prepare plasminogen and to remove plasminogen from other components. Plasminogen plays a central role in the dissolution of blood clots[4] and it is used as a component of other fibrinolytic assays, such as plasminogen activator. Plasminogen itself is studied from a structural point of view and the multiple forms of this proenzyme have now been explained, as described in a recent review article."[5]

1. Alkjaersig N, Fletcher A P & Sherry S. ε-Aminocaproic acid: an inhibitor of plasminogen activation. *J. Biol. Chem.* 234:832-7, 1959.
2. Cuatrecasas P, Wilchek M & Anfinsen C B. Selective enzyme purification by affinity chromatography. *Proc. Nat. Acad. Sci. US* 61:636-43, 1968.
3. Porath J, Axen R & Ernback S. Chemical coupling of proteins to agarose. *Nature* 215:1491-2, 1967. (835 cites.)
4. Deutsch D G. An orthomolecular approach to thrombolysis. *Perspect. Biol. Med.* 20:307-9, 1977.
5. Castellino F J & Powell J R. Human plasminogen. *Meth. Enzymology* 80:365-78, 1981.

# This Week's Citation Classic

CC/NUMBER 39
SEPTEMBER 29, 1980

Huisman T H J & Dozy A M. Studies on the heterogeneity of hemoglobin. IX. The use of tris(hydroxymethyl)aminomethane–HCl buffers in the anion-exchange chromatography of hemoglobins. *J. Chromatography* 19:160-9, 1965.
[Dept. Biochem., Medical College of Georgia, Augusta, GA]

A modified procedure for the separation of various hemoglobin types by anion exchange chromatography has been presented. DEAE-Sephadex, A-50 medium, was preferred over DEAE-cellulose as chromatographic medium. Complete separations of many hemoglobin fractions were obtained by applying a single pH gradient to the columns using TRIS-HCl buffers of reasonably high concentrations (0.05 M). The method was applicable both for analytical and preparative purposes. [The *SCI*® indicates that this paper has been cited over 460 times since 1965.]

Titus H. J. Huisman
Department of Cell and
Molecular Biology
Medical College of Georgia
Augusta, GA 30901

July 15, 1980

"This paper was one in a long series devoted to the development of chromatographic procedures to quantitate and isolate variants of human and animal hemoglobins (Hb). My technical assistant, Andrée Dozy, played a key role in these studies. She worked with me on these and related problems for some ten years, and is presently a key member of the research team of Y. W. Kan in San Francisco, California.

"Numerous human Hb variants had been discovered, mainly after 1960, and a great need existed for a reliable procedure to isolate these proteins for further structural analyses. The DEAE-Sephadex procedure is ideally suited for this purpose; it is an easy method which provides a pure isolated hemoglobin variant. The method has been modified by various investigators. Dozy, Enno Kleihauer (presently professor of pediatrics at the University of Ulm, Germany), and I introduced the use of the beaded DEAE-Sephadex anion exchanger which gave sharp Hb zones.[1] Presently the method has been replaced in several laboratories (including my own) by one which uses DEAE-cellulose and glycine-KCN-NaCl developers and was developed by Edathara C. Abraham, Alice Reese, Mamie Stallings, and myself.[2] Over the years I had numerous discussions and conducted collaborative studies with my friend Walter Schroeder, who has worked at Caltech in Pasadena for over 30 years. Together we developed and modified techniques for Hb separations which are presently used all over the world. We summarized our combined efforts in a book which just came off the press.[3]

"Anion exchange chromatography similar to the one described in this 'Citation Classic' will continue to be important in the analyses of Hb and its variants. Important new developments are modifications allowing analyses of micro quantities of Hbs in mixtures (numerous examples are listed in ref. 3). In the very near future we will observe a major advance when these anion exchangers are used in high performance liquid chromatographic methods. It seems likely that the long lasting (2-4 days) Hb analyses of the mid-1960s will be replaced by highly automated analyses which will be completed in less than 2 hours. It is gratifying to know that Dozy, Kleihauer, and I were able to develop methodology which forms the basis for so many approaches in column chromatography of Hbs."

1. Dozy A M, Kleihauer E F & Huisman T H J. Studies on the heterogeneity of hemoglobin. XIII. Chromatography of various human and animal hemoglobin types on DEAE-Sephadex. *J. Chromatography* 32:723-7, 1968.
2. Abraham E C, Reese A, Stallings M & Huisman T H J. Separation of human hemoglobins by DEAE-cellulose chromatography using glycine-KCN-NaCl developers. *Hemoglobin* 1:27-44, 1976.
3. Schroeder W A & Huisman T H J. *Chromatography of hemoglobin.* New York: Marcel Dekker, 1980. 272 p.

# This Week's Citation Classic

CC/NUMBER 2
JANUARY 11, 1982

Clegg J B, Naughton M A & Weatherall D J. Abnormal human haemoglobins: separation and characterization of the $\alpha$ and $\beta$ chains by chromatography, and the determination of two new variants, Hb Chesapeake and Hb J (Bangkok).
*J. Mol. Biol.* 19:91-108, 1966.
[Depts. Biophys. and Med., Johns Hopkins Univ. Sch. Med., Baltimore, MD]

The paper described a chromatographic technique for the quantitative fractionation of the peptide chains of human hemoglobins. The high resolution allowed separation on a preparative scale of globins differing by only a single charged residue. [The *SCI*® indicates that this paper has been cited over 890 times since 1966.]

John B. Clegg
Nuffield Department of Clinical Medicine
John Radcliffe Hospital
University of Oxford
Headington, Oxford OX3 9DU
England

October 5, 1981

"My first postdoctoral job in the US turned out to be a disaster which culminated after six months in a simultaneous resignation/firing ceremony and a hectic search for alternative employment. Eventually, relief came in the form of Howard Dintzis and Mike Naughton of the department of biophysics, Johns Hopkins University School of Medicine. I joined them in studying the biosynthesis of insulin by pulse-labelling, but nine months later the project was in the doldrums. There seemed to be no rhyme or reason for the labelling patterns of the peptides we had so laboriously isolated from our pancreas incubations. We subsequently discovered that our troubles had been caused by a fluorescent whitener in the blotting paper used to dry the papers for high voltage electrophoresis!

"By the way of a palliative, Naughton suggested that perhaps as a sideline I should turn back to protein work; in particular he had long been interested in hemoglobin and had tried unsuccessfully over some years, as had many others, to devise a way of easily fractionating human globin chains. It was clear from the amino acid compositions that the most likely pHs for optimum resolution by ion-exchange chromatography or electrophoresis would be in the neutral range, where the relative net charges on the globins were minimal—precisely the pHs where globin was virtually insoluble in aqueous solution. Here I was on home ground, for I had spent most of my PhD devising a method for fractionating the peptide chains of fibrin.[1] This too was insoluble, a complication that I had overcome by doing all the experiments in 8M urea. It seemed natural, then, to try on globin the method that had finally worked for fibrin, chromatography on CM-cellulose in urea.

"The first attempts were disappointing, with poor recoveries and what can only be described as a splurge instead of peaks, but we had ignored the free sulphydryl groups in globin; neutral pH in 8M urea would be ideal for forming mixed disulphides of any oxidised cysteines. A few mls of mercapthethanol in the next experiment produced a beautiful separation, and by the end of a week we had settled the conditions, essentially as they were subsequently published. Shortly afterward, David Weatherall came over from the department of medicine to try out the method on some of his radioactive thalassemic globin samples,[2] thus beginning a friendship and collaboration that has continued ever since.

"The method appeared at an appropriate time. Interest in human hemoglobin genetics, and the hereditary disorders like thalassemia, was rapidly gathering pace in the 1960s and there was a need for a good quantitative preparative technique, hence its appeal.

"This story seems to me to illustrate the absurdity of some of the pure-versus-applied science arguments. In recent years the method has been widely used for the antenatal diagnosis of the thalassemias.[3] It arose out of a personality clash and fluorescent blotting paper, and was paid for out of an insulin project. What politician would have ordained it that way?"

1. Clegg J B & Bailey K. The separation and isolation of the peptide chains of fibrin.
    *Biochim. Biophys. Acta* 63:525-7, 1962.
2. Weatherall D J, Clegg J B & Naughton M A. Globin synthesis in thalassaemia: an in vitro study.
    *Nature* 208:1061-5, 1965.
3. Alter B P & Nathan D G. Antenatal diagnosis of haematological disorders.
    *Clin. Haematol.* 7:195-216, 1978.

## This Week's Citation Classic

CC/NUMBER 32
AUGUST 6, 1979

Benesch R & Benesch R E. The effect of organic phosphates from the human erythrocyte on the allosteric properties of hemoglobin.
*Biochem. Biophys. Res. Commun.* 26:162-7, 1967. [Columbia University, College of Physicians and Surgeons, Department of Biochemistry, New York, NY]

2,3 Diphosphoglycerate (DPG) was found to be a powerful allosteric effector of oxygen binding by hemoglobin. The dramatic decrease in oxygen affinity in the presence of this compound suggested its role in regulating oxygen release under physiological conditions. [The *SCI®* indicates that this paper has been cited over 585 times since 1967.]

Reinhold Benesch
Department of Biochemistry
College of Physicians &
Surgeons of Columbia University
New York, NY 10032

May 9, 1979

"Our discovery in 1966 of the effect of DPG on hemoglobin was quite surprising, since so much of the relevant information had been around for a long time. The compound had been isolated in large quantities from red cells by Greenwald in 1925.[1] During the next 40 years biochemists vainly searched for a metabolic function and its possible relation to oxygen transport was ignored, although it was suspected in some quarters, e.g., by Barcroft in 1921, that there might be 'some third substance present...which forms an integral part of the oxygen-hemoglobin complex.'[2]

"We were led to DPG by the casual question of a graduate student, Peter Model, who is now a professor at Rockefeller. He wanted to know why we always measured oxygen equilibrium curves in phosphate buffer and how much inorganic phosphate there was in human erythrocytes. A standard textbook of biochemistry showed right away that while there was precious little inorganic phosphate, the organic phosphate, 2,3 DPG, loomed very large, i.e., in almost equimolar amounts with hemoglobin. Since the compound was even commercially available, two and two were quickly put together. It became obvious that DPG combines preferentially with the deoxy form of hemoglobin, thereby lowering the oxygen affinity and facilitating the unloading of oxygen. It does this by prevailing on the hemoglobin molecule to return more readily to the state in which it was before it picked up oxygen.

"The interaction of DPG with hemoglobin quickly became a 'growth industry' since it had important implications in so many fields. The finding that only one DPG molecule acted allosterically on four oxygen finding hemes introduced a novel element into classical symmetry-dominated allosteric theory. It also added another dimension to the multibly-linked interactions of hemoglobin with ligands, e.g., the competition of $CO_2$ and DPG for the same site.

"On the physiological side quite a few loose ends were tied up. The much lower oxygen affinity of avian blood could be accounted for by the replacement of DPG by more powerful cofactors, i.e., inositol polyphosphates. In fact, the chicken switches from DPG to inositol pentaphosphate after hatching to cope with the altered oxygen requirement. In man, on the other hand, the higher oxygen affinity of fetal as compared with maternal blood is due to the lower affinity of fetal hemoglobin for DPG.

"A practical application of the DPG story was the realization that the increased oxygen affinity of blood during storage is due to destruction of DPG. As a result new blood storage media have been introduced. Perhaps the most striking physiological lesson has been the response of the red cell DPG level to oxygen need. Thus, ascent to high altitude rapidly raises DPG levels to 50% above normal and a similar increase is seen in a variety of diseases where too little oxygen reaches the tissues. In many anemias, such as sickle cell disease, patients with very low hemoglobin levels can often deliver normal amounts of oxygen because of their increased DPG concentration.

"It has been very satisfying to see how a relatively simple experiment has made it possible to find answers to so many old puzzles."

1. Greenwald I. A new type of phosphoric acid compound isolated from blood, with some remarks on the effect of substitution on the rotation of λ-glyceric acid. *J. Biol. Chem.* 63:339-49, 1925.
2. Adair G S, Barcroft J & Bock A V. The identity of haemoglobin in human beings. *J. Physiol.* 55:332-8, 1921.

# This Week's Citation Classic

CC/NUMBER 46
NOVEMBER 16, 1981

Smith D W & Williams R J P. The spectra of ferric haems and haemoproteins.
*Struct. Bond.* 7:1-45, 1970.
[Dept. Chemistry, University of Sheffield, Sheffield, and Inorganic Chemistry Lab., Oxford, England]

The electronic spectra of ferric haems and haemoproteins are reviewed. Their analysis requires explicit consideration of mixing between $\pi$-$\pi^*$ and charge transfer states, as well as the effects of axial coordination and the recognition of spin equilibria. The probe properties of spin equilibria are discussed. [The SCI® indicates that this paper has been cited over 120 times since 1970.]

Derek W. Smith
Department of Chemistry
School of Science
University of Waikato
Hamilton, New Zealand

July 9, 1981

"In mid-1965, I began my doctoral work in Bob Williams's group in Oxford. I was then mainly interested in ligand field spectra, and I inherited from Andy Thomson a homemade microspectrophotometer for making single crystal measurements. It was a temperamental brute, driven by a battery of lead accumulators, but capable of good spectra given due care and attention. Measurements were best made in the dead of night, and the hours I kept caused my landlady to harbour doubts about my morals and sobriety. Bob's infectious enthusiasm soon seduced me into bio-inorganic chemistry, and we decided upon a study of ferrimyoglobin crystals. We began by making an expedition to 'that other place' (i.e., Cambridge) to pick the protein experts' brains at the MRC laboratory. I began to have second thoughts about the whole thing when Chris Nobbs asked about our supplies of whale meat, and explained the elaborate extraction and crystallisation procedure. But in the end, he kindly donated to us a generous supply of beautiful crystals. These provided us with useful information about the much-neglected near infrared spectra of ferrimyoglobin and its derivatives; we were also able to make some measurements on cytochrome-c. We then turned to a general interpretation of the electronic spectra of ferric haems and haemoproteins. Bands could be identified as arising from the high-spin and low-spin forms, both present in equilibrium, and could be assigned as porphyrin $\pi$-$\pi^*$ transitions or as ligand-to-metal charge transfer transitions, with extensive configuration interaction. Peter Day's understanding of charge transfer spectra came in useful here. Gaussian analysis of the spectra made it possible to determine the relative amounts of the two spin states in several axially-substituted derivatives.

"The work on ferrimyoglobin was relevant to studies of other haemoproteins, and after completing my D. Phil. thesis in 1968 I suggested to Bob that we publish a review together. The first 30 pages of the article as it appeared in print were largely abstracted from my thesis, while the last section, dealing with the probe properties of spin equilibria, was entirely Bob's work.

"I would like to think that the article has been extensively cited because it presents a comprehensive analysis of Fe(III)-porphyrin spectra, useful in the study of spin equilibria and axial coordination in haemoproteins. But I suspect that the article's popularity owes more to the provocative speculations and penetrating insights in Bob's contribution. The use of electronic spectra in studies of haemoproteins has, to a great extent, been overtaken by the rapid developments in other techniques in the 1970s. These are reviewed in A.S. Brill's admirable monograph."[1]

1. Brill A S. *Transition metals in biochemistry.* Berlin: Springer-Verlag, 1977. p. 81-117.

# This Week's Citation Classic

CC/NUMBER 19
MAY 12, 1980

Wetlaufer D B. Ultraviolet spectra of proteins and amino acids.
*Advan. Prot. Chem.* 17:303-90, 1962.
[Dept. Biochemistry, Indiana Univ. Sch. Med., Indianapolis, IN]

Practical and interpretive aspects of protein absorption spectroscopy were reviewed, especially with a view to extracting protein structural information. [The *SCI®* indicates that this paper has been cited over 390 times since 1962.]

Donald B. Wetlaufer
Department of Chemistry
University of Delaware
Newark, DE 19711

December 4, 1979

"My first work in protein ultraviolet spectroscopy was at the Carlsberg Laboratory where I was a postdoctoral visitor with K. Linderstrøm-Lang. I applied the then rather new technique of difference spectroscopy in model compound studies to test Crammer and Neuberger's suggestion that some of ovalbumin's tyrosyl residues were H-bonded to carboxylate sidechains in the protein.[1] My results appeared to rule out all but very weak associations.

"When I moved on to work in John Edsall's laboratory at Harvard it was inevitable that, in addition to my primary work on myosin, I became involved in studies already under way on tyrosyl peptides. In the late 1950s most biochemical labs had only limited physical chemical instrumentation, most likely pH meters and spectrophotometers. The attraction of using any plausible and available method to get protein fine-structure information struck several of a then young group of investigators at about the same time, and we were shortly in the middle of a minor publication explosion. Difference spectroscopy and refractive index perturbations became firmly rooted, and it was clear from synthetic polypeptide studies that the peptide absorption is dependent on conformation.

"I agreed to review the field, and started writing shortly after taking a new position in the mid-west. It was a very busy time, moving and settling a young family and equipping and staffing a new laboratory. Critical readings by Walter Gratzer and Edsall helped me smooth out a first draft, and by the time the review was actually published we were deep into a new vein of protein work.

"Protein spectroscopy matured with further studies, particularly in the peptide absorption region. These have been skillfully reviewed by Gratzer.[2]

"It's something of a surprise to learn that my review has continued to be useful. Perhaps the fact that it was pitched at an audience with limited physical chemical sophistication has contributed to its longevity. It also seems sensible that some familiarity with absorption spectroscopy, both theoretical and practical, is a useful preface to ORD, CD, and fluorescence studies."

1. Crammer J L & Neuberger A. The state of tyrosine in egg albumin and in insulin as determined by spectrophotometric titration. *Biochemical J.* 37:302-10, 1943.
2. Gratzer W B. Ultraviolet absorption spectra of polypeptides. (Fasman G D, ed.) *Poly-α-amino acids: protein models for conformational studies.* New York: Marcel Dekker, 1967. p. 177-238.

# This Week's Citation Classic

CC/NUMBER 25
JUNE 22, 1981

McKenzie H A. Milk proteins. *Advan. Prot. Chem.* 22:55-234, 1967.
[Dept. Physical Biochemistry, Institute of Advanced Studies, Australian National Univ., Canberra, ACT, Australia]

The occurrence, isolation, and properties of milk proteins are reviewed. The individual proteins exhibit an array of interactions. The problems that these pose in the study of the proteins are discussed. Attention is given to milk proteins in allergenicity. Present knowledge is critically assessed and predictions are made concerning future research. [The *SCI*® indicates that this paper has been cited over 140 times since 1967.]

Hugh A. McKenzie
Department of Physical Biochemistry
John Curtin School of Medical Research
Institute of Advanced Studies
Australian National University
Canberra City, ACT 2601
Australia

April 25, 1981

"No major review of the chemistry of milk proteins had appeared since that of McMeekin in 1954,[1] just prior to the commencement of a new era in the methods of protein chemistry. Soon after his review, two major discoveries did much to stimulate new work: that by von Hippel and Waugh[2] of a new casein component, κ-casein, considered responsible for the stabilization of milk micelles and that by Aschaffenburg and Drewry[3] of genetic variants. Thus the field was ripe for review.

"I believed that the unique significance of milk proteins in nutrition and protection of the newborn alone justified their study. However, they possess physico-chemical properties that make them of great importance in protein chemistry. Their interactions pose particular problems in their study. Hence special attention was given in the review to the pitfalls involved, and the precautions necessary for proper study.

"While I endeavoured to give a clear overall picture, I made no apology for emphasising certain aspects, and giving examples from my own work. This policy brought criticism from two readers, but the majority of the opinions received in the past 14 years have been laudatory. Thus, I believe, the approach has proved valuable, and it is this which accounts for the article's frequent citation. Another feature was that predictions were made concerning future discoveries as well as suggestions for future research. A particularly pleasing aspect has been that a number of these have proved to be valid, e.g., that the 'proteose peptones' are essentially caseins[4] and that variants without charge differences would be found.[3,5]

"Since 1967 there have been some remarkable advances in our knowledge of proteins, and of milk proteins in particular. The amino acid sequences of the major bovine milk proteins are now known. There is considerable knowledge of the proteins of other species. Nevertheless, we still do not know the detailed conformation of any of these proteins[6] (with the possible exception of α-lactalbumin). Also in 1967 it was stated in the last sentence of the review, 'Only when we can understand the many reactions of whole milk will we be able to say that we understand the proteins.' This has still not been attained in 1981.

"The article is probably unique among *Citation Classics* in that the idea of it was conceived in a prison. During the early 1960s Kevin Bell of the University of Queensland and I were studying the new lactoglobulin variant, β-lactoglobulin C. At that time the only homozygous animal located conveniently close to the university was at the Wacol prison. I often had to wait for considerable periods before the formalities of movement were completed and had time to contemplate: thus the idea was born (another example of Alyea's relevance of irrelevance). The real impetus to writing the article was provided by sabbatical leave spent during 1965 at the Frick Laboratory of Princeton University in the stimulating environment of a longtime friend, Walter Kauzmann. It provides an example of how important such sabbaticals are to those of us who are located in the antipodes."

1. McMeekin T L. Milk proteins. (Neurath H & Bailey K, eds.) *The proteins: chemistry, biological activity, and methods.* New York: Academic Press, 1954. Vol. 2A. p. 389-434.
2. Waugh D F & von Hippel P H. κ-casein and the stabilization of casein micelles. *J. Amer. Chem. Soc.* 78:4576-82, 1956.
3. Aschaffenburg R. Genetic variants of milk protein: their breed distribution. *J. Dairy Res.* 35:447-60, 1968.
4. Andrews A T. The composition, structure and origin of proteose-peptone component 5 of bovine milk. *Eur. J. Biochem.* 90:59-65, 1978.
5. Bell K, McKenzie H A & Shaw D C. Bovine β-lactoglobulins E, F and G of Bali (banteng) cattle. *Bos (Bibos) javanicus. Aust. J. Biol. Sci.* 34:133-47, 1981.
6. Green D W, Aschaffenburg R, Camerman A, Coppola J C, Dunnill P, Simmons R M, Komorowski E S, Sawyer L, Turner E M C & Woods K F. Structure of bovine β-lactoglobulin at 6 Å resolution. *J. Mol. Biol.* 131:375-97, 1979.

# This Week's Citation Classic

Ramachandran G N & Sasisekharan V. Conformation of polypeptides and proteins. *Advan. Prot. Chem.* 23:283-437, 1968.
[Centre of Advanced Study in Biophysics, Univ. Madras, India and Dept. Biophysics, Univ. Chicago, IL]

The three-dimensional architecture of a molecule is very important in deciding its chemical properties and biological activity. The article deals with the rules governing the shapes of proteins, polypeptides and related biomolecules, their internal structure and the nature and origin of the variations which occur, both in solution and the solid state. [The *SCI®* indicates that this paper has been cited over 580 times since 1968.]

---

G.N. Ramachandran
Mathematical Philosophy Group
Indian Institute of Science
Bangalore 560 012
India

January 22, 1981

"When I joined the University of Madras as a research professor in 1952, I looked out for a new line of work to follow in my future career. The beautiful papers by Pauling and Corey[1] on the helical structure of proteins and polypeptides, which had appeared in 1951, influenced me to take up the field of biomolecular structure as my life interest. Our first studies[2] were on the fibrous protein, collagen, and these led to the first proposal, in 1954, of the chain configuration in the triple helical structure of collagen, which is now universally accepted. The next year, several groups, Kartha and I[3] in Madras; Rich and Crick[4] in Cambridge; and Cowan, McGavin, and North[5] in London, improved the structure to have 3.3 units per turn for collagen, and also showed that the prototype helix occurs for related polypeptides like polyglycine and polyproline. This was extended to polyhydroxyproline by my colleague, Sasisekharan,[6] in Madras in 1959.

"Having been interested in several such structures in specific examples, we worked out a general theory in the early 1960s of the conditions for the conformation of a pair of peptide units to be stable, in terms of 'contact criteria,' and a short note was published in the *Journal of Molecular Biology* in 1963.[7] This paper is also a widely quoted reference, and as a result of this, we held an International Conference on the Conformation of Biopolymers in Madras in 1967, the first of its kind, and one attended by all the leading workers of that time in the field.

"In 1964, I was elected an honorary member of the American Society of Biological Chemists, and soon after that, J.T. Edsall invited me to write the article under discussion. It contains a detailed account of the description of peptide and polypeptide conformations arising from variations of the dihedral angles phi and psi, and of the consequent phi-psi map, which has now come to be known as the Ramachandran Map. The mathematical technique of making such variations and calculating atomic coordinates, helical parameters, and so on, was presented systematically in that review. It was also one of the earliest to discuss techniques of energy minimization, which has now come to be widely used.

"It is worthwhile commenting on the fact that my colleague, Ramakrishnan, spent some six months with a desk calculator for obtaining the data for the first phi-psi-plot, while this can be accomplished in a few minutes on the present day electronic computers. Thanks to such sophisticated aids, the subject of biopolymer conformation has become explosive in recent years, and conformational analysis (both theory and experiment) has now been accepted as the most important technique for the understanding of biology in terms of physics and chemistry.

"There have been any number of reviews after 1970 on this subject, but two may be mentioned in particular—namely the books by Fraser and MacRae[8] on fibrous proteins and Schulz and Schirmer[9] on globular proteins."

1. **Pauling L & Corey R B.** Series of papers on the structure and helical configurations of polypeptide chains. *Proc. Nat. Acad. Sci. US* 37:235, 241, 251, 256, 261, 272, 1951.
2. **Ramachandran G N & Kartha G.** Structure of collagen. *Nature* (London) 174:269-70, 1954.
3. ————————. Structure of collagen. *Nature* (London) 176:593-5, 1955.
4. **Rich A & Crick F H C.** The structure of collagen. *Nature* (London) 176:915-6, 1955.
5. **Cowan P M, McGavin S & North A C T.** The polypeptide chain configuration of collagen. *Nature* (London) 176:1062-4, 1955.
6. **Sasisekharan V.** Structure of poly-L-hydroxyproline A. *Acta Crystallogr.* 12:903-9, 1959.
7. **Ramachandran G N, Ramakrishnan C & Sasisekharan V.** Stereochemistry of polypeptide chain configurations. *J. Mol. Biol.* 7:95-9, 1963. [The *SCI®* indicates that this paper has been cited over 200 times since 1963.]
8. **Fraser R D B & MacRae T P.** *Conformation in fibrous proteins.* New York: Academic Press, 1973. 628 p.
9. **Schulz G E & Schirmer R H.** *Principles of protein structure.* New York: Springer-Verlag, 1979. 314 p.

## This Week's Citation Classic

CC/NUMBER 5
FEBRUARY 2, 1981

Traub W & Piez K A. The chemistry and structure of collagen.
*Advan. Prot. Chem.* 25:243-352, 1971.
[Dept. Chemistry, Weizmann Inst. Science, Rehovot, Israel and Natl. Inst. Dental Res., NIH, Bethesda, MD]

This article reviews chemical and conformational aspects of the structure of collagen, the main protein of connective tissue. [The *SCI®* indicates that this paper has been cited over 275 times since 1971.]

Wolfie Traub
Department of Structural Chemistry
Weizmann Institute of Science
Rehovot
Israel

October 28, 1980

"Karl Piez and I wrote this article in the last three months of 1970, while we were 6,000 miles apart. Over the previous decade, Karl's laboratory at the NIH had become the outstanding centre of collagen chemistry, and he had attracted a number of exceptionally able colleagues. Together they had elucidated the chain compositions of collagens, identified cross-links between them, developed methods for specific chemical and enzymatic cleavage, and initiated determination of the amino acid sequences.

"My contribution to the article was based largely on x-ray diffraction and physicochemical structural investigations, in Rehovot, of synthetic polypeptide models for collagen. Arieh Berger and Ephraim Katchalski first introduced me to these compounds, which resembled collagen in having every third residue glycine and high contents of imino acids. Since 1955, this protein was known to have a rope-like triple-helix structure,[1] but three distinct models of this type differing substantially in conformation had been proposed. Because of collagen's complex intermolecular packing and amino acid sequence a clear choice proved difficult. We hoped that simpler polytripeptide sequences would lead to collagen-like structures which could be determined unambiguously.

"The first polytripeptide whose structure I investigated, (Gly-Gly-Pro)$_n$, turned out to have a helical conformation like the individual chains of the collagen molecule, but in parallel array rather than twisted into three-stranded ropes. However, our next model, (Gly-Pro-Pro)$_n$, showed evidence of trimeric association in solution and gave a sharp x-ray pattern including the triple-helix features of collagen but indicating a simple hexagonal molecular packing.[2] Ada Yonath, then a graduate student, and I were able to make a systematic conformational analysis and show that the structure was close to the collagen II model, with one interchain hydrogen bond per tripeptide. However, to really test the alternative two-bond model we needed an amino acid with an NH group following glycine. Dave Segal, who had now joined us as a post-doc, synthesised (Gly-Ala-Pro)$_n$, but to our disappointment we found that its structure resembled that of (Gly-Gly-Pro)$_n$ rather than collagen. We decided to try to force the conformation by adding triple-helix-favouring tripeptides, and Dave synthesised the hexapeptides (Gly-Ala-Pro-Gly-Pro-Pro)$_n$, (Gly-Ala-Pro-Gly-Pro-Ala)$_n$ and (Gly-Ala-Ala-Gly-Pro-Pro)$_n$. To our delight, all showed collagen-like x-ray patterns and behaviour in solution, and detailed conformational analyses revealed structures almost identical with that of (Gly-Pro-Pro)$_n$. It seemed clear that collagen itself must have a conformation close to that formed by all these diverse sequences.

"This conclusion has been supported by various more recent investigations and this conformation still represents a good model for collagen. The chemical methods and results described by Karl have led to further great advances, notably in the identification of a considerable variety of genetically distinct tissue-specific collagens and complete amino acids sequence determinations for several of these. There has, in fact, been a great burgeoning of interest in collagen, much of it directed towards elucidating the biosynthesis and various newly discovered biological properties of the protein, but nevertheless related to its chemistry and structure. In these circumstances, our article appears to have been timely and useful, though the recent publication of an updated, more comprehensive review[3] may diminish its impact in the future."

1. Ramachandran G N & Kartha G. Structure of collagen. *Nature* 176:593-5, 1955.
2. Engel J, Kurtz J, Traub W, Berger A & Katchalski E. On the mechanism of collagen denaturation and renaturation and on conformational changes in related polypeptides. (Fitton Jackson S, Harkness R D, Partridge S M & Tristram G R, eds.) *Structure and function of skeletal and connective tissue.* London: Butterworths, 1965. p. 241-9.
3. Bornstein P & Traub W. The chemistry and biology of collagen. (Neurath H & Hill R L, eds.) *The proteins.* New York: Academic Press, 1979. Vol. 4. p. 411-632.

# This Week's Citation Classic

**Margoliash E & Schejter A.** Cytochrome *c*. *Advan. Prot. Chem.* 21:113-286, 1966.
[Biochemical Res. Dept., Abbott Labs., North Chicago, IL and Dept. Chemistry, Univ. Pennsylvania, Philadelphia, PA]

This article is a review, as complete and thorough as we could make it, of all the information concerning cytochrome c that had been gathered from the time of its rediscovery in 1924 and preparation in 1930 by David Keilin,[1] until the early months of 1966. [The *SCI*® indicates that this paper has been cited over 270 times since 1966.]

Emanuel Margoliash
Department of Biochemistry
and Molecular Biology
Northwestern University
Evanston, IL 60201

March 14, 1980

"Cytochrome c functions in the mitochondrial respiratory chain to transfer electrons between the reductase segment and cytochrome c oxidase. As it occurs in all eukaryotes, is small, and is the only respiratory chain heme protein that can readily be purified in water-soluble form, it has been studied extensively. An impressively large amount of work has been done on its structure, function, evolution and on the myriad other aspects of its biochemistry and molecular biology. As more information accumulates—for example, we now know the amino acid sequences of close to 100 eukaryotic cytochromes c—the protein presents an ever more attractive model for experimental approaches that are far more difficult in less well documented cases. This may be why this review has been cited so often. Even though apparently exhaustive, with 174 pages and 674 references, the flow of science since 1966 has left it far behind. Clearly, the review needs to be rewritten, even though as thorough an article is now likely to be several times longer.

"For such an enterprise to be really successful, it is best for the author to have spent much of his research life working on the subject, to have treated its delicate and uncertain growth with tender loving care and, if possible, to have been an actor or a spectator at the crucial turning points in its development. In this regard I have been very fortunate. I shall never forget how on arriving on September 10, 1951 for a sabbatical leave at the Molteno Institute, University of Cambridge, intending to study a complex phenomenon involving liver catalase in rats, Keilin gently dissuaded me from that unrewarding enterprise saying: 'Why don't you start by doing something simpler? Why don't you make some cytochrome c? It is such a pleasant preparation.' This conversation, *inter alia*, decided the course of my scientific life. It soon led to a simple preparation of pure and fully active cytochrome c and, after agonizingly slow progress, to the complete amino acid sequence of horse cytochrome c in the laboratory of Emil Smith in Salt Lake City in 1958 1960. Shortly thereafter (1962-1963) followed the realization that there was an obvious correlation between the primary structures of the cytochromes c of various species and their phylogenetic relationships. This last started a train of work that is still continuing.

"My coauthor, Abel Schejter, came in contact with cytochrome c while working with me on his PhD thesis in the mid-1950s at the Hebrew University in Jerusalem. The taste for it he acquired then has not left him, as he has continued to contribute, most effectively, to our knowledge of the protein.

"Following involvement in early work on the genetics, immunology, and X-ray crystallography of cytochrome c, the emphasis in more recent years has been on structure-function relations, physiological mechanisms of action, and the molecular biology of the protein. This last may soon provide cytochromes c of any desired amino acid sequence, and thus present a really practical approach to both structure-function and protein evolution problems.

"As the reader will notice, with some determination it is not too difficult to make a lifelong living out of a single protein, if one applies some imagination to the problem."

1. **Keilin D.** *The history of cell respiration and cytochrome.* Cambridge: Cambridge University Press, 1966. 416 p.

# This Week's Citation Classic

CC/NUMBER 8
FEBRUARY 25, 1980

Spiro R G. Glycoproteins. *Adv. Prot. Chem.* 27:349-467, 1973.
[Depts. Biological Chem. and Med., Harvard Med. Sch.; and Elliott P. Joslin Res. Lab., Boston, MA]

By 1973 it had been recognized that the attachment of sugar residues represents one of the major posttranslational modifications which a protein may undergo. This article reviews the information available on glycoproteins at that time in regard to their distribution, biological properties, composition, structural features, and metabolism. [The *SCI®* indicates that this paper has been cited over 185 times since 1973.]

Robert G. Spiro
Department of Biological Chemistry
Harvard Medical School
Joslin Research Laboratory
Boston, MA 02215

August 30, 1979

"It was as a medical student that I first became interested in glycoproteins after being introduced to the pathological manifestations of diabetic microvascular disease. The nodular lesions in the kidney glomerulus particularly fascinated me as they were known to react intensely with the periodic acid-Schiff stain and were presumed to consist of glycoprotein material.

"When I started my postdoctoral biochemical training under A. Baird Hastings at the Harvard Medical School in 1956, I became aware that very little was known about glycoproteins at that time and indeed I remember that all the pertinent reprints could comfortably fit into one file folder. It therefore became evident to me that, as so often happens in science, I would have to take a long detour in order to understand the diabetic glomerular deposits. I soon developed a keen interest and commitment to the study of the biochemistry and biology of glycoproteins which has stayed with me to the present day.

"The 1960s proved to be an exciting period in which work from a small number of laboratories defined the structural features of carbohydrate-containing proteins and characterized some of the enzymes involved in their assembly. My laboratory concentrated at that time on fetuin, $\alpha_2$-macroglobulin, thyroglobulin, collagens and basement membranes and eventually showed that the diabetic glomerular lesions were made up of basement membrane material. Moreover, during this period sophisticated analytical and structural techniques were developed which permitted the study of glycoproteins at high levels of sensitivity and which revealed that they are widely distributed in nature as biologically important molecules, including enzymes, hormones, immunoglobulins, lectins, transport and clotting proteins, lubricants, and membrane components.

"When I reviewed this rapidly expanding field for the *New England Journal of Medicine* and the *Annual Review of Biochemistry*, I discovered that investigators in quite diverse fields had developed a desire to know more about these molecules. The decision to write an article for *Advances in Protein Chemistry* was based on a perception that this interest in glycoproteins would further escalate and that a comprehensive review would serve a useful function. Indeed in recent years there has been a veritable explosion of activity in the area of glycoproteins.

"I believe that my review in *Advances in Protein Chemistry* may be cited by investigators who are attempting to correlate structure with function after becoming aware of the biological role of glycoproteins, particularly at the cell surface, where the sugar components appear to be specific determinants involved in recognition phenomena. Another group of researchers who may refer to this review are cell biologists interested in the assembly of membrane and secretory proteins, since carbohydrate attachment can occur at an early stage through the mediation of lipid-saccharide donors, followed by processing reactions to yield the mature carbohydrate units."

# This Week's Citation Classic
CC/NUMBER 6
FEBRUARY 11, 1980

Tanford C. Protein denaturation. *Advan. Prot. Chem.* 23:121-282, 1968.
[Dept. Biochemistry, Duke Univ. Medical Center, Durham, NC]

This paper reviews work that had been done in my laboratory during the preceding several years on the characterization of proteins denatured by urea and guanidine hydrochloride, and similar work done elsewhere on other protein denaturants. [The *SCI®* indicates that this paper has been cited over 680 times since 1968.]

---

Charles Tanford
Department of Biochemistry
Duke University Medical Center
Durham, NC 27710

August 22, 1979

"Proteins, alive and functioning, carry within their structure the secret of virtually all life processes. But when they are denatured, they are dead, like a car in a junkyard. Who today would want to study denatured proteins? I don't know, and don't know why there should be more than scattered citations of my review in the literature.

"In the 1960s there was a valid biological reason for investigating denatured proteins. There was a growing conviction that the three-dimensional structure and biological activity of proteins are uniquely determined by the amino acid sequence of their constituent polypeptide chains. The conviction was based on the spontaneous recovery of structure and function after denaturation by guanidine hydrochloride and subsequent removal of the denaturing agent. But that experiment actually does not prove anything, unless it can be demonstrated that all secondary and tertiary structure had been lost when the protein was transiently in the denatured state. (In fact, had the argument been based on recovery from denaturation by sodium dodecyl sulfate, it would have been invalid. As Jacqueline Reynolds showed later, this denaturing agent does not cause loss of protein secondary structure.[1])

"Purely by chance, I had in my laboratory at the time two exceptionally good polymer chemists, Kazuo Kawahara from the University of Osaka and Savo Lapanje from the University of Ljubljana, and they agreed to help me study the state of a number of proteins in guanidine hydrochloride solution. Our approach was rigorous and objective, and we had no vested interest in what the result would be. As it turned out, the denatured proteins did in fact behave as truly structureless polymer chains ('random coils'), and the sequence/structure/function hypothesis was proved. The original work (which, I should mention, ultimately involved several students and postdoctoral fellows in addition to Kawahara and Lapanje) was published in about a dozen papers. I summarized what we had learned in the review that is the subject of this article, and added to it what was then known about other denaturing agents, such as alcohols, heat, etc.

"Live proteins are more interesting than dead ones, and I have not pursued the subject of denaturation. Looking through the papers published from my laboratory during the last six years, I find that we ourselves have cited the denaturation review only twice, both times in connection with demonstrating that a protein we were studying resisted denaturation by guanidine hydrochloride."

---

1. **Reynolds J A & Tanford C.** The gross conformation of protein-sodium dodecyl sulfate complexes. *J. Biol. Chem.* 245:5161-5, 1970.

# This Week's Citation Classic
CC/NUMBER 16
APRIL 20, 1981

Blundell T, Dodson G, Hodgkin D & Mercola D. Insulin: the structure in the crystal and its reflection in chemistry and biology. *Advan. Prot. Chem.* 26:279-402, 1972.
[Lab. Molecular Biophysics, South Parks Rd., Oxford, England]

This review described the relation of the tertiary structure of insulin determined by X-ray analysis to the solution structure. It demonstrated the role of zinc insulin hexamers in storage and the importance of certain residues on the surface of the monomer to receptor binding and full biological potency. [The *SCI*® indicates that this paper has been cited over 205 times since 1972.]

---

T.L. Blundell
Department of Crystallography
Birkbeck College
University of London
London WC1E 7HX
England

March 24, 1981

"Dorothy [Crowfoot] Hodgkin started work on insulin in 1934 when she reported the first X-ray patterns of wet crystals and demonstrated that insulin was arranged as three identical 12,000 molecular weight units.[1] I well remember her introductory comments 35 years later to my lecture reporting our high resolution results, when she reminded us that she had started work before I was born. She could have said the same of my colleagues Guy and Eleanor Dodson, Margaret Adams, and M. Vijayan, and it was typical that she should ask one of us to give the first lecture rather than give it herself.

"The importance of the review three years later was that it was the first detailed attempt to relate the crystal structure analysis to the chemistry and biology of insulin. We began by showing that the crystal structure was relevant to the conformation in aqueous solutions. In this task we were fortunate to have Dan Mercola join us in 1970 from Ed Arquilla's laboratory, where he had gained experience in many chemical, biophysical, and immunological techniques. Together we reviewed the many hundreds of papers concerning the chemistry, spectroscopy, and biology of insulin. We showed that the hydrophobic core of insulin was remarkably conserved in evolution and suggested that an invariant surface area was a good candidate for the receptor binding region. We were very fortunate that our work was paralleled by dramatic advances in receptor binding technology which helped us firm up our hypothesis.

"Some of our conclusions were unexpected, and proved to have radical implications. For instance, Don Steiner had shown that proinsulin, which allows insulin to fold, had a connecting peptide of about 30 amino acid residues.[2] As a result, most of us had imagined that B30 and A1 would be widely separated in the three-dimensional structure. In fact they were only 10 Å apart, a distance spanned by three residues. One consequence of this was the suggestion that insulin might be synthesized with a much smaller link which could easily be removed, an idea that has been developed in Helmut Zahn's and other laboratories.

"Most of our suggestions have stood the test of time and experiment. However, in 1972 we were very uncertain of our conclusions. The manuscript went through many versions until at last, several months after the deadline for submission, Dodson, Mercola, and I reached agreement. We received polite but firm letters from Fred Richards and Chris Anfinsen requesting the article promptly. But Dorothy was adamant that it needed more improvement. Telegrams began to arrive from our anxious editors. Dorothy recounted how she had, 30 years before, worked hard to produce an article for an editor on whose desk it was still to be found one year later! Finally when we sent off the review, Dorothy took this with good humour saying that at least the third editor, John Edsall, would return the manuscript allowing the necessary modifications. But Edsall proved quite pragmatic: he wrote to say how much he had enjoyed reading it! And although in retrospect I agree with Dorothy, I am pleased that others continue to cite it. More recent reviews have appeared in both *Trends in Biochemical Science* and *Nature*."[3,4]

1. Crowfoot D. X-ray single crystal photographs of insulin. *Nature* 125:591-2, 1935.
2. Steiner D F, Clark J L, Nolan C, Rubenstein A H, Margoliash E, Aten B & Oyer P E. Proinsulin and the biosynthesis of insulin. *Recent Progr. Hormone Res.* 25:207-40, 1969.
3. Blundell T L. Insulin conformation and molecular biology of polypeptide hormones. I. Insulin, insulin-like growth factor and relaxin. *Trends Biochem. Sci.* 4:51-4, 1979.
4. Blundell T L & Humbell R E. Hormone families: pancreatic hormones and homologous growth factors. *Nature* 287:781-7, 1980.

# This Week's Citation Classic

CC/NUMBER 46
NOVEMBER 14, 1983

Havel R J, Eder H A & Bragdon J H. The distribution and chemical composition of ultracentrifugally separated lipoproteins in human serum.
*J. Clin. Invest.* 34:1345-53, 1955.
[Lab. Metabolism, Natl. Heart Inst., Natl. Insts. Health, Dept. Health, Education, and Welfare, Bethesda, MD]

A versatile method for isolating lipoprotein fractions from blood serum according to their hydrated density is presented, enabling them to be characterized and quantified by ordinary procedures. Results are given for normal humans and animals and human hyperlipidemics. [The *SCI*® indicates that this paper has been cited in over 2,160 publications since 1961.]

Richard J. Havel
Cardiovascular Research Institute
University of California
San Francisco, CA 94143

July 28, 1983

"When the National Institutes of Health Clinical Center opened in 1953, I came from Cornell Medical School to the National Heart Institute with a group of young physicians recruited by James Shannon. I joined a small group organized by Christian Anfinsen to work on lipids and chose to work in the laboratory of Joseph Bragdon, a pathologist who was interested in the pathogenesis of atherosclerosis. One year later, Howard Eder, also an internist from Cornell, joined the group. He had worked with David Barr and Ella Russ on the distribution of cholesterol and phospholipids in two lipoprotein fractions separated from human blood plasma by 'Cohn fractionation'; they had made a number of interesting observations on alterations of lipoproteins in several diseases, including atherosclerosis.[1] I was by then separating lipoproteins for metabolic studies by preparative ultracentrifugation, based upon the pioneering work of Gofman, Lindgren, and their associates at the University of California at Berkeley.[2] They had used analytical ultracentrifugation extensively to characterize human plasma lipoproteins and their alterations in patients with atherosclerosis and other diseases. What appeared to be needed was a straightforward method of isolating and characterizing a larger number of fractions than provided by Cohn fractionation. A combination of preparative ultracentrifugation and chemical analysis appeared to have promise as a reproducible, quantitative procedure. A preliminary study by Eder and others demonstrated that the ultracentrifugal fraction containing high density lipoproteins corresponded closely to the Cohn fraction that contains alpha lipoproteins.[3] The procedure that we designed, now known as 'sequential preparative ultracentrifugation,' yielded clean lipoprotein fractions that corresponded to those defined by the work of Gofman and Lindgren.[2] To illustrate the usefulness of the procedure, we provided limited normative data and some values for patients with various forms of hyperlipoproteinemia. We also tested the method in six other mammals and demonstrated large variations in lipoprotein concentrations and distributions which are now widely appreciated. We separately reported the detailed composition of those lipoprotein fractions that are now analyzed routinely in clinical and epidemiological research.[4]

"Although we referred in the laboratory to the fraction that we separated at a nonprotein ('background') solvent density of 1.019 g/ml as 'very low density lipoproteins' to distinguish them from those subsequently separated at 1.063 g/ml (low density lipoproteins), such terminology was not permitted by the editors of the *Journal of Clinical Investigation* until 1957.[5] Since then, this terminology has 'stuck,' and 'VLDL,' 'LDL,' and 'HDL' have become the standard jargon of the field.

"Our description of this procedure evidently helped to make the study of lipoproteins readily accessible to both clinical and basic scientists. Several modifications have been published, some of which are also cited frequently, and the phenomenon of 'extinction' is increasingly evident. I was among the first to point out that ultracentrifugation dissociates certain proteins from lipoproteins[6] and I have advocated gentler methods that avoid ultracentrifugal artifacts! At the University of California, San Francisco, we now make extensive use of chromatographic and immunoadsorption methods to characterize lipoprotein particles that are modified or even destroyed by ultracentrifugation.

"None of the authors of this *Citation Classic* felt that it represented a major conceptual advance and we were initially surprised by the wide attention that it received. Evidently, even rather straightforward methodological efforts can sometimes help to open up a fruitful field of research."

1. Barr D P, Russ E M & Eder H A. Protein-lipid relationships in human plasma. II. In atherosclerosis and related conditions. *Amer. J. Med.* 11:480-93, 1951. (Cited 245 times.)
2. Lindgren F T, Elliott H A & Gofman J W. The ultracentrifugal characterization and isolation of human blood lipids and lipoproteins, with applications to the study of atherosclerosis.
*J. Phys. Colloid Chem.* 55:80-93, 1951.
3. Eder H A, Russ E M, Pritchett R R A, Wilber M A & Barr D P. Protein-lipid relationships in human plasma: in biliary cirrhosis, obstructive jaundice, and acute hepatitis. *J. Clin. Invest.* 34:1147-62, 1955. (Cited 90 times.)
4. Bragdon J, Havel R J & Boyle E. Human serum lipoproteins. I. Chemical composition of four fractions. *J. Lab. Clin. Med.* 48:36-42, 1956. (Cited 135 times.)
5. Havel R J. Early effects of fat ingestion on lipids and lipoproteins of serum in man. *J. Clin. Invest.* 36:848-54, 1957. (Cited 70 times.)
6. Fainaru M, Havel R J & Imaizumi K. Apoprotein content of plasma lipoproteins of the rat separated by gel chromatography or ultracentrifugation. *Biochem. Med.* 17:347-55, 1977. (Cited 60 times.)

# This Week's Citation Classic

CC/NUMBER 10
MARCH 10, 1980

Hatch F T & Lees R S. Practical methods for plasma lipoprotein analysis.
*Advan. Lipid Res.* 6:1-68, 1968.
[Bio-Medical Div., Lawrence Radiation Lab., Univ. Calif., Livermore, CA
and Rockefeller Univ., New York, NY]

In this review we presented full details for using some relatively simple and only moderately expensive procedures of lipoprotein analysis. Because the review was intended to be every 'person's' lipoprotein manual, we included comparisons of results obtained with older or more complex methods, so that an appropriate selection of methods could be made by the readers. [The *SCI®* indicates that this paper has been cited over 315 times since 1968.]

Frederick T. Hatch
Biomedical Sciences Division
Lawrence Livermore Laboratory
University of California
Livermore, CA 94550

October 12, 1979

"In 1960 I organized a new Arteriosclerosis Unit at the Massachusetts General Hospital. This unit was to do both basic biochemical and clinical research on atherosclerosis and coronary heart disease, with emphasis on lipoprotein metabolism. During the 1950s the work of John Gofman and associates at the Donner Laboratory in Berkeley established that the plasma lipids are transported in the form of a series of macromolecular complexes of lipids and proteins, the lipoproteins.[1] They also found that patients with certain hereditary disorders exhibited striking abnormalities of the lipoprotein spectrum. The foregoing work utilized new and complex techniques of ultracentrifugation that were available in only a handful of laboratories in the world.

"One of our earliest objectives was to adapt or develop a suite of relatively simple methods that would enable quantitative or semiquantitative measurements of lipoprotein patterns in animals and humans to be performed in our laboratory and other biophysically unsophisticated laboratories. Methods involving zone electrophoresis in paper and other media soon took on a major role in this suite of methods. N. Zöllner of Munich suggested that we add albumin during electrophoresis, since this had previously been used to improve the resolution of radioactively labeled insulin. Robert Lees and I indeed found that adding one percent serum albumin to the buffer for paper electrophoresis greatly improved the resolution of lipoprotein fractions, and we showed that these corresponded to the fractions separated with the ultracentrifuge.[2] Our method contributed heavily to the development of a typing system for the hereditary lipoprotein disorders, which is still widely used today.[3] Subsequently, there was an explosive growth of clinical research on lipoprotein disorders and of interest in the metabolic and other risk factors for coronary heart disease. Our electrophoretic method became a widely used standard, later to be replaced by superior agarose gel methods. This is undoubtedly responsible for the frequent citation of the review 'handbook' on methods that Lees and I wrote, together with the fact that we aimed it at clinical researchers. With various other collaborators I later made the paper and agarose gel electrophoretic techniques as quantitative and standardized as was feasible. The current state of this art was published by Frank Lindgren and a Donner Laboratory group.[4]

"Although I have now reoriented my research toward chromatin biochemistry and genetic toxicology, it is very gratifying to have participated for a long time in the assault on heart disease, which recently is showing clear signs of success. Lees remains active in the clinical investigation and treatment of patients with atherosclerosis."

1. Gofman J W, Glazier F, Tamplin A, Strisower B & de Lalla O. Lipoproteins, coronary heart disease, and atherosclerosis. *Physiol. Rev.* 34:589-607, 1954.
2. Lees R S & Hatch F T. Sharper separation of lipoprotein species by paper electrophoresis in albumin-containing buffer. *J. Lab. Clin. Med.* 61:518-28, 1963.
3. Fredrickson D S, Levy R I & Lees R S. Fat transport in lipoproteins: an integrated approach to mechanisms and disorders. *N. Eng. J. Med.* 276:32-44, 94-103, 148-56, 215-26, 273-81, 1967.
4. Wong R A, Banchero R G, Jensen L C, Pan S S, Adamson G A & Lindgren F T. Automated microdensitometry and quantification of lipoproteins by agarose gel electrophoresis. *J. Lab. Clin. Med.* 89:1341-8, 1977.

# Citation Classics

**Number 3** — January 16, 1978

Fredrickson D S, Levy R I & Lees R S. Fat transport in lipoproteins; an integrated approach to mechanisms and disorders.
*N. Engl. J. Med.* 276:34-44, 94-103, 148-56, 215-25, 273-81; 1967.

The authors review the structure and function of the plasma lipoproteins, with particular reference to abnormal lipoprotein metabolism associated with certain clinical disorders, and discuss two methods for separating lipoproteins—paper electrophoresis and lipoprotein quantification. By identifying the specific lipoprotein pattern, hyperlipidemia may be translated to hyperlipoproteinemia. [The *SCI*® indicates that the five parts of this paper were cited a total of 3973 times in the period 1967-1976.]

Dr. Robert I. Levy, Director
National Heart, Lung and Blood Institute
National Institutes of Health
Bethesda, Maryland 20014

April 11, 1977

"Perhaps one of the intrinsic reasons for the volume of citation received by our articles is that they drew attention to an important group of diseases that are common and often potentially fatal. Up to the time the articles were published, specialists in the field who attempted to treat patients with hyperlipidemia had to grapple with complicated classifications that often proved contradictory and misleading in clinical practice. The system of classifying blood lipid disorders that we introduced was based on a combination of three things: the rich clinical material we collected at the NIH Clinical Center, a great deal of experience (much then unpublished) and relatively simple methodology.

"We focused principally on two methods of separating lipoproteins—paper electrophoresis and lipoprotein quantification. Although the former has been de-emphasized with the passage of time, its merit at that stage was in highlighting the lipoproteins, rather than lipids, as a basis for classifying and treating hyperlipidemia. The second method—quantification of lipoproteins—allowed for the separation of several disorders with similar plasma lipid levels. Used within the clinical context, it called for a degree of precision which was often absent in the previous management of such patients. A drawback of this method—then and now—is that it is based on systems which are unavailable to many clinicians, although they are accessible to many researchers.

"Our studies at NIH dealt primarily with the genetic or familial forms of hyperlipoproteinemia. But our work was also relevant to the general problem of lipids, lipoproteins and coronary heart disease, introducing concepts that applied to subjects within the so-called 'normal' range of lipid and lipoprotein distribution.

"The typing system for hyperlipoproteinemia that we introduced was a simpler, more convenient code than the existing classifications, but it contains a number of limitations which we recognized from the outset. The major deficiency is that it is not based on the specific metabolic defects which underlie each form of HLP. The precise abnormality associated with a minority of these disorders is only now beginning to be comprehended. There has also been some misapplication of the system, particularly in the steps that have to be taken between recognizing the presence of a hyperlipoproteinemia and in deciding whether it is primary or secondary, familial or non-familial, and in determining the appropriate therapy. Later findings, especially with respect to familial combined hyperlipidemia, highlight some of the present gaps in our knowledge of the lipid transport disorders including their genetic characteristics.

"Despite the deficiencies of our classification (of which, with hindsight, there are many), it did provide, and continues to do so, a useful and practicable system for the classification, investigation and treatment of hyperlipoproteinemic patients. Our objective was to break down some of the conventional cliches and approaches to the management of these patients by providing a more rational and workable alternative. Perhaps the frequency with which our work is cited is proof that in some measure we succeeded."

## This Week's Citation Classic

CC/NUMBER 11
MARCH 16, 1981

Fleischman J B, Porter R R & Press E M. The arrangement of the peptide chains in γ-globulin. *Biochemical J.* **88**:220-8, 1963.
[Department of Immunology, St. Mary's Hospital Medical School, London, England]

This paper describes the isolation and biochemical properties of the polypeptide chains of rabbit IgG antibodies. Based on these data, the paper proposed a structural model for the antibody molecule made up of four polypeptide chains (two heavy and two light). [The *SCI®* indicates that this paper has been cited over 465 times since 1963.]

Julian B. Fleischman
Department of Microbiology
and Immunology
Washington University
School of Medicine
St. Louis, MO 63110

February 13, 1981

"It is a pleasure to learn that this paper emerges as one of the most frequently cited according to the *Science Citation Index®*. The paper proposed a structure for the antibody molecule; it was based on work in R.R. Porter's laboratory in the early 1960s. I think the main reason that this paper was the one most cited is that it unified the known structural features of antibodies into a consistent model.

"The background for the work was (a) Porter's 1959 discovery that rabbit antibody could be split by papain into three fragments (two Fab plus one Fc), with an intact antigen-combining site in each Fab fragment,[1] and (b) Edelman's simultaneous proof that the molecule was made up of more than one polypeptide chain.[2] I started a postdoctoral fellowship in Porter's laboratory at St. Mary's in London in the autumn of 1961. The objective at that stage was to separate and recover the antibody polypeptide chains in a soluble form so that they could be properly characterized. Porter had exploited Cecil and Wake's observation that interchain disulfide bonds were reducible under mild conditions.[3] This proved the key to separating the heavy and light chains of rabbit antibody and recovering them in soluble form. I had brought to the lab some goat antisera to the Fab and Fc fragments which I had helped prepare in Melvin Cohn's lab at Stanford. (Our most pungent scrub goats made by far the best antisera—the titers were so high that Boris, the biggest and ripest goat, keeled over in anaphylactic shock after his last boost.) The antisera proved most useful in demonstrating that both heavy and light chains were present in Fab, but only heavy chains were in Fc—an important clue to the architecture of the molecule. St. Mary's had no goats but propionic, butyric, and valeric acids (which we used to separate chains), laced with smog from Paddington Station, provided the olfactory stimulation.

"Figuring out the four-chain structure of the antibody molecule had a trace of the 'double-helix' element to it. We had to arrange the chains according to their distribution in Fab and Fc, and we also knew that similar efforts were under way at the Rockefeller Institute. We each promised to spend the weekend independently trying to work out a model. I had a couple of head-scratching days in Hampstead, and we met in Porter's office on Monday.

"The model we came up with is the four-chain structure illustrated in our paper. I remember stressing Nisonoff's critical observation that pepsin digestion of the antibody molecule left the two Fabs intact and linked by a single disulfide bond. This forced us to line up the two Fabs, each with a single antigen-combining site, next to each other, giving us the now-familiar Y-shaped molecule. We reluctantly discarded the then-popular cigar-shape with combining sites at either end, a favorite among both immunologists and textbook illustrators at the time. I think that the main strength of our model was that it provided a basic pattern for the many different classes of vertebrate antibodies studied since that time."[4]

1. Porter R R. The hydrolysis of rabbit gamma-globulin and antibodies with crystalline papain.
   *Biochemical J.* 73:119-27, 1959.
2. Edelman G M. Dissociation of gamma-globulin. *J. Amer. Chem. Soc.* 81:3155-6, 1959.
3. Cecil R & Wake R G. The reactions of inter- and intra-chain disulphide bonds in proteins with sulphite.
   *Biochemical J.* 82:401-6, 1962.
4. Nisonoff A, Hopper J E & Spring S S. *The antibody molecule.* New York: Academic Press, 1975. 542 p.

# This Week's Citation Classic

CC/NUMBER 25
JUNE 18, 1979

Boyer P D. Spectrophotometric study of the reaction of protein sulfhydryl groups with organic mercurials. *J. Amer. Chem. Soc.* 76:4331-47, 1954.
[University of Minnesota, Department of Agricultural Biochemistry, Minneapolis, MN]

As stated in the summary of the paper, 'A procedure is described for the sensitive and rapid spectrophotometric measurement of the extent and rate of reaction of various organic mercury compounds, particularly *p*-mercuribenzoate, with sulfhydryl groups.' Included with the paper are examples of differing reactivities of sulfhydryl groups in proteins, and the effect of ionic composition and pH on rate and extent of reaction. [The *SCI*® indiates that this paper has been cited over 1315 times since 1961.]

Paul D. Boyer
Molecular Biology Institute
University of California
Los Angeles, CA 90024

March 30, 1978

"In common with a number of its companion 'Citation Classics' the principal reason my paper has been frequently cited is probably because it fulfills a definite methodological need. Another reason is that the results demonstrate simply and quantitatively differing reactivities of sulfhydryl groups in proteins. This paper was a direct outgrowth of a need for quantitation of the number and reactivity of sulfhydryl groups in enzymes.

"H. L. Segal, as a graduate student in my laboratory, had obtained some of the first convincing evidence for acyl-S-enzyme formation as part of the coupling mechanism of glyceraldehyde 3-phosphate dehydrogenase.[1] We wanted more information about the number and reactivity of the sulfhydryl groups.

"The reaction of mercurials with sulfhydryl groups of proteins had been widely studied prior to this paper, but methods available were cumbersome. In considering the need and the problem, it seemed plausible that the orbital electrons in the benzene ring of *p*-mercuribenzoate would be sufficiently perturbed when a hydroxide ligand on the mercury is replaced by sulfur to allow detection by change in ultraviolet absorption. Experimental measurements showed that this was indeed the case; there was sufficient increase in absorption accompanying mercaptide formation to give a sensitive measure.

"It is fortunate for the application of the procedure that the change in ultraviolet spectrum is considerable in the 250-260 nm region where there is a minimum in the absorption of proteins. Since this publication a number of other measures have been found for sulfhydryl groups, but the spectrophotometric procedure with *p*-mercuribenzoate remains useful for measurement of mercaptide formation.

"I recall another interesting aspect of this work. It was supported by a modest grant from the National Science Foundation, and included in the approved budget were sufficient funds to allow me to travel to a scientific meeting. The research was scheduled for presentation at the 123rd Meeting of the American Chemical Society in Los Angeles, but my use of the National Science Foundation funds was blocked by an administrator of the University of Minnesota on the basis that this very large university had already approved attendance of another staff member from another department at this meeting.

"My objections were of no avail, and the paper was presented at my personal expense. This is an example of poor administration at a fine university, which in general has been administered much better than most. More importantly, such action was a forerunner of the present condition of science support where the time expended and restrictions required for conformity to regulation often stifle progress."

1. Segal H L & Boyer P D. The role of sulfhydryl groups in the activity of glyceraldehyde 3-phosphate dehydrogenase. *J. Biol. Chem.* 204:264-81, 1953.

# This Week's Citation Classic

CC/NUMBER 28
JULY 13, 1981

Goodwin T W & Morton R A. The spectrophotometric determination of tyrosine and tryptophan in proteins. *Biochemical J.* 40:628-32, 1946.
[Johnston Labs., Dept. Biochemistry, Univ. Liverpool, England]

The availability for the first time of a photoelectric spectrophotometer allowed the accurate determination of tyrosine and tryptophan in solutions containing mixtures of the two. The method was adapted to assay these amino acids in unhydrolysed protein solutions. [The *SCI*® indicates that this paper has been cited over 1,175 times since 1961.]

T.W. Goodwin
Department of Biochemistry
University of Liverpool
Liverpool L69 3BX
England

June 11, 1981

"It is somewhat ironic that after I have spent over 30 years investigating carotenoid and sterol biochemistry, I should receive an accolade of a *Citation Classic* for the very first paper I published in the *Biochemical Journal* on a topic remote from terpenoids. A moment's thought, however, reveals that this selection is purely an alphabetical accident. The work reported in this paper began when I was a research student and it was conceived and directed by my supervisor, the late R.A. Morton. It was completed during the war in spare time during the tenure of a research assistantship in Morton's laboratory. Morton can be considered one of the true founders of biochemical spectroscopy and, like me, never returned to amino acids after this investigation.

"The early spectrometric apparatus used to analyse binary mixtures of substances with overlapping absorption spectra had neither the sensitivity nor accuracy required for satisfactory results. The appearance of the Beckman photoelectric spectrophotometer changed this situation overnight and the availability to us of one of the first of these machines to be produced allowed us to develop an accurate method of tryptophan and tyrosine analysis. The knowledge that tyrosine and tryptophan are the only amino acids in proteins which significantly absorbed ultraviolet light above 220 nm and that their spectra were essentially unaltered in peptide linkage (only later did more sophisticated methods reveal small but significant differences related to the micro-environment of these amino acids in proteins) allowed us to propose a reasonably accurate method of assay after a means of correcting for scattered light caused by the opalescence of protein solutions had been worked out.

"It is interesting to record why Morton's laboratory was the first academic department in the UK to obtain a photoelectric spectrophotometer (Shell Research had the first). Morton, because of his expertise in vitamins and spectroscopy, had been asked by the British Ministry of Food to control a programme of vitaminization of margarine and other products, and to take part in a large collaborative investigation on determining the vitamin A requirement of humans. I was the research assistant employed on these projects. After much discussion it was agreed that in order to forward this work, Morton should be provided with the Beckman spectrophotometer under the Lend-Lease agreement between the US and UK governments.

"The main reason why the method became popular and is still quoted is that it is simple; it can be carried out directly on a solution of a protein and no hydrolysis is necessary, a particularly important advantage when the instability of tryptophan to alkaline hydrolysis is remembered; furthermore, no derivatization was necessary. The second main reason at the time the paper was published was the relatively small amount of protein required for an assay—25 mg! This requirement would certainly be a major disadvantage today.

"The method has survived into the present time probably because of the correctness of its basis, its simplicity, and the fact that with modern instrumentation it can be scaled down to use acceptable amounts of protein. Further information on the subject has been reported by T.T. Herskovits."[1]

1. **Herskovits T T.** Difference spectroscopy. *Methods Enzymol.* 11:748-75, 1967.

# This Week's Citation Classic

Bencze W L & Schmid K. Determination of tyrosine and tryptophan in proteins.
*Anal. Chem.* 29:1193-6, 1957.
[Dept. Medicine, Harvard Med. Sch., and Massachusetts Gen. Hosp., Boston, MA]

A spectrophotometric method is described which relies on the pattern of tyrosine and tryptophan ultraviolet absorption peaks rather than on the exact location of the absorption maxima. By addition of tyr and try stock solutions to the native protein the degree of the inherent background or extraneous absorption of the protein can be estimated. [The *SCI*® indicates that this paper has been cited over 695 times since 1961.]

William L. Bencze
Pharmaceuticals Division
CIBA-GEIGY Ltd.
CH 4000 Basel
Switzerland

January 13, 1981

"Upon my immigration to the US in 1954 my task was to contribute new data towards the characterization of human acid glycoprotein. I arrived in Boston as a postdoctoral fellow coming from the University of Zürich, Switzerland, where I had earned my PhD in organic chemistry under the guidance of the late Hans Schmid. My new master at Massachusetts General Hospital was Karl Schmid. Although both Schmids were born in the same county of Switzerland, they were no kin. Both of them were research oriented teachers, endowed with the exactitude of a Swiss watchmaker, assiduously implanting their meticulous techniques in their students and co-workers. I had been duly impressed with the painstaking care which afforded the pure glycoprotein and pledged that I would devote the same heaping measure of care to each milligram of the protein that was entrusted to me.

"Recalling the skilled eagle eyes of my teacher in Switzerland, who would cast a glance on the ultraviolet absorption spectrum of a plant extract and then proclaim that it contained a coumarin but no chromone, I commenced to record the spectra of varying mixtures of tyr and try. It was then convincingly clear that the pattern of the ultraviolet absorption curve could reliably tell whether the ratio of tyr:try was 3:7 or 2:8, respectively. Consequently, a method was at hand that was independent from the bathochromic shift of the absorption maxima. The only problem awaiting to be tackled was the imponderable extraneous absorption. Again, the stock solutions of tyr and try offered their invaluable aid. Addition of known amounts of tyr and try to the solution of the native protein will furnish a new ultraviolet absorption pattern. Repeated additions of tyr and/or try and calculations should lead to an approximate assay of the extraneous absorption of the protein. We set up plans to study the tyr and try content of large proteins and to investigate light scattering as a part of the extraneous ultraviolet absorption.

"However, the printers ink on this publication wasn't even dry when I abandoned my engagement with proteins. I have turned to medicinal chemistry. Karl Schmid remained true and loyal to the proteins and keeps on turning out a steady stream of new data on his favorite glycoproteins.

"Perhaps it is the enticing simplicity of the method that turned the publication into a *Citation Classic*. However, the task to unravel the nature of the extraneous absorption remains a difficult and a disturbing problem. It may very well be the case that a substantial portion of the citations are cursing the method and its advocators in an outright fashion. Luckily, the computer merely counts the citations and will keep silent about the complaints."

# Citation Classics

Number 25 — June 20, 1977

Spies, Joseph R & Chambers, Dorris C. Chemical determination of tryptophan in proteins. *Analytical Chemistry* 21:1249-66, 1949.

Several variations of a method for colorimetric analysis of unhydrolyzed proteins are described. The basic method was based on fundamental studies of the behavior of free and peptide-linked tryptophan. These studies included a method of alkaline hydrolysis which protects tryptophan from external destruction at temperatures up to 185°C without addition of antioxidants to the solution. Dr. Spies, now retired, conducted his research in the Allergens Research Division of the U.S. Department of Agriculture. [The *SCI*® indicates that this paper was cited 739 times in the period 1961-1975.]

---

Dr. Joseph Reuben Spies
507 North Monroe Street
Arlington, Virginia 22201

January 24, 1977

"It is most gratifying that our paper is 'one of the most cited articles ever published.'

"Dissatisfaction, in the early 1940s, in the use of existing methods for the determination of tryptophan prompted our studies. The first of our several papers on the subject was published in 1948 and the last in 1967. Originally we never intended to get so involved. We started out by simply trying to substitute sulfuric acid for the concentrated hydrochloric acid used in some existing methods to avoid the corrosive fumes of the latter acid. But, one thing led to another, so that this small beginning led to over four years of full-time research with the excellent technical assistance of Dorris Chambers part of this time.

"The experimental research was just plain pleasure. But getting the results published was something else. Originally I attempted to publish the contents of this paper as four separate articles, little suspecting the grueling road that lay ahead. Three of these articles were submitted to the late Walter J. Murphy, Editor of *Analytical Chemistry*, on May 10, 1948, and the fourth on October 6, 1948. A total of 135 pages of manuscript was thus being considered at one time. The reviewers were hard and in one case prejudiced to a point where I asked for and received his disqualification.

"During the ensuing months I answered the reviewers' many comments with a general statement and specific reply to each point raised. My responses totaled 62 pages over all with numerous revisions of the manuscripts. The final result was the combination of the four articles into one consisting of 71 pages of manuscript. Exactly one year to the day after the first submission, the revised manuscript was tentatively accepted for publication subject to more editorial revisions. The review process was onerous and a strong will was required to keep from giving it all up somewhere along the way. However, with one exception, I am deeply indebted to the anonymous reviewers who performed a difficult and seemingly thankless task. I am also indebted to Walter Murphy for his patience and fairness. Thanks are also due to Stella Anderson, Assistant Editor, and Miss Gordon for encouragement and editorial help in preparation of the manuscript. The resulting paper was much better than the original because of the cooperation of all of these persons. A final, as far as I am concerned, revision of the original method was published in 1967.[1] [Cited 76 times, 1967-1975.]

"Although many other papers on the determination of tryptophan have appeared, especially in the last decade, our method filled a considerable need by many researchers following its publication in 1949, and indeed the original method and its subsequent modification still enjoys considerable usage."

---

1. **Spies JR.** Deterimination of tryptophan in proteins. *Analytical Chemistry* 39:1412-16, 1967.

# This Week's Citation Classic

CC/NUMBER 9
MARCH 2, 1981

Adams C W M. A p-dimethylaminobenzaldehyde-nitrite method for the histochemical demonstration of tryptophane and related compounds.
*J. Clin. Pathol.* 10:56-62, 1957.
[Bernhard Baron Inst. Pathol., London Hosp., London University, England]

This paper described a new method for the histochemical detection of tryptophan. Staining depends on a reaction between tryptophan and p-dimethylaminobenzaldehyde (DMAB) and a subsequent stage of nitrosation with nitrous acid. Tryptophan-containing proteins in tissue-sections are stained in an intense blue colour. [The $SCI^®$ indicates that this paper has been cited over 250 times since 1961.]

C.W.M. Adams
Department of Pathology
Guy's Hospital Medical School
London University
London SE1 9RT
England

January 27, 1981

"This paper was written at the beginning of a career in pathology when I was a Freedom Research Fellow at Bernhard Baron Institute at London Hospital. It was stimulated by the need for better histochemical methods for identifying tissue proteins. I am happy to remember that this enthusiasm was shared with John Sloper and Ken Swettenham at the London Hospital and with George Glenner and Ralph Lillie at the NIH.[1] Other colleagues took a friendly but sometimes less committed attitude to this work, particularly when the fumes of strong acids penetrated their offices! In the mid-1950s exhaust ventilation was not as effective as now and largely depended upon a strong draft and an open window!

"The method itself is remarkably simple and takes only two minutes to complete. The reagent is DMAB dissolved in concentrated hydrochloric acid. It was adapted from spot tests used in organic chemistry, coupled with the substitution of nitric acid by nitrous acid in the second stage. It is this last partly serendipitous addition that results in the intense blue colour and makes the method of practical use. The reaction is between tryptophan and the aldehyde dissolved in a strong mineral acid. Similar but weaker results can be obtained with glyceraldehyde and even formaldehyde, hence the need to avoid prolonged fixation with formalin or other aldehydes. Unfortunately, it is still not known how nitrosation or diazotization enhances the pigment properties of the tryptophan-aldehyde adduct. However, since the tryptophan molecule is an integral part of the final pigment complex, the specificity of the method is virtually absolute.

"In pathology, this DMAB-nitrite method proved to be particularly useful for demonstrating amyloid, fibrin, immuno- and other globulins, intestinal and salivary zymogen granules, and the structural proteins of the myelin sheath in the PNS. By contrast, connective tissues contain relatively little tryptophan and are unstained. At the time, the method added a useful amino acid stain to back up those for tyrosine, cystine, and arginine.[2] The relative staining intensity for these various amino acids provided a rough 'fingerprint' to distinguish different proteins in tissue-sections. However, this application has been largely superseded by the well-known immunohistochemical methods using a specific antibody to a particular protein, which is then detected by a fluorescein- or peroxidase-labelled antiimmunoglobulin antibody.[2]

"The possible reason for the interest that has been shown in the DMAB method is that it is remarkably quick and easy to perform. Moreover, the method is quite reliable and the reaction rapidly reaches completion. Nevertheless, I must admit to be more than a little surprised at this interest. However, it could also be that the method has come to be used as a spray-reagent in chromatography; it would certainly be appropriate for this purpose."

1. Glenner G G. The histochemical demonstration of indole derivatives by the rosindole reaction of E. Fischer. *J. Histochem. Cytochem.* 5:297-304, 1957.
2. Pearse A G E. *Histochemistry: theoretical and applied.* London: Churchill, 1968. Vol. 1. p. 106-247.

# This Week's Citation Classic

CC/NUMBER 2
JANUARY 14, 1980

Woessner J F, Jr. The determination of hydroxyproline in tissue and protein samples containing small proportions of this imino acid.
Arch. Biochem. Biophys. 93:440-7, 1961. [Labs. Biochem., Howard Hughes Med. Inst., and Dept. Biochem., Univ. Miami Sch. Med., Miami, FL]

This paper presents an improvement on the method of Stegemann[1] for the determination of hydroxyproline and an extension of that method for the determination of hydroxyproline in samples containing as little as one part of hydroxyproline in 4,000 parts of other amino acids. [The *SCI*® indicates that this paper has been cited over 645 times since 1961.]

J. Frederick Woessner
Department of Biochemistry
University of Miami
School of Medicine
PO Box 016960
Miami, FL 33101

July 19, 1979

"Collagen is one of the most widely distributed proteins in the animal kingdom. It is found in most tissues and is intimately involved in many important disease processes including atherosclerosis, arthritis, cirrhosis, and tumor invasion. It is not surprising, therefore, that there is great interest in the measurement of hydroxyproline, an amino acid which is found almost exclusively in collagen and which provides a direct measure of collagen content. There have been at least five other published methods that, in my estimation, could be considered 'Citation Classics.'[1-5]

"My interest in hydroxyproline measurement began in the early fifties during my doctoral work at the Massachusetts Institute of Technology involving the measurement of small amounts of collagen produced in tissue culture. The best method at that time was barely adequate for the purpose.[4] It was based on oxidation of hydroxyproline with hydrogen peroxide, followed by coupling of the resultant pyrrole to dimethylaminobenzaldehyde. The color development was variable due to difficulty in removing the last traces of peroxide. High levels of amino acids tended to depress color formation, while other compounds such as tyrosine gave extraneous color. Fortunately, these errors tended to balance one another.

"In 1958 a classic method was published by Stegemann.[1] Almost all current methods of hydroxyproline determination trace their roots to this work. Stegemann introduced the use of chloramine-T as an oxidizing agent in place of peroxide. Unfortunately, this paper did not enjoy the popularity it deserved because it was written in German. Part of my contribution was to present the method in an English version. At the same time, I was able to increase the intensity and stability of the final color. More important was the modification of the method so that hydroxyproline could be measured in the presence of large amounts of other amino acids. Recalling earlier problems caused by peroxide, I turned this around and used peroxide to destroy the final chromagen. The remaining color was taken as a blank attributable to interfering substances.

"At that time, and continuing to the present, it has been my practice to have one or two high school students in the laboratory in the afternoons and summers to become acquainted with scientific research. When I had written the method in such a form that the current students were able to carry through the assay successfully on their own, it was deemed suitable for publication.

"In 1976, while preparing a chapter on hydroxyproline determination, I found that over 60 methods and modifications had appeared in the literature since 1961, with a current rate of 5-7 per year.[6] Although several of these offer advantages over my original method, none has improved on the determination of hydroxyproline in the presence of large amounts of other amino acids. This feature continues to be important for the assay of small amounts of hydroxyproline in serum, culture media, and tissues of low collagen content. However, the continuing outpouring of methods gives eloquent testimony to the fact that the ideal hydroxyproline assay method has yet to be published."

1. **Stegemann H.** Mikrobestimmung von Hydroxyprolin mit Chloramin-T und p-Dimethylamino-benzaldehyd. (Microdetermination of hydroxyproline with chloramine-T and p-dimethlamino-benzaldehyde.)
   *Hoppe-Seylers Z. Physiol. Chem.* **311**:41-5, 1958.
2. **Bergman I & Loxley R.** Two improved and simplified methods for the spectrophotometric determination of hydroxyproline. *Anal. Chem.* **35**:1961-5, 1963.
3. **Prockop D J & Udenfriend S.** A specific method for the analysis of hydroxyproline in tissues and urine.
   *Anal. Biochem.* **1**:228-39, 1960.
4. **Neuman R E & Logan M A.** The determination of hydroxyproline. *J. Biol. Chem.* **184**:299-306, 1950.
5. **Juva K & Prockop D J.** Modified procedure for the assay of H$^3$- or C$^{14}$- labeled hydroxyproline.
   *Anal. Biochem.* **15**:77-83, 1966.
6. **Woessner J F, Jr.** Determination of hydroxyproline in connective tissues. *The methodology of connective tissue research.*
   (Hall D A. ed.) Oxford, England: Johnson-Bruvvers Ltd., 1976. p. 235-45.

# Citation Classics

Number 23 — June 6, 1977

Piez K A & Morris L. A modified procedure for the automatic analysis of amino acid. *Analytical Biochemistry* 1:187-201, 1960.

The Sparkman, Moore, and Stein procedure for amino acid analysis[1] was modified by using a continuous salt and pH gradient for elution and separation on a single column of all the amino acids in protein hydrolysates, including hydroxyproline and hydroxylysine. [The *SCI*® indicates that this paper was cited 690 times in the period 1961-1975.]

Karl A. Piez, Chief
Laboratory of Biochemistry
National Institutes of Health
Bethesda, Maryland 20014

January 18, 1977

"As the title of the article by me and Louise Morris indicates, it reports a modified method based on the earlier Moore and Stein procedure for the ion exchange analysis of amino acids which had been automated by Spackman, Moore and Stein. Our reason for modifying what was already a highly developed method and one of the most important procedures in modern biochemistry was twofold. First, it seemed to be an advantage to be able to do a complete analysis on a single column in one run rather than two runs on separate columns as required by the Spackman procedure. Second, we wanted to analyze for hydroxyproline, hydroxylysine and other amino acids not resolved by the two-column method. This was done by using a continuous gradient, which has greater flexibility, rather than a step gradient.

"We were helped greatly in construction by Stanford Moore who kindly provided us with the plans for his instrument. Our instrument was built in the NIH shops and was used continuously for more than ten years. As is well known the Spackman procedure was quickly adopted by instrument manufacturers. Our modification came into commercial use more slowly because it was technically more difficult. With improved instrumentation and computer operation, the single-column method is now usually the method of choice. The continuous gradient is no longer necessary since instruments now allow multiple small steps which serve the same purpose. I don't believe they serve as well, but automatic instrumentation to produce the precise gradients required has not developed.

"Perhaps the most important aspect of the publication to me is that the method was my first step into the collagen and connective tissue biochemistry field. That has resulted in publications which I would like to think are more important contributions than a method, even though individually they may be cited less frequently. Still, all advances depend on methods."

1. Spackman D H, Stein W H & Moore S. Automatic recording apparatus for use in the chromatography of amino acids. *Analytical Chemistry* 30(7):1190-1206, 1958.

# This Week's Citation Classic
CC/NUMBER 52
DECEMBER 27, 1982

König W & Geiger R. Eine neue Methode zur Synthese Von Peptiden: Aktivierung der Carboxylgruppe mit Dicyclohexylcarbodiimid unter Zusatz von 1-Hydroxy-benzotriazolen. (A new method for synthesis of peptides: activation of the carboxyl group with dicyclohexylcarbodiimide using 1-hydroxybenzotriazoles as additives.) *Chem. Ber.* **103**:788-98, 1970.
[Hoechst AG, Frankfurt am Main, Federal Republic of Germany]

1-Hydroxybenzotriazole and a number of substituted 1-hydroxybenzotriazoles are suitable additives in the synthesis of peptides using the dicyclohexylcarbodiimide method.[1] These additives considerably reduce racemisation as well as the formation of by-products and provide peptides in excellent yield and a high state of purity. [The *SCI*® indicates that this paper has been cited in over 680 publications since 1970.]

Wolfgang König
Pharma Synthese G838
Hoechst Aktiengesellschaft
6230 Frankfurt am Main
Federal Republic of Germany

November 4, 1982

"R. Geiger and I started our careers as peptide chemists in the laboratories of Friedrich Weygand (Geiger in 1954 at Tübingen and Berlin and I in 1962 at Munich). There I investigated the rate of racemisation during peptide synthesis. For this purpose I developed, in collaboration with A. Prox, a sensitive and versatile gas chromatographic racemisation test.[2] In 1965, I joined the department of pharmaceutical research at Hoechst AG, where I met Geiger, who had made that step eight years before. One of our tasks was to develop new methods in peptide chemistry, thus accompanying the work of our admired teacher, Weygand, which we continued after his deplorable death in 1969. Concerning the rate of racemisation, Weygand already observed the beneficial effect of the addition of several N-hydroxy compounds to dicyclohexylcarbodiimide, the convenient peptide-forming reagent. We also tested many compounds as potential additives to dicyclohexylcarbodiimide, but the results were discouraging. One day I glanced by chance at an advertisement of my principal, Hoechst AG, showing the formula of benzotriazole. I was immediately fascinated by the idea that the 1-hydroxy-derivative should comprise structural elements most favourable for our purposes. In a matter of days we showed in preliminary tests that 1-hydroxybenzotriazole was indeed the additive we had looked for the whole time.

"In 1969, I reported this result at the European Peptide Symposium in Abano Terme. It was well accepted by the audience and broadly applied immediately. In 1973, an outline of 1-hydroxybenzotriazole in peptide synthesis appeared in Japanese.[3]

"The simple synthesis of 1-hydroxybenzotriazole and its availability from suppliers of fine chemicals resulted in a reserve of another compound, 3-hydroxy-4-oxo-3,4-dihydro-1,2,3-benzotriazine, the use of which is sometimes more advantageous than 1-hydroxybenzotriazole. We frequently use this additive in segment condensation, but most peptide chemists are apparently not aware of the advantages of this compound, possibly also because they overvalue a minor side reaction, which we described in another paper.[4]

"I think that the report about 1-hydroxybenzotriazole in peptide synthesis became a Citation Classic because the reagent is easily accessible and application ensures optimal conditions for peptide synthesis in most cases. The method is so well established that citation is often omitted or secondary literature is cited as, e.g., in the *Pierce Handbook and General Catalog 1979-80*.[5] It is thus the more surprising that such a high citation rate is found."

1. Sheehan J C & Hess G P. Letter to editor. (A new method of forming peptide bonds.)
    *J. Amer. Chem. Soc.* **77**:1067-8, 1955.
2. Weygand F, Prox A & König W. Racemisierung bei Peptidsynthesen. *Chem. Ber.* **99**:1451-60, 1966.
3. Munekata E & Sakakibara S. Use of 1-hydroxybenzotriazole in peptide synthesis.
    *J. Syn. Org. Chem. Jpn.* **31**:853-8, 1973.
4. König W & Geiger R. Eine neue Methode zur Synthese von Peptiden: Aktivierung der Carboxylgruppe mit Dicyclohexylcarbodiimid und 3-Hydroxy-4-oxo-3,4-dihydro-1,2,3-benzotriazin. *Chem. Ber.* **103**:2034-40, 1970.
    [The *SCI* indicates that this paper has been cited in over 65 publications since 1970.]
5. 24460. 1-Hydroxybenzotriazole monohydrate (HBT). *Pierce handbook and general catalog 1979-80*.
    Rockford, IL: Pierce Chemical Co., 1979. p. 333.

# This Week's Citation Classic

Yarmolinsky M B & de la Haba G L. Inhibition by puromycin of amino acid incorporation into protein. *Proc. Nat. Acad. Sci. US* 45:1721-9, 1959.
[McCollum-Pratt Inst., and Mergenthaler Lab. Biol., Johns Hopkins Univ., Baltimore, MD]

---

The authors suggest that the antibiotic, puromycin, is an analog of the terminal aminoacyl-adenosine portion of aminoacyl-transfer RNA. They show that puromycin is an inhibitor of protein synthesis and, by a study of separate steps in the *in vitro* process of radioactive amino acid incorporation into protein, provide evidence that the inhibition occurs at a step following the charging of transfer RNAs with amino acids. [The *SCI*® indicates that this paper has been cited over 560 times since 1961.]

Michael Yarmolinsky
Molecular Genetics
Cancer Biology Program
Frederick Cancer Research Center
Frederick, MD 21701

October 22, 1979

"Having been asked to comment on or explain why an article I coauthored with Gabriel de la Haba has been so often cited, I am beset with conflicting desires: to express irritation at the computer-assisted trivialization of renown and to express pleasure at receiving an honor, however dubious.

"The work came about as a consequence of a review, which I prepared for a few colleagues in 1957, of the then meager literature on protein synthesis. The antibiotic, puromycin, had been shown (by a team at Lederle Laboratories in 1953) to consist of a methylated adenosine linked by a peptide bond from an amino group replacing the hydroxyl group on carbon 3 of the ribose to a methylated tyrosine.[1] In 1955 E. H. Creaser had noted that puromycin inhibited the induction of β-galactosidase in staphylococcus.[2] I called attention to these results in a seminar delivered to my colleagues in the Laboratory of Cellular Pharmacology at the National Institute of Mental Health, suggesting that puromycin might specifically inhibit protein synthesis at a reaction involving both a nucleic acid and an amino acid component. The proposal was perhaps too vague to awaken much interest at a time when the role of transfer RNAs as 'adaptor molecules' was not recognized. Among those present at this seminar were Gabriel de la Haba and a visitor, Edward Reich.

"It was de la Haba who, two years later, encouraged me to take my proposal seriously. We had both returned to our alma mater, Johns Hopkins University, as postdoctoral fellows. The structure of the business end of 'charged' (aminoacyl) transfer RNA had just been published; the resemblance to puromycin was striking. De la Haba had promising ideas of his own concerning the possible sensitivity of protein synthesis to esterase inhibitors such as diisopropyl fluorophosphate (DFP). On a part-time basis the two of us embarked on a collaboration to study the action of DFP and puromycin. The DFP experiments were disappointing, but the puromycin experiments justified our weekend labors. After our results were published, we learned that similar efforts were being planned at the Rockefeller Institute. The unfortunate investigator was Ed Reich.

"Our findings provided a glimmer of rationality in a murkily empirical field. But it is not this achievement that accounts for the frequent citations. It is rather our demonstration that puromycin can be a convenient tool. It has been used (often with skill and effectiveness) to dissect molecular events in protein synthesis in prokaryotes and in eukaryotes. It has been used (often with recklessness and subsequent regret) to dissect the role of protein synthesis in more complex physiological phenomena.

"The puromycin article summarizes an isolated undertaking. Other articles of mine, closer to the line of my subsequent scientific development, mean more to me. One such article, coauthored with Max Gottesman, is about integration-deficient mutants of bacteriophage lambda.[3] Its citation success is unknown to me, but I would not be surprised if its rating is largely a reflection of a trivial happenstance; the article provides the first published mention of a now widely used recombination-deficient mutant of *E. coli* kindly sent to us by a scientist whose renown requires no certification by computer, Matthew Meselson."

1. Waller C W, Fryth P W, Hutchings B L & Williams J H. Achromycin. The structure of the antibiotic puromycin. I. *J. Amer. Chem. Soc.* 75:2025, 1953.
2. Creaser E H. The induced (adaptive) synthesis of β-galactosidase in *Staphylococcus aureus*. *J. Gen. Microbiol.* 12:288-97, 1955.
3. Gottesman M E & Yarmolinsky M B. Integration-negative mutants of bacteriophage lambda. *J. Mol. Biol.* 31:487-505, 1968. [The *SCI*® indicates that this paper has been cited over 205 times since 1968.]

# Chapter 5
# Enzymes

The fifty-five Citation Classics grouped here represent one of the largest units in this collection. Because of this large number and the variety of subjects included under the Enzyme banner—methods, kinetics, enzyme inducers, etc.—it may be informative first to consider some generalizations.

At the time of accounting, eleven of these Classics had been cited more than 1,000 times and fourteen had a citation record greater than 750. The eleven top scorers form a heterogeneous group. Astrup and Müllertz found their fibrin plate assay for determining fibrinolytic activity popular because it led to the discovery of plasminogen activator and because the method is also adaptable to other proteases, to lipases, and to other enzymes. Kalckar's Classic is based on a change in ultraviolet absorption of certain nucleosides or nucleotides when they serve as substrates for purine enzymes. Since UV spectrophotometry had not been used in the United States for enzymatic analysis prior to his reports and because several of the purine metabolizing enzymes gained clinical importance, his report became unusually popular. Skou's review on ATPase-activated membrane transport of $Na^+$ and $K^+$ is popular because it is an authoritative review appropriate to the needs of contemporary researchers in membrane energetics and transport. The dominance of adenylate cyclase and cyclic AMP in the biochemistry literature of the past two decades is well known and is based largely on the work of Sutherland and his associates, which is discussed further below. Influential in the cyclic AMP area was the reporting by Salomon and his colleagues of a sensitive adenylate cyclase assay, an assay that was cited 1,105 times in a period of 8 years.

One of the most cited articles is particularly pertinent to contemporary recombinant DNA technology. The restriction endonucleases I and II from *Haemophilus parainfluenzae* are important and useful enzymes of this class, and their discovery made use of the sensitive ethidium bromide fluorescence of DNA fragments (p. 161). Tsou describes the use of nitroblue tetrazolium as an indicator of succinic dehydrogenase in tissue, but the dye is also applicable to the cytochemical location of many dehydrogenases and therein lies its popularity (1,150 citations). The Classic described by Schenkman (p. 171), cited over 1,025 times, was one of the early reports on cytochrome P-450 monooxygenase, and the ease with which this method measures binding to this cytochrome by a shift in its absorption spectrum popularized this method as a means to examine drug interaction with microsomes. Conney reviewed the status of enzyme induction, particularly of the microsomal oxygenases, which when once activated accelerated the generation of carcinogens from their precursors. He also reviewed the effect of drugs on the induction of other microsomal enzyme systems that influenced electron transport, steroid metabolism, and other enzyme-related activities.

The famous discovery by de Duve that many liver enzymes were packaged in a distinct cell organelle, the lysosome, led to many important publications on this topic. The Classic described here contains the first use of the word lysosome in a scientific paper. Moncada describes his oft-cited report with Vane (among other coauthors) on the discovery of prostacyclin, an inhibitor of platelet aggregation derived from arachidonic acid. The prostacyclins are formed by an enzyme pathway separate from that used for the production of thromboxane $A_2$, a platelet aggregator. Thus arachidonic acid is a precursor of both a platelet stabilizer and aggregator. And last in the 1,000-fold citation list is the article by McCord and Fridovich on superoxide dismutase. Their demonstration that erythrocuprein had this dismutase activity was interesting in itself, but this article became popular because it clearly described the isolation and assay of the enzyme. This enzyme is now believed to be important to many bacteria for their protection against the toxic superoxide anion.

Biochemists will find many familiar names as authors of Classics in this section. A partial list would include Knox, Udenfriend, Kalckar, Potter, Sutherland, Temin, Spiegelman, Theorell, de Duve, Samuelsson, Vane, Lardy, and some who were already mentioned. No fewer than seven of these are Nobel Prize recipients. Sutherland's exposure of adenylate cyclase and cyclic AMP as part of the messenger sequence that converted external stimuli into a secretory response (in cells capable of a secretory function) is a concept applicable to cells in many tissues and organs. This is intimately tied to the mobility of $Ca^{2+}$ across the outer membranes of these cells. Temin's Nobel Prize, shared with David Baltimore, was awarded for their unexpected finding that RNA could serve as the source of genomic information when reverse transcriptase (RNA-dependent DNA polymerase) generated DNA from this RNA. This "upstream" conversion of RNA to DNA is a necessity for the successful replication of the RNA viruses that are devoid of DNA. Theorell's Nobel Prize was based on his extensive studies of enzyme reactions and enzyme kinetics. De Duve received his Nobel award for descriptions of the condensed enzyme organelles, the lysosomes, first found in liver cells but later found to be abundant in other cells that have a specialized degradative function. And more recently, Samuelsson and Vane shared the 1982 Nobel Prize in Medicine and Physiology with Bergström for their informative studies of arachidonic acid metabolites, which has so significantly altered thinking about inflammation, analgesics, allergy, cell membrane structure, thromboembolic disease, and other unresolved problems in medicine. Mitchell became a Nobel Laureate for advancing his theory of proton transport across the mitochondrial membrane.

But all is not science here. Several of these "This Week's Citation Classic" (TWCC) narratives reveal much about the human nature of scientists and their approach to problems. For example, Erlanger, Fahrney, Salomon, Lindell, and Ellman all refer to the initial rejection of their manuscripts, which later became Classics. Each of them overcame this rejection: Erlanger selected a second journal, Fahrney pressed successfully for additional editorial review, Salomon applied force upon the editor, Lindell took the second-journal route, and Ellman overcame the rejection by an expansion of the data reported. Since it is generally agreed that most research articles submitted are eventually published, these five incidents are not unusual in the end result, only in the avenues taken to achieve these results and for the eventual status of these publications among scientists.

Invocation of serendipity as a key to success is also noticeable in a few of these TWCCs. Skou cites the unintended replacement of $Na^+$ with $K^+$ to balance the ionic strength as instrumental in his studies of ATPase-dependent cell transport of these cations. Dahlqvist found a shift in buffer choice essential to his development of a reliable disaccharidase assay. Other TWCCs contain important comments, including Erlanger's, written in 1981 and now a refrain among scientists, that "It is time that we correct this situation" of scant funding of young scientists.

# This Week's Citation Classic

CC/NUMBER 41
OCTOBER 8, 1984

Astrup T & Müllertz S. The fibrin plate method for estimating fibrinolytic activity. *Arch. Biochem. Biophys.* 40:346-51, 1952.
[Biological Institute, Carlsberg Foundation, Copenhagen, Denmark]

This paper gives a description of the optimum conditions of the fibrin plate method, thereby making possible the accurate assessment of small quantities of fibrinolytic agents. [The *SCI®* indicates that this paper has been cited in over 1,060 publications since 1955.]

---

Tage Astrup
Section of Coagulation and Fibrinolysis
Department of Clinical Chemistry
Ribe County Hospital
and
Section of Thrombosis Research
South Jutland University Centre
6700 Esbjerg
Denmark

June 13, 1984

"The roots of this paper go back in time to the mid-1930s. The study of mammalian cells grown *in vitro* was in its adolescent stage with numerous unsolved problems. The chicken plasma clot providing the solid matrix for the growing cells was often seen to undergo a process of liquefaction causing the cell culture to collapse. This occurred most frequently when the cultivation of explants from certain epithelia and tumors was attempted. It was a major goal of the Copenhagen Institute (headed by Albert Fischer) to elucidate the particular interactions between the cells and their substrates causing this liquefaction.

"In those days, this was not an easy task. Methods for the production of sufficient quantities of purified fibrinogen and thrombin had to be worked out first. To emulate conditions in tissue culture, solutions of bovine fibrinogen were clotted in a petri dish with bovine thrombin forming a layer of fibrin, on the surface of which samples of tissue or drops of lytic solutions could be placed causing the formation of areas of lysis. Thus was born the fibrin plate method. The first major result of its application was the discovery of the tissue plasminogen activator.[1] Subsequently, the method was standardized, resulting in the publication cited here. Among important, early findings were the demonstration by Sten Müllertz[2] of a plasminogen activator in blood, and the observation by Olesen[3] that a plasminogen activator is generated from a humoral precursor by acid polysaccharides (now called the intrinsic system of fibrinolysis). The method made possible the quantitative assay of the small amounts of plasminogen activator usually present in human and animal tissues.[4] The method became popular because of its simplicity and sensitivity, and because it simulated conditions in the body, but many investigators encountered difficulties in mastering the technique. This was chiefly caused by the use of inferior grades of plasminogen-rich fibrinogen. Conditions have been worked out in detail,[5] the latest test of precision appearing recently.[6]

"Müllertz, now professor and head of the department of clinical chemistry at the Hvidovre Hospital and the University of Copenhagen, has retained his interest in fibrinolysis. I transferred my research activities to Washington, DC, in 1961, supported by a grant from the National Heart, Lung, and Blood Institute, National Institutes of Health. Having retired in 1976 to my native country, I am happily continuing research in blood coagulation, fibrinolysis, and thrombosis in my home town, Esbjerg, aided by dedicated and highly qualified colleagues and associates. For this I owe a debt of gratitude to good fortune."

---

1. **Astrup T & Permin P M.** Fibrinolysis in the animal organism. *Nature* 159:681-2, 1947. (Cited 175 times since 1955.)
2. **Müllertz S.** A plasminogen activator in spontaneously active blood. *Proc. Soc. Exp. Biol. Med.* 82:291-5, 1953.
3. **Olesen E S.** Peptone activation of a serum fibrinolytic system. *Acta Pharmacol. Toxicol.* 15:197-206, 1959.
4. **Astrup T & Albrechtsen O K.** Estimation of the plasminogen activator and the trypsin inhibitor in animal and human tissues. *Scand. J. Clin. Lab. Invest.* 9:233-43, 1957. (Cited 170 times since 1955.)
5. **Astrup T & Kok P.** Assay and preparation of tissue plasminogen activator. *Meth. Enzymology* 19:821-34, 1970.
6. **Jespersen J & Astrup T.** A study of the fibrin plate assay of fibrinolytic agents: optimal conditions, reproducibility and precision. *Haemostasis* 13:301-15, 1983.

## This Week's Citation Classic

Alkjaersig N, Fletcher A P & Sherry S. The mechanism of clot dissolution by plasmin. *J. Clin. Invest.* **38**:1086-95, 1959.
[Dept. Med., Washington Univ. Sch. Med., St. Louis, MO]

In this paper we described in vitro experiments and in vivo observations concerned with thrombolytic mechanisms. Since plasminogen is found in plasma and also is a constituent of thrombi, clot lysis occurs by a dual mechanism. The primary mechanism of thrombolysis involves the diffusion or adsorption of plasminogen activator to the thrombus, activation of intrinsic clot plasminogen, and thrombolysis. The secondary mechanism involving digestion of the thrombus by extrinsic plasmin action appears to be of negligible importance. [The *SCI®* indicates that this paper has been cited in over 550 publications since 1959.]

Norma Alkjaersig
Geriatric Research, Education, and Clinical Center
Veterans Administration Medical Center
St. Louis, MO 63125

May 16, 1984

"This paper, published simultaneously with two clinical investigative studies,[1,2] appeared at a time of considerable controversy in the developing field of thrombolytic therapy. Essentially, it had been shown that clot lysis (thrombolysis) could be produced in the experimental animal by infusion of several proteolytic enzymes including 'plasmin' preparations (subsequently shown to contain high concentrations of streptokinase), or by streptokinase (a plasminogen activator) alone. Fortunately, one of my colleagues had been invited on an extensive tour of US nuclear facilities during the preparatory phase of the 1955 Geneva conference on 'Atoms for Peace' and had returned convinced of the potential inherent in radiochemical assay methods. The developing of an assay for thrombolytic activity based on the use of $^{131}$I labeled fibrin greatly facilitated studies showing the major importance of plasminogen activator and the relatively minor effect of enzymes such as plasmin on clot lysis.

"At that time, it was difficult to see how an *in vivo* fibrinolytic state could be achieved, since it was already known that plasma contained greater inhibitory capacity than the potentially available plasmin. *In vitro* experiment, however, showed that much greater fibrinolysis than fibrinogenolysis occurred when a plasminogen activator was introduced into a plasma-clot system; consequently a mechanism existed which favored lysis of thrombi and at the same time protected the blood coagulation system. Further experiments indicated that plasminogen was adsorbed onto the clot, where, when activated, it was in close proximity to its substrate, fibrin, and in a relatively inhibitor-free environment. Initially, it was difficult to accept that plasminogen was adsorbed to the clot, since there was little difference in plasminogen content of plasma and serum; however, washed clots could be ground up and extracted and did indeed release measurable plasminogen. Later, it was shown that approximately four percent of the plasminogen is bound to fibrin.[3]

"These *in vitro* studies, together with earlier studies on the clearance rate of streptokinase, formed the basis for determining doses and infusion rate, and for defining the necessary laboratory measurements to ensure an active fibrinolytic state and an adequate blood coagulation system.

"Our original concern over introducing a hemorrhagic diathesis by the infusion of streptokinase has been obviated by newer plasminogen activators, first urokinase, which has a more favorable ratio of fibrinolysis to fibrinogenolysis than streptokinase, and most recently by the development of tissue plasminogen activator, which is capable of lysing clots *in vivo* with little, if any, effect on the coagulation system. Collen and others have been especially active in this area, and a review by Collen[3] outlines the development of this activator, and summarizes the more recent developments in the biochemistry of fibrinolysis.

"When I joined Sol Sherry, he was chief of the Medical Service at the Jewish Hospital of St. Louis, where Tony Fletcher soon joined as well. The early part of these studies took place there but, shortly after, we moved to Barnes Hospital; throughout, all three of us held appointments at Washington University. My clinical investigator colleagues were excited by the avenues opened by the plasminogen activators as thrombolytic agents and were instrumental in designing the *in vitro* experiments. We were all overjoyed to find the *in vivo* studies correlating so well with projections.

"I think that this paper has been referred to frequently because, with the accompanying clinical studies, it offered a rational basis for the use of plasminogen activators as thrombolytic agents and also because of interest in the methodology employed."

1. Fletcher A P, Alkjaersig N & Sherry S. The maintenance of a sustained thrombolytic state in man. I. Induction and effect. *J. Clin. Invest.* **38**:1096-110, 1959. (Cited 225 times since 1959.)
2. Fletcher A P, Sherry S, Alkjaersig N, Smyrniotis F E & Jick S. The maintenance of a sustained thrombolytic state in man. II. Clinical observations on patients with myocardial infarction and other thromboembolic disorders. *J. Clin. Invest.* **38**:1111-19, 1959. (Cited 100 times since 1959.)
3. Collen D. On the regulation and control of fibrinolysis. Edward Kowalski Memorial Lecture. *Thromb. Haemost.* **43**:77-89, 1980.

# This Week's Citation Classic

CC/NUMBER 4
JANUARY 26, 1981

Erlanger B F, Kokowsky N & Cohen W. The preparation and properties of two new chromogenic substrates of trypsin. *Arch. Biochem. Biophys.* 95:271-8, 1961.
[Dept. Microbiology, Coll. Physicians and Surgeons, Columbia Univ., New York, NY]

This paper describes the synthesis and properties of two new trypsin substrates, L-lysine p-nitroanilide and benzoyl-DL-arginine p-nitroanilide (BAPA). BAPA was found to be an excellent substrate, highly sensitive and specific and stable in the absence of enzyme. A novel method of synthesis was devised that was simple, economical, and which gave excellent yields. [The *SCI®* indicates that this paper has been cited over 775 times since 1961.]

Bernard F. Erlanger
Department of Microbiology
Health Sciences Center
College of Physicians and Surgeons
Columbia University
New York, NY 10032

December 24, 1980

"At the time of this research Bill Cohen was a postdoctoral fellow, fresh from F.F. Nord's laboratory. He is now a professor of biochemistry at Tulane. Nick Kokowsky had been a research assistant in my laboratory for several years and is now running his own chemical business. Bill was working on the catalytic mechanism of chymotrypsin by studying the reactivation of DEP-chymotrypsin by various nucleophiles. Our assay used acetyl-DL-phenylalanine B-naphthyl ester as the substrate, and included a color development with a diazonium salt.[1] The naphthyl ester was unstable, giving high blanks that differed in magnitude for each nucleophile.

"When we started studies on trypsin, we sought a substrate that would be stable and colorless and would yield a colored product upon tryptic hydrolysis. We prepared benzoyl-DL-arginine p-nitroanilide (BAPA) and L-lysine p-nitroanilide. BAPA was found to be an excellent substrate, highly sensitive, very specific and stable in the absence of trypsin. The product of tryptic hydrolysis, p-nitroaniline, is orange in color and can be measured directly, even in a colorimeter.[2]

"The synthesis we developed was esthetically satisfying in that (a) it was a one-step process followed by only one crystallization, (b) the yield was excellent, and (c) it was an original procedure that used $P_2O_5$ as a condensing agent (for which there was no precedent in the literature as far as I knew). The problem was to find a solvent and I happened to have a sample of diethylphosphite. It worked.

"We have used this assay as an easy way to teach students how to measure $k_{cat}$ and $K_m$ of enzyme-substrate reactions. We also prepared p-nitroanilide substrates for chymotrypsin,[3] and there now exist nitroanilide substrates for many peptidases.

"The BAPA assay did not take hold for many years, a surprise to me since we found it more reproducible than titrimetric procedures with ester substrates. Its popularity grew suddenly when it was used in an indirect assay of trypsin inhibitor in the serum of patients with cystic fibrosis. Its ability to be automated was also probably a contributing factor. Although I cannot find the correspondence, it is my recollection that the paper was found not to be of general interest by the first journal to which we sent it.

"Although we cite both the Office of Naval Research (ONR) and National Institutes of Health (NIH) for support in the paper, the former was my major source of encouragement. It was through this agency that I was paid during my dissertation research and they gave me my first grant as an independent researcher. ONR, at that time, helped many young researchers, thanks to people like Bill Consolazio and Orr E. Reynolds. Unfortunately, the Mansfield Amendment severely limited this kind of support on the part of ONR. Coupled with the trend of other agencies toward contracts, program grants, and institutional grants, less money is now available to fund promising young scientists who can do a great deal with relatively small grants. It is time that we correct this situation."

1. **Cohen W & Erlanger B F.** Studies on the reactivation of diethylphosphorylchymotrypsin.
   *J. Amer. Chem. Soc.* 82:3928-34, 1960.
2. **Cohen W, Lache M & Erlanger B F.** The reactivation of diethylphosphoryltrypsin. *Biochemistry* 1:686-93, 1962.
3. **Erlanger B F, Edel F & Cooper A G.** The action of chymotrypsin on two new chromogenic substrates.
   *Arch. Biochem. Biophys.* 115:206-10, 1966.

# This Week's Citation Classic

CC/NUMBER 42
OCTOBER 17, 1983

Fahrney D E & Gold A M. Sulfonyl fluorides as inhibitors of esterases. I. Rates of reaction with acetylcholinesterase, α-chymotrypsin, and trypsin.
*J. Amer. Chem. Soc.* 85:997-1000, 1963.
[Departments of Biochemistry and Neurology, Columbia University, New York, NY]

This paper reported the remarkable reactivity of phenylmethanesulfonyl fluoride (PMSF) toward chymotrypsin. PMSF is 10,000 times more reactive than the methane analog. We suggested that correct binding at the active site provides the driving force for the PMSF reaction. [The *SCI*® indicates that this paper has been cited in over 485 publications since 1963.]

David E. Fahrney
Department of Biochemistry
Colorado State University
Fort Collins, CO 80523

August 2, 1983

"Landmark papers in the early 1950s established that certain enzymes react chemically with substrates to form covalently bonded enzyme-substrate intermediates.[1-3] Thus, the stage was set for the design of 'active-site directed' irreversible inhibitors as probes for enzyme mechanisms in the 1960s.

"As a graduate student with David Nachmansohn at Columbia University, I was expected to work on acetylcholinesterase. But my interest turned to a paper on the inactivation of chymotrypsin by dansyl chloride.[4] This sulfonyl chloride reacts with a histidine residue at the active site, instead of serine. Although evidence that the Nazi nerve gas DFP reacted at a serine residue seemed irrefutable to some, others argued for nucleophilic attack by a histidine residue, followed by transfer of the phosphoryl group to a nearby serine hydroxyl group. In contrast, model studies indicated that transfer of a sulfonyl group from histidine to serine was unlikely. Since sulfonyl fluorides are much less reactive than phosphoryl fluorides, I thought that a sulfonyl fluoride would react at the active site only if it were juxtaposed against the attacking nucleophile. If the R group were phenylmethyl instead of DFP's biologically irrelevant isopropyl group, the sulfonyl fluoride might mimic phenylalanine substrates and react with the correct nucleophilic group. Nachmansohn liked the idea and a research plan was sent to the National Science Foundation. But he did not let me start in the lab until I had filled two notebooks with handwritten abstracts—not Xerox copies—of papers on chymotrypsin. The work proceeded smoothly, due largely to the superb guidance of Allen Gold, then a postdoctoral fellow in Nachmansohn's group. When the time to submit the manuscript arrived, Nachmansohn felt Al and I should be the authors. First author was decided by a toss of a coin. The manuscript received two diametrically opposed reviews and was initially rejected by the editor. Nachmansohn demanded a third reviewer and the manuscript was accepted verbatim.

"Although cited often by researchers designing new enzyme inhibitors, citations also occur in papers on the isolation and purification of proteins from a wide variety of biological systems. Phenylmethanesulfonyl fluoride (PMSF) is added to buffers in place of the highly toxic DFP to block proteases. Unfortunately, PMSF doesn't always work: after all, it was designed to fool chymotrypsin.

"PMSF was a new compound. Twenty years later the field of designing new enzyme inactivators is still very active. Perhaps the most ingenious are the recent 'suicide substrates.'"[5]

1. Wilson I B, Bergmann F & Nachmansohn D. Acetylcholinesterase. X. Mechanism of the catalysis of acylation reactions. *J. Biol. Chem.* 186:781-90, 1950. (Cited 80 times.)
2. Segal H L & Boyer P D. The role of sulfhydryl groups in the activity of D-glyceraldehyde 3-phosphate dehydrogenase. *J. Biol. Chem.* 204:265-81, 1953. (Cited 55 times.)
3. Hartley B S & Kilby B A. The reaction of p-nitrophenyl esters with chymotrypsin and insulin. *Biochemical J.* 56:288-97, 1954. (Cited 200 times.)
4. Hartley B S & Massey V. The active centre of chymotrypsin. I. Labelling with a fluorescent dye. *Biochim. Biophys. Acta* 21:58-70, 1956. (Cited 155 times.)
5. Walsh C T. Suicide substrates: mechanism-based enzyme inactivators with therapeutic potential. *Trends Biochem. Sci.* 8:254-7, 1983.

## This Week's Citation Classic

CC/NUMBER 11
MARCH 17, 1980

Knox W E. Two mechanisms which increase *in vivo* the liver tryptophan peroxidase activity: specific enzyme adaptation and stimulation of the pituitary-adrenal system. *Brit. J. Exp. Pathol.* 32:462-9, 1951.
[Molteno Institute, University of Cambridge, Cambridge, England]

The amount of an enzyme, now called L-tryptophan oxygenase, was greatly increased in liver by treatment of living rats with the enzyme's specific substrate. Some unrelated compounds caused lesser increases by a different mechanism, but only if the adrenal glands were present. The first mechanism was like the substrate induction of enzymes known in microorganisms. The second mechanism was a way hormones could act to affect metabolism by altering the amounts of specific enzymes in cells. [The *SCI*® indicates that this paper has been cited over 250 times since 1961.]

W. Eugene Knox
Department of Biological Chemistry
Harvard Medical School
and Cancer Research Institute
New England Deaconess Hospital
Boston, MA 02215

January 18, 1980

"This paper was often cited, perhaps, because it left so much more to be done in the field of metabolic regulation that it started. It described the first inductions of increased amounts of an enzyme in animal tissues by a substrate or by glucocorticoid hormones. It was a big surprise in 1951 that enzyme concentrations in cells did change.

"How hormones in trace amounts could influence metabolism was then a major question for enzyme physiology. Most experimentation envisioned them as a kind of participant in the enzyme reaction, something like the role of coenzymes that had recently explained the actions of vitamins. Typical experiments were always additions of hormones *in vitro* to cell-free enzyme preparations.

"Hormones offered one way to alter metabolism, and parallel alterations of appropriate enzyme reactions could be expected. Our approach differed from the fashionable one only in that we produced the alterations in the animal and not in the test tube.

"We had sought and found something analogous to the enzyme adaptation (induction) of microorganisms: a great elevation in the amount of an enzyme that degraded tryptophan whenever excess tryptophan was administered to an animal. However, control experiments showed that administration of certain nonsubstrate compounds also increased the enzyme. Such nonspecific noxious stimuli were known to release ACTH from the pituitary and cause the adrenal glands to pour out cortisone (the 'stress response'). Cortisone was not yet available, so a biological test was necessary to find whether the nonsubstrate elevators of the liver enzyme acted through the pituitary-adrenal system to release cortisone. Adrenalectomy interrupted the pituitary-adrenal system and proved both mechanisms: it eliminated the effects of the nonsubstrate compounds but preserved that of tryptophan. In addition to the biologically primitive regulation of an enzyme by its substrate, the higher animals also regulated tissue enzyme amounts by their hormones.

"Some citations were inevitable during the cleanup of numerous problematic aspects left from the discovery in this paper. Did the adrenal hormone itself actually cause the enzyme change? Was the metabolic machinery of animal cells so plastic that environmental conditions could alter the proportions of its enzymes? If so, could other examples of regulated enzymes be found? Several years had already been expended on the properties and assay of this admittedly complex enzyme, the first of the oxygenases to be recognized, and one whose mechanism is still unclear. We could distinguish between merely more activity of the unchanged enzyme and an increased amount of the enzyme. But, inexplicably, for a decade biochemists did not readily assimilate the fact that enzyme concentrations in cells might change. We plumped for a straightforward regulation by hormones of the then still mysterious synthesis of specific proteins.

"Protein synthesis by living cells was apparently necessary for hormones to act 'not by affecting the enzyme reaction itself, but by altering the amount of the enzyme.' In spite of the paper's apparently large readership, fruitless experiments seeking *in vitro* actions of hormones in cell-free systems that could not synthesize proteins continued to be published for a decade. Then the problem of hormone actions evaporated. Suddenly it had become a part of common sense, known even to schoolboys, that hormones changed enzyme concentrations in cells."

# This Week's Citation Classic

CC/NUMBER 5
FEBRUARY 4, 1980

Nagatsu T, Levitt M & Udenfriend S. Tyrosine hydroxylase: the initial step in norepinephrine biosynthesis. *J. Biol. Chem.* **239**:2910-17, 1964.
[Lab. Clin. Biochem., National Heart Institute, National Institutes of Health, Bethesda, MD]

This paper describes the presence and properties of a new pteridine-requiring enzyme tyrosine hydroxylase in brain, adrenal medulla, and sympathetically innervated tissues. [The *SCI*® indicates that this paper has been cited over 645 times since 1964.]

Toshiharu Nagatsu
Department of Life Chemistry
Graduate School at Nagatsuta
Tokyo Institute of Technology
Yokohama 227
Japan

September 4, 1979

"It is most gratifying to learn that our tyrosine hydroxylase paper has been so frequently cited. The work in this paper was started in 1963 at the laboratory of Sidney Udenfriend in the NIH, where I was a NIH international postdoctoral fellow and Morton Levitt was a graduate student. At that time, among the four enzymes involved in the catecholamine biosynthesis, only the enzyme responsible for converting tyrosine to dopa was elusive. Therefore, assuming that such an enzyme may exist in catecholamine-containing tissues, we first developed a highly sensitive isotopic assay for the enzyme activity. L-($^{14}$C) tyrosine was used as a substrate, and L-($^{14}$C) dopa, enzymatically formed, was isolated on an alumina column and assayed. We started our initial work to discover the enzyme in tissue slices and minces which should contain necessary cofactors with the enzyme, and we found a substantial formation of dopa from L-($^{14}$C) tyrosine. However, we also found a significant nonenzymatic hydroxylation of tyrosine to dopa in heated tissue slices and minces. This may be the reason why the presence of this enzyme went unnoticed for so long. Fortunately, we found the absolute stereospecificity of this enzyme which permitted the use of D-($^{14}$C) tyrosine as a control, and we became convinced that we were really detecting a new enzyme.

"We found bovine adrenal medulla contained a large amount of the enzyme activity in the soluble fraction, and we could isolate the enzyme by ammonium sulfate fractionation. After testing many probable cofactor substances, the preparations were shown to require for activity a tetrahydropteridine and molecular oxygen. The requirement of a tetrahydropteridine as a cofactor of the hydroxylating process was first discovered with rat liver phenylalanine hydroxylase by Seymour Kaufman in the NIH in 1959.[1]

"Another significant finding in this work was the inhibition of the enzyme by the products catecholamines, and we proposed the possibility of the feedback inhibition *in vivo*, which is now of great interest for the short term regulation of catecholamine biosynthesis.

"The reasons why our publication is so frequently cited may be the great physiological significance of this enzyme, its widely-applicable assay procedure, and its interesting properties as a pteridine-dependent monooxygenase.

"Sidney Udenfriend is the director of Roche Institute of Molecular Biology, and Morton Levitt is at the New York State Psychiatric Institute. We had another chance to collaborate again in 1972 at Roche Institute.[2]

"I recall distinctly how much we were pleased to see the enzyme activity of tyrosine hydroxylase in front of the liquid scintillation counter at midnight. The best reward for scientists may be the pleasure and excitement of new findings."

1. Kaufman S. Studies on the mechanism of the enzymatic conversion of phenylalanine to tyrosine. *J. Biol. Chem.* **234**:2677-88, 1959.
2. Nagatsu T & Udenfriend S. Photometric assay of dopamine β-hydroxylase activity in human blood. *Clin. Chem.* **18**:980-3, 1972.

# This Week's Citation Classic

CC/NUMBER 43
OCTOBER 24, 1983

Russell D & Snyder S H. Amine synthesis in rapidly growing tissues: ornithine decarboxylase activity in regenerating rat liver, chick embryo, and various tumors. *Proc. Nat. Acad. Sci. US* **60**:1420-7, 1968.
[Depts. Pharmacol. and Exp. Therapeutics, and Psychiat. and Behavioral Sci., Johns Hopkins Univ. Sch. Med., Baltimore, MD]

A decarboxylase, specific for ornithine decarboxylation, was shown to be rapidly and dramatically elevated in regenerating rat liver and in developing chick embryos. Ornithine decarboxylase activity was elevated threefold within one hour and 25-fold within 16 hours in rat liver after partial hepatectomy. In chick embryo, activity correlated with embryonic growth rate. It was suggested that ornithine decarboxylase was involved in the initiation of the growth process. This report implicated ornithine decarboxylase and polyamine biosynthesis as important parameters of mammalian cell growth regulation. [The *SCI*® indicates that this paper has been cited in over 630 publications since 1968.]

Diane H. Russell
Department of Pharmacology
College of Medicine
University of Arizona
Tucson, AZ 85724

August 4, 1983

"In 1967, I finished my PhD degree under the tutelage of Donald S. Farner, a renowned avian physiologist, and embarked on a postdoctoral fellowship with Solomon H. Snyder at Johns Hopkins University School of Medicine. Evidence was accumulating that polyamines, organic cations implicated in the regulation of protein and RNA synthesis in bacteria and other microorganisms, also might be of importance in similar physiological regulation processes in mammals. The stigma to the study of these compounds in higher animals can be attributed, in retrospect, to the common names assigned to these important nitrogen-rich, ubiquitously occurring, short-chain hydrocarbons. Spermine, a polyamine containing four amine groups, was described first in human semen by van Leeuwenhoek,[1] the inventor of the microscope. Spermidine, a triamine-containing polyamine, was found later to be present also in high concentrations in seminal fluid. Putrescine, the diamine precursor of spermidine and spermine, was isolated by Brieger in 1887 from the cholera-producing bacterium, *Vibrio cholerae*.[2] The possibility that these misnamed compounds might be important in mammalian metabolism was suggested by Dykstra and Herbst[3] who demonstrated in regenerating rat liver the rapid and extensive uptake of [$^3$H]putrescine and its conversion to spermidine in parallel with ribosomal RNA synthesis. These data coupled with the ubiquitous presence of putrescine, spermidine, and spermine in plant and animal tissues led us to determine whether ornithine decarboxylase, the enzyme which catalyzes putrescine formation, might be important in the initiation of rapid growth processes. We chose to measure ornithine decarboxylase activity as a function of time after partial hepatectomy in the rat, after fertilization in chick embryos, and in rat hepatomas and sarcomas. In regenerating rat liver, the activity of ornithine decarboxylase was threefold of control within one hour and was 25-fold elevated within 16 hours. The peak activity in chick embryos occurred at five days of age, the time of limb-bud formation and the most rapid growth rate. The very early and striking increase in ornithine decarboxylase activity in regenerating rat liver and in chick embryo, which preceded increased nucleic acid synthesis, suggested a physiological function for the enzyme in the initiation of the growth process.

"At this time, ornithine decarboxylase is an established biochemical marker of growth initiation and has been shown to increase in a dose-dependent manner in target tissues after exposure to trophic hormones, growth factors, and steroid hormones. Its extent of increase (up to 1,000-fold) and rapid half-life have indicated further its unique position as an internal marker of cell surface receptor-mediated activity.

"The large number of citations may be due to: 1) the establishment of elevated ornithine decarboxylase activity as an early pronounced event in a variety of rapidly growing animal tissues; 2) the presence in the paper of a useful, simple method for the measurement of ornithine decarboxylase activity in avian and mammalian tissues; 3) the demonstration in regenerating rat liver that the enzyme was specific for ornithine as a substrate; and 4) widespread interest generated by the paper for the measurement of ornithine decarboxylase as a biochemical marker of hormone action.

"A recent monograph of the field was published in 1978[4] and sketches from my point of view some important physiological aspects of polyamines and of polyamine biosynthesis as biochemical markers of normal and malignant growth."

1. **van Leeuwenhoek A.** Observationes D. Anthonii Lewenhoeck de natis e semine genitali animalculus. *Phil. Trans. Roy. Soc. London* **12**:1040-3, 1678.
2. **Cohen S S.** *Introduction to the polyamines.* Englewood Cliffs, NJ: Prentice-Hall, 1971. p. 4.
3. **Dykstra W G, Jr. & Herbst E J.** Spermidine in regenerating liver: relation to rapid synthesis of ribonucleic acid. *Science* **149**:428-9, 1965.
4. **Russell D H & Durie B G M.** *Polyamines as biochemical markers of normal and malignant growth.* New York: Raven Press, 1978. 178 p.

# This Week's Citation Classic™
CC/NUMBER 26
JUNE 25, 1984

Kalckar H M. Differential spectrophotometry of purine compounds by means of specific enzymes. III. Studies of the enzymes of purine metabolism.
*J. Biol. Chem.* **167**:461-75, 1947.
[Div. Nutrit. and Physiol., Public Health Res. Inst. of the City of New York, Inc., NY]

A description is given for the preparation of adenosine deaminase, adenylic deaminase, adenylpyrophosphatase, xanthine oxidase, uricase, nucleoside phosphorylase, and guanase in sufficient purity to be used as analytical reagents for the measurement of the purines on which they act. [The *SCI*® indicates that this paper has been cited in over 1,195 publications since 1955.]

---

Herman M. Kalckar
Department of Chemistry
Boston University
Boston, MA 02215

April 30, 1984

"The principle of determining enzymatic reactions by following optical changes at specific wavelengths had not, as far as I knew in 1944, been used in the US, at least not in the ultraviolet region. In Berlin, Otto Warburg, who built ultraviolet spectrophotometers in his biochemistry lab, introduced the principle in his studies of redox enzymes in 1935.[1]

"In 1943, working at the Public Health Institute of the City of New York, Inc., I had available to me for the first time a Beckman ultraviolet spectrophotometer. I now explored some known spectral changes in the more shortwaved ultraviolet spectra of purines, purine nucleosides, or nucleotides which had been observed by changing the pH from the acid to the alkaline range. It soon dawned upon me that these sensitive spectral changes could be used in the field of enzymology of nucleosides and nucleotides.

"This was first published in *Federation Proceedings*,[2] where I also introduced a new active ester ribose-1-phosphate and its enzymic reaction with purines to form nucleosides (in surprisingly good yields). The details of this work appeared in 1947 in the *Journal of Biological Chemistry*. This appeared in four papers entitled 'Differential spectrophotometry of purine compounds by means of specific enzymes, I-IV' by myself (with the technical assistance of Manya Shafran).[3-5] The third paper is the one being classified as one of the most-cited items in the biochemistry literature. This paper focuses on the determination of a variety of enzymes of purine, nucleoside, and nucleotide metabolism. Some of these enzymes, like uricase or xanthine oxidase, as well as their respective substrates, also happen to have clinical interest. Hence, when ultraviolet spectrophotometers became a common item in clinical labs, the methods were probably deemed of additional interest.

"A clinical derivation of the principle of ultraviolet spectrophotometry of purines was presented in the 1947 paper, through the ultraviolet determination of *uric acid* in serum (see, for instance, the Sigma catalog[6]). This is a sensitive method for use in cases like arthritic urica, or in infants, to spot the serious inborn error, the Lesch-Nyhan syndrome (lack of the enzyme of HGPRT).

"For recent references, see Murphy et al.[7] and Tritsch."[8]

1. **Warburg O & Christian W.** Pyridine, the hydrogen transferring constituent of fermentation enzymes. (Translated from German by Kalckar H M.) *Biological phosphorylations*. (Kalckar H M.) Englewood Cliffs, NJ: Prentice-Hall, 1969. p. 86-97.
2. **Kalckar H M.** Enzymatic synthesis of nucleosides. *Fed. Proc.* **4**:248-52, 1945.
3. ─────. Differential spectrophotometry of purine compounds by means of specific enzymes. I. Determination of hydroxypurine compounds. *J. Biol. Chem.* **167**:429-43, 1947. (Cited 780 times since 1955.)
4. ─────. Differential spectrophotometry of purine compounds by means of specific enzymes. II. Determination of adenine compounds. *J. Biol. Chem.* **167**:445-59, 1947. (Cited 365 times since 1955.)
5. ─────. The enzymatic synthesis of purine riboside. *J. Biol. Chem.* **167**:477-86, 1947. (Cited 155 times since 1955.)
6. Reagents for the ultraviolet determination of uric acid in serum or urine at 292 nm per procedure. *Biochemical and organic compounds for research and diagnostic clinical agents*. St. Louis: Sigma Chemical Co., 1984. p. 1093. No. 292-UV.
7. **Murphy J, Baker D C, Behling C & Turner R A.** A critical reexamination of the continuous spectrophotometric assay for adenosine deaminase. *Anal. Biochem.* **122**:328-37, 1982.
8. **Tritsch G L.** Validity of the continuous spectrophotometric assay of Kalckar for adenosine deaminase activity. *Anal. Biochem.* **129**:207-9, 1983.

# This Week's Citation Classic
NUMBER 7
FEBRUARY 12, 1979

Bollum F J & Potter V R. Nucleic acid metabolism in regenerating rat liver. 6. Soluble enzymes which convert thymidine to thymidine phosphates and DNA. *Cancer Research* 19: 561-5, 1959.

DNA polymerase and thymidine kinase are described as soluble enzyme systems and correlated with DNA replication *in vivo*. [The *SCI®* indicated that this paper has been cited 291 times since 1961.]

F.J. Bollum
Uniformed Services University
of the Health Sciences
Bethesda, MD 20014

December 22, 1977

"The work Van Potter and I did on DNA synthesis in regenerating rat liver contained findings useful to scientists interested in DNA replication in eukaryotic cells. The primary focus of this paper was measurement of the actual increase in DNA polymerase in regenerating rat liver. Regenerating liver had already been well studied for *in vivo* incorporation of DNA precursors, so it was of interest to compare our new findings with the earlier work. The final result was a mixture of new findings correlated with old findings; what editors call a 'timely' piece of research. I also sense that many investigators had been unsuccessful in demonstrating enzymatic DNA synthesis in eukaryotic systems at that time and our work (and that from E. S. Cannellakis' laboratory at Yale) opened the door on this subject.

"In proper perspective it should be mentioned that the work was done in Van Potter's laboratory, University of Wisconsin, where I was a (rather fresh) USPHS postdoctoral fellow. The clever experiment in the paper was his idea. The problem was to correlate *in vivo* results with *in vitro* results without embarking on a tedious statistical study. Simple enough to Van Potter! Just inject $^{14}$C-Orotic acid two hours before sacrificing animals for enzyme assay. Then isolate $^{14}$C-DNA from nuclei to estimate *in vivo* rate of synthesis and measure $^{3}$H-deoxynucleotide incorporation with the cytoplasmic extract and exogenous DNA template. The same rat was used for *in vivo* and *in vitro* measurements of capacity for DNA replication. The relation was remarkably good during the induction phase and is called the 'Correlation Curve' in the paper. The correlation was not perfect, however, and the deviations still require explanation. I think Potter's clever experiment and our unorthodox use of cytoplasmic extracts in these may have excaped the comprehension of many readers.

"I clearly remember the day Van Potter called me into his office, drew his conception of the 'Correlation Curve' out of thin air and chalk on a very small blackboard he kept there and said, 'If what we already know is true, this will be true. If we make this demonstration, and publish it, I think anyone will understand that our findings correlating DNA enzymology and DNA replication are correct.' We made the demonstration, I think his perception was accurate, and I learned more than DNA enzymology from this experiment."

149

## This Week's Citation Classic

CC/NUMBER 13
MARCH 30, 1981

Post R L, Merritt C R, Kinsolving C R & Albright C D. Membrane adenosine triphosphatase as a participant in the active transport of sodium and potassium in the human erythrocyte. *J. Biol. Chem.* 235:1796-802, 1960.
[Dept. Physiology, Vanderbilt Medical School, Nashville, TN]

A correspondence between the kinetics of energy-dependent transport of sodium and potassium ions across intact human erythrocyte membranes and that of sodium plus potassium ion-dependent adenosine triphosphatase activity of fragments of the same membranes suggests that a single entity performs both transport and hydrolase functions. [The *SCI*® indicates that this paper has been cited over 865 times since 1961.]

Robert L. Post
Department of Physiology
School of Medicine
Vanderbilt University
Nashville, TN 37232

March 9, 1981

"In 1954 when I started on the kinetics of active sodium and potassium transport in human erythrocytes, I began as an observer. The transports of the two ions turned out to be linked so that transport of each required transport of the other. There was one pump for both. Then in 1957 Jens Skou published his now classic paper[1] on an ATPase activity in particles derived from crab nerves. This ATPase activity required both sodium and potassium ions simultaneously. Skou wrote, 'Characteristics of the system suggest that the ATPase studied here may be involved in the active extrusion of sodium from the nerve fiber.'

"Skou was, and still is, at the University of Aarhus in Denmark. I had become acquainted with him in 1953 when he visited the United States. In Vienna in 1958 at the International Biochemical Congress we met again and compared notes. As I listened to him, I became convinced that he had the pumping enzyme. In turn I told him about linkage in transport and about the cardioactive steroid inhibitors, which are specific for this pump. I began to appreciate that work on the erythrocyte could identify a sodium plus potassium ATPase activity and active transport of sodium and potassium as distinct functions of a single system. I began work on the ATPase in erythrocytes when I got back home.

"I am pleased to learn that this paper became a *Citation Classic*. I believe this happened for two reasons. The first reason was practical. The paper provided a method for identifying a transport system in a preparation of broken membranes. This made it easy to search for this system in tissues in which transport studies were difficult. The second reason was conceptual. The paper built a bridge between the then mysterious process of active transport and familiar concepts of enzymology. It confirmed to transportologists that ATP was a direct source of energy for an ion pump and it showed enzymologists that ions which might have been considered only as catalysts could in fact be substrates for translocation. Also the sodium and potassium ion pump of the erythrocyte turned out to be representative of a unique transport system found in almost all animal cells.

"The paper initiated an abbreviation with the expression '$(Na^+ + K^+)$-dependent ATPase.' Nowadays some authors keep the parentheses, (Na,K)ATPase, and others keep the hyphen, Na,K-ATPase.

"My collaborators were students. Cullen Merritt and Charlie Albright are now practicing internal medicine, the former in Nashville and the latter in Tyler, Texas. Richard Kinsolving is director of research at Pennwalt Pharmaceutical Co., Rochester, NY. An account of our adventures has been published.[2]

"Of many recent reviews on the (Na,K) ATPase[3] perhaps the broadest is that of Hobbs and Albers."[4]

1. **Skou J C.** The influence of some cations on an adenosine triphosphatase from peripheral nerves. *Biochim. Biophys. Acta* 23:394-401, 1957.
2. **Post R L.** A reminiscence about sodium, potassium-ATPase. *Ann. NY Acad. Sci.* 242:6-11, 1974.
3. ──────. A perspective on sodium and potassium ion transport adenosine triphosphatase. (Mukohata Y & Packer L, eds.) *Cation flux across biomembranes.* New York: Academic Press, 1979. p. 3-19.
4. **Hobbs A S & Albers R W.** The structure of proteins involved in active membrane transport. *Annu. Rev. Biophys. Bioeng.* 9:259-91, 1980.

# This Week's Citation Classic

CC/NUMBER 20
MAY 18, 1981

Skou J C. Enzymatic basis for active transport of Na$^+$ and K$^+$ across cell membrane. *Physiol. Rev.* 45:596-617, 1965.
[Inst. Biophysics, Univ. Aarhus, Aarhus, Denmark]

The paper is a review on the characteristics of a membrane bound Na$^+$ + K$^+$ activated ATPase which shows that the system is responsible for the energy requiring transport of Na$^+$ and K$^+$ across the cell membrane. [The *SCI*® indicates that this paper has been cited over 1,580 times since 1965.]

Jens C. Skou
Institute of Biophysics
University of Aarhus
DK-8000 Aarhus C
Denmark

April 13, 1981

"In the beginning of the 1950s I was interested in the effect of local anaesthetics (l.a.) and had found that the increase in surface pressure by penetration of local anaesthetics into a monolayer of lipids extracted from nerves correlated to their local anaesthetic effect. This raised the question: Can an increase in surface pressure from penetration of l.a. into the lipid part of a nerve membrane influence the configuration of proteins in the membrane and thereby block the trigger mechanism in the Na$^+$-channels? Experiments showed that the activity of surface spread enzymes (catalase and acetylcholin esterase) was surface pressure dependent which was taken as an indication of an effect of surface pressure on protein configuration. I then needed a monolayer of a lipoprotein with enzyme activity to test the effect of penetration of l.a. Libet[1] had shown that there is an ATPase in the sheath part of giant axons, and being membrane bound it was likely that it was a lipoprotein. I had no access to giant axons but looked for and found an Mg-ATPase in the microsomal fraction from a homogenate of crab nerves. However, activity varied from preparation to preparation and with no explanation. Finally after three months of work it was observed that K$^+$ in the test solution increased activity. I went on a summer holiday to forget about ATPases and crab nerves. After returning, the experiment was repeated but no effect of K$^+$ was found. However, addition of Na$^+$, which had little or no effect in the presence of Na-ATP, increased the activity when K-ATP was used, i.e., the activity was Na$^+$ + K$^+$ dependent. This explained the varying results. Sometimes Na$^+$ and sometimes K$^+$ had been used for ionic strength effect in the buffer and as counter ion for ATP. The characteristics of the system suggested that it was involved in active transport of Na$^+$ and K$^+$ across the cell membrane. This shifted my interest to active transport of cations. The results were published in 1957.[2]

"Crab nerves were a lucky choice and later experiments showed that it is one of the few tissues where the Na,K-ATPase activity is revealed without use of detergents. A problem was, however, to kill the crabs, 25,000 shore crabs—200,000 nerves. The only usable way was to put them in boiling water immediately after having cut the legs—but the smell! 'Couldn't you use another tissue,' was a standing remark in the department.

"The often cited paper is a review on the following eight years of research on the Na,K-ATPase by many different authors which gave the evidence that the enzyme is found in the membrane of most cells and is responsible for the active transport of Na$^+$ and K$^+$ across the cell membrane. It appeared at a time when membranes had come into focus and there was a lot of interest in cations. For more recent information and references see, 'Isolation and characterization of the components of the Na$^+$ pump' and *Na,K-ATPase, Structure and Kinetics*."[3,4]

1. Libet B. Adenosinetriphosphatase (ATP-ase) in nerve. *Fed. Proc.* 7:72-3, 1948.
2. Skou J C. The influence of some cations on the adenosine triphosphatase from peripheral nerves. *Biochim. Biophys. Acta* 23:394-401, 1957.
3. Jørgensen P L. Isolation and characterization of the components of the Na$^+$ pump. *Quart. Rev. Biophys.* 7:239-74, 1975.
4. Skou J C & Nørby J G, eds. *Na,K-ATPase, structure and kinetics.* London: Academic Press, 1979. 549 p.

# This Week's Citation Classic™

CC/NUMBER 23
JUNE 4, 1984

Bárány M. ATPase activity of myosin correlated with speed of muscle shortening.
*J. Gen. Physiol.* **50**:197-218, 1967.
[Institute for Muscle Disease, New York, NY]

Pure myosin was isolated from 25 different muscles whose contraction time varied 250-fold. The ATPase activity of the myosins was correlated with the contraction time of their respective muscles, suggesting that the myosin ATPase determines the speed of muscle contraction. [The *SCI*® indicates that this paper has been cited in over 725 publications. It is one of the five most-cited papers for this journal.]

---

Michael Bárány
Department of Biological Chemistry
University of Illinois
College of Medicine
Chicago, IL 60612

May 4, 1984

"The nature of the muscle engine was one of the primary interests of biochemists in the first half of this century. Engelhardt and Ljubimowa discovered that myosin, one of the contractile proteins of muscle, hydrolyzes ATP, the compound which provides the energy for muscular contraction.[1] Subsequently, Engelhardt introduced the term mechanochemistry, i.e., the protein which performs the mechanical work is also an enzyme capable of liberating the energy necessary for the work. This idea was challenged by Straub and Feuer a decade later, who found that the globular to fibrous transformation of actin, the other contractile protein of muscle, is correlated with an ATP to ADP transformation of the nucleotide bound to actin.[2] Straub postulated the term mechano-chemical coupling, i.e., the protein which does the mechanical work carries a prosthetic group capable of changing its chemical energy. In Straub's concept, the energy is built in the structure of the muscle engine. In contrast, in Engelhardt's concept, the energy flows from an outside reservoir to the engine similar to that of a car.

"I, as a student of Straub, had spent many years in his institute in Budapest, Hungary, attempting to prove his theory, and continued this research on my own in America. My failure, as well as that of others, to find a role for the actin-bound nucleotide in contraction forced me to return to myosin. In collaboration with my wife, Kate Bárány, we isolated pure myosin from the three major types of muscle — skeletal, heart, and smooth — and compared these myosins for their size and shape, digestibility by proteolytic enzymes, sulfhydryl content, and ATPase activity. The only significant difference found was in the ATPase activity. Following the suggestion of the late Ernest Gutmann, we prepared myosin from the slow and fast muscles of various animals and have found much higher ATPase activity in myosin from fast muscle than that from slow muscle. The relationship between speed of muscle contraction and ATPase activity of myosin became clear when myosin from extremely slow muscles, like sloth or turtle, exhibited very low ATPase activity.

"This work was scheduled for presentation at a New York Heart Association symposium. Just before my lecture, fire broke out in the hotel and both the audience and lecturers left the scene. For our results to be known, we had to wait for the publication of the symposium's proceedings. In the meantime, we have shown that cross-innervation of rat muscle, which transforms fast muscle to slow muscle and vice versa, also transforms the ATPase activity of myosin.[3] It was also observed that the ATP-induced conformational change of fast muscle myosin was much larger than that of slow muscle myosin.[3] Finally, by introducing radioactive reagents, capable of forming covalent linkages with myosin, into intact muscle, we have shown that, during contraction, myosin undergoes a conformational change, which is the driving force for generating the tension and the movement.[4] Accordingly, in the muscle engine, the fuel, ATP, changes the structure of the engine, myosin.

"This paper is frequently cited because, in all muscles, the ATPase activity of myosin is related to the speed of contraction. This paper was also the first demonstration of a relationship between the enzymic activity of a pure protein and a basic biological phenomenon, motion."

---

1. Engelhardt W A & Ljubimowa M N. Myosine and adenosinetriphosphatase. *Nature* **144**:668-9, 1939.
   (Cited 220 times since 1955.)
2. Straub F B & Feuer G. Adenosinetriphosphate, the functional group of actin. *Biochim. Biophys. Acta* **4**:455-70, 1950.
   (Cited 185 times since 1955.)
3. Bárány M & Close R I. The transformation of myosin in cross-innervated rat muscles. *J. Physiology* **213**:455-74, 1971.
   (Cited 210 times.)
4. Bárány M & Bárány K. A proposal for the mechanism of contraction in intact frog muscle.
   *Cold Spring Harbor Symp.* **37**:157-68, 1972. (Cited 30 times.)

# This Week's Citation Classic

CC/NUMBER 25
JUNE 21, 1982

Walsh D A, Perkins J P & Krebs E G. An adenosine 3',5'-monophosphate-dependent protein kinase from rabbit skeletal muscle. *J. Biol. Chem.* 243:3763-5, 1968.
[Department of Biochemistry, University of Washington, Seattle, WA]

A protein kinase that catalyzes an adenosine 3', 5'-monophosphate (cyclic AMP)-dependent phosphorylation of casein and protamine was purified from rabbit skeletal muscle. The $K_m$ values of cAMP for these reactions are $1 \times 10^{-7}$ and $6 \times 10^{-8}$, respectively. The protein kinase markedly increases the rate of the cAMP-dependent activation and phosphorylation of phosphorylase kinase by ATP. [The *SCI®* indicates that this paper has been cited over 675 times since 1968.]

Donal A. Walsh
Department of Biological Chemistry
School of Medicine
University of California
Davis, CA 95616

February 25, 1982

"I joined the laboratories of Ed Krebs and F.H. Fischer as a postdoctoral fellow in the autumn of 1966. Krebs was the past master of having fellows 'choose' the project that he thought best for them. So I 'selected' to work on the activation of phosphorylase kinase. Possibly, I was spurred on by Krebs's comment that he did not think a 'phosphorylase kinase kinase' existed. At that time, phosphorylase kinase was already highly purified; the stimulation by cAMP of the ATP-dependent activation was well recognized. As is the axiom of biochemistry, the conditions of an experiment should be optimized; so since the activation and phosphorylation of phosphorylase kinase were faster at higher concentrations of ATP-$Mg^{2+}$, it was examined under those conditions. After the fact, we understand that high ATP, in fact, masked the specificity of cAMP-dependent activation.

"Four observations led to the discovery of the cAMP-dependent protein kinase. The first was the conviction that at high ATP there were two catalytic processes to activate phosphorylase kinase. The second was the unpublished observation of Bob Kemp that there was far less than a stoichiometric binding of cAMP to phosphorylase kinase. The third, albeit unrecognized by the investigators,[1] was that anomalously phosphorylase kinase was reactivated at a much faster rate after dephosphorylation by a crude protein phosphatase preparation than it was in the initial activation.

"The fourth observation was the most crucial. One night, while 'watching' a column, I picked up Carmen Gonzalez's MS thesis. This was work done in 1962, unpublished because it was difficult to interpret, and somewhat forgotten. Carmen had described how a heat stable factor, first described because it interfered with the assay of cAMP,[2] modified phosphorylase kinase activation. What became apparent was that if phosphorylase kinase activation involved two catalysts, Carmen's inhibitor only blocked one of them. Could it be that this inhibitor would block the activator that was present, but unrecognized, in the crude extracts in the phosphatase experiment? Could it be that the activating factor in crude extracts was a mediator of cAMP? Could it be that phosphorylase kinase preparations were contaminated with this mediator of cAMP? Well, the history now is well known and the answer to all three questions was yes. The first definitive experiment was designed and from that single experiment came the identification of both the cAMP-dependent protein kinase and the heat-stable inhibitor protein of that enzyme. Ironically, this experiment was not published until later,[3] and the first paper was built upon experiments that were possible only after this cAMP-dependent 'phosphorylase kinase kinase' was purified.

"The identification of the cAMP-dependent protein kinase was an experiment whose time had come and I was fortunate to have been at the right place at the right time. Its discovery marked the end of an era of investigation with all the links in the chain of the glycogenolytic cascade now known.[4] Krebs had chosen to investigate this system recognizing that it served as a model for cAMP-mediated hormonal action. Because, with the final link in, Krebs thought it not erudite to point out the obvious, he has not been given the unique recognition that he deserves."

1. **Riley W D, DeLange R J, Bratvold G E & Krebs E G.** Reversal of phosphorylase kinase activation.
   *J. Biol. Chem.* 243:2209-15, 1968.
2. **Posner J B, Hammermeister K H, Bratvold G E & Krebs E G.** The assay of adenosine-3',5'-phosphate in skeletal muscle.
   *Biochemistry* 3:1040-4, 1964.
3. **Walsh D A, Perkins J P, Brostrom C O, Ho E S & Krebs E G.** Catalysis of the phosphorylase kinase activation reaction.
   *J. Biol. Chem.* 246:1968-76, 1971.
4. **Walsh D A & Cooper R H.** The physiological regulation and function of cAMP-dependent protein kinases.
   (Litwack G, ed.) *Biochemical actions of hormones.* New York: Academic Press, 1979. Vol. 6. p. 1-75.

# This Week's Citation Classic

Kuo J F & Greengard P. Cyclic nucleotide-dependent protein kinases. IV.
Widespread occurrence of adenosine 3',5'-monophosphate-dependent protein
kinase in various tissues and phyla of the animal kingdom.
*Proc. Nat. Acad. Sci. US* **64**:1349-55, 1969.
[Dept. Pharmacology, Yale Univ. Sch. Med., New Haven, CT]

Cyclic AMP-dependent protein kinase was found in every one of about 30 sources examined, which included many mammalian tissues as well as species representative of eight invertebrate phyla. The data support a unifying theory for the mechanism of action of cyclic AMP, namely, that its many and diverse effects are mediated through activation of cyclic AMP-dependent protein kinase, with the resultant phosphorylation of various cellular proteins. [The *SCI*® indicates that this paper has been cited in over 750 publications since 1969.]

---

J.F. Kuo
Department of Pharmacology
Emory University School of Medicine
Atlanta, GA 30322

February 1, 1983

"Although a great diversity of actions that involve cyclic AMP in many tissues and cells had already been documented,[1] the molecular mechanism for the actions of this intracellular second messenger for hormones remained unknown. The paper by Walsh, Perkins, and Krebs[2] published in 1968 describing a skeletal muscle protein kinase stimulated by cyclic AMP provided the first clue as to how cyclic AMP worked. At that time, Paul Greengard, after spending some time at Vanderbilt University, had just moved from Geigy to Yale University as professor of pharmacology. I, assuming the similar 'industrial shunt,' left Lederle to join him as assistant professor in the department of pharmacology. We teamed up because of our common interest in the cyclic nucleotide systems; thus began our four-year association which was extremely important to the earlier stage of my career. We had evidence showing the occurrence of cyclic AMP-dependent protein kinase in several tissues, including the brain. Greengard suggested (in retrospect, I am glad he did) that we ought to systematically survey the enzyme in mammalian tissues as well as tissues from various invertebrate phyla in order to test its widespread occurrence in the animal kingdom. We found that the enzyme was readily detectable even in the crude extracts of many tissues, but that its presence in many other tissues can be unequivocally established only after the crude extracts were further treated with isoelectric precipitation, ammonium sulfate fractionation, and/or DEAE-cellulose chromatography. Without exception, we found the enzyme in every one of the diverse tissues examined. Based upon the apparently ubiquitous occurrence of cyclic AMP-dependent protein kinase activity, we proposed a unifying hypothesis that the actions of cyclic AMP are mediated through activation of this enzyme. This paper was published in 1969 in the *Proceedings of the National Academy of Sciences of the USA*, most fortunately and appropriately communicated by the late Nobelist Earl W. Sutherland, who, in 1957 with Theodore W. Rall, discovered cyclic AMP.[3] This hypothesis has been tested over the years; it is still considered correct today in all eukaryotic systems.

"In 1970, Greengard and I discovered cyclic GMP-dependent protein kinase, and we further suggested that this enzyme may serve as a mediator for cyclic GMP actions.[4] One obvious determinant for the tissue-specific effects of cyclic AMP and cyclic GMP resides in the nature of tissue-specific substrates that the two cyclic nucleotide-dependent protein kinases phosphorylate. This point of view has been dealt with in Greengard's review article[5] entitled 'Phosphorylated proteins as physiological effectors.'

"The reason for the frequent citation of our 1969 paper probably is the particularly attractive feature of the unifying hypothesis which provides a single reaction mechanism by which cyclic AMP can bring about its diverse effects."

---

1. **Sutherland E W & Robison G A.** The role of cyclic 3',5'-AMP in responses to catecholamines and other hormones. *Pharmacol. Rev.* **18**:145-61, 1966.
2. **Walsh D A, Perkins J P & Krebs E G.** An adenosine 3',5'-monophosphate-dependent protein kinase from rabbit skeletal muscle. *J. Biol. Chem.* **243**:3763-5, 1968.
   [Citation Classic. *Current Contents/Life Sciences* **25**(25):16, 21 June 1982.]
3. **Rall T W, Sutherland E W & Berthet J.** The relationship of epinephrine and glucagon to liver phosphorylase. IV. Effects of epinephrine and glucagon on the activation of phosphorylase in liver homogenates. *J. Biol. Chem.* **224**:463-75, 1957.
4. **Kuo J F & Greengard P.** Cyclic nucleotide-dependent protein kinases. VI. Isolation and partial purification of a protein kinase specific for guanosine 3',5'-monophosphate. *J. Biol. Chem.* **245**:2493-8, 1970.
5. **Greengard P.** Phosphorylated proteins as physiological effectors. *Science* **199**:146-52, 1978.

# This Week's Citation Classic

Robison G A, Butcher R W & Sutherland E W. Cyclic AMP.
*Annu. Rev. Biochem.* 37:149-74, 1968. [Departments of Pharmacology and Physiology, Vanderbilt University, Nashville, TN]

The article reviews work on cyclic AMP, which had by 1968 been recognized as a second messenger mediating many of the actions of a number of hormones. [The *SCI®* indicates that this paper has been cited over 860 times since 1968.]

G. Alan Robison
Department of Pharmacology
University of Texas
Houston, TX 77025

"This article and another review article published around the same time[1] were by-products of a monograph which we were working on at the time, and which was finally published three years later.[2] Sutherland and T.W. Rall and their colleagues had first published on cyclic AMP in 1957. The literature on this subject grew slowly at first, but, by the time I joined the group in 1962 (Butcher had been already there, having been a graduate student with Sutherland), it had begun to grow more rapidly. By the end of 1965, by which time he had moved from Western Reserve to Vanderbilt, we were still able to keep up with the literature but it was getting to be a problem.

The decision to write a monograph on the subject was to some extent based on the feeling that if we were having trouble in this regard, others might be having even more trouble. It struck us that we could perhaps do our colleagues a favor by summarizing everything that was known about the subject. With this in mind, we prepared a list of 100 references for the Institute for Scientific Information® so that we could get weekly ASCA® reports on what was being published.

"That was the start of a rather hectic period in our lives, since we soon realized that the literature was growing even faster than we had thought. It began taking a substantial amount of our time just to read these papers, which to an increasing extent were leading us into areas of research about which we knew nothing. As we wrote in the preface to our 1971 monograph, it was a difficult book to write 'because the subject refused to sit still. It was not simply that it behaved like a naughty child at the photographer's, for that would be expected in any viable field of scientific research. Rather, it has seemed to us more like an imaginary child who, in the course of having his picture taken, suddenly grew to adult proportions and than left the studio badly in need of a shave.'

"Three years earlier it had not seemed quite that bad, and we were well prepared when the request came to write an article for the *Annual Review of Biochemistry*. We simply took the various outlines and drafts we had been working on for the monograph and turned them into a review article of the requested number of pages. With an appropriate shift of emphasis, we did the same thing for the *Circulation* article.

"I think the main reason these articles were cited so often is that they appeared at a time when a large number of investigators were just starting to write papers about cyclic AMP, which was then sufficiently unfamiliar that most authors felt the need to refer to an earlier article whenever they first mentioned it. Our reviews were particularly handy for this purpose because they were among the first to deal with the subject as a whole, most earlier reviews having dealt with only one or a few aspects of it. Our monograph was later used for the same purpose, until cyclic AMP gradually became so familiar that the automatic citation of a review article no longer seemed necessary. I think another factor, since there were other review articles that could have been cited instead, is that these reviews were written for the express purpose of being useful to the reader. I think they were appreciated for that reason.

"More satisfying than the frequency with which these reviews were cited was the experience, years later, of meeting young investigators who told us they had been stimulated to enter the field by reading one or another of these reviews, and who had subsequently gone on to make important discoveries of their own. Of course it was the subject itself that was exciting, rather than our writing about it, but it was good to know that we hadn't completely prevented the excitement from showing through."

1. **Sutherland E W, Robison G A & Butcher R W.** Some aspects of the biological role of adenosine 3',5'-monophosphate cyclic AMP. *Circulation* 37:279-306, 1968.
2. **Robison G A, Butcher R W & Sutherland E W.** *Cyclic AMP.* New York: Academic Press, 1971, 531 p.

## This Week's Citation Classic

CC/NUMBER 50
DECEMBER 13, 1982

Thompson W J & Appleman M M. Multiple cyclic nucleotide phosphodiesterase activities from rat brain. *Biochemistry—USA* 10:311-16, 1971.
[Depts. Biological Sciences and Biochemistry, Univ. Southern California, Los Angeles, CA]

Physical and kinetic criteria defined the multiplicity of cyclic nucleotide phosphodiesterase(s), the enzyme system solely responsible for cyclic AMP and cyclic GMP catabolism. A theoretical basis for the anomalous kinetic behavior of the enzyme was derived and a new assay procedure described in this paper. [The *SCI®* indicates that this paper has been cited in over 700 publications since 1971.]

---

W. Joseph Thompson
Department of Pharmacology
University of Texas Medical School
Houston, TX 77025

September 20, 1982

"As a graduate student in the late-1960s interested in the hormonal regulation of carbohydrate or lipid metabolism, I was led inexorably to cAMP (this despite the fact that one of my thesis advisers swore that cAMP was an artifact). Ultimately, this translated in the laboratory to having to watch Dowex columns drip. I was okay with big columns, but little Pasteur pipettes or glass columns of Dowex resin dripping seemed like the world's biggest waste of time and drove me up the wall. I was struggling with my columns in Mike Appleman's laboratory at the University of Southern California in order to use $^3$H-cAMP (just then commercially available) as a substrate for PDE as a way to increase the sensitivity of the published procedure. This work was based on that of Bob Kemp and Mike published a year earlier.[1] Their studies also prompted Gary Brooker to develop an isotope dilution assay for tissue cAMP[2] during his graduate studies with Mike. Gary's cAMP assay was never used for much because the radioimmunoassay for cAMP came out about the same time. However, his work provided us with the basis for a PDE assay because the success of the isotope dilution assay required the existence of high affinity PDE. Mercifully, that assay required no columns since the exchange resin quenches bound tritium which allowed the whole thing to be done in a scintillation vial or centrifuged and an aliquot measured. It is ironic that several years later we discovered Dow had changed its resin processing procedure such that too much nonspecific reaction product binding occurred and we ultimately had to go back to little columns again. This time we designed a vacuum approach for a hundred columns so separation on the little monsters only takes about two minutes and is thus tolerable.[3]

"It is sort of remarkable to me that this assay is still being used by so many labs 11 years after it was published. I think the paper is cited because the method is cheap, easy to do, and works. It is also cited as a reference to the initial characterization of the multiple forms of the enzyme and the logic of why a mixture of two separate catalytic sites of varying affinity may display apparent negative cooperativity. I was surprised to learn that only 25 percent of the *Citation Classics* are methods papers.[4] This paper does not fit that category solely, but its method component is certainly a major cause of its popularity. I plead guilty to standing in the right place at the right time.

"I suggest to graduate students that in addition to dedication and hard work, anticipation of where fields are going is a useful practical attribute for a scientist.

"I believe that this paper is successful because my thesis adviser, Mike Appleman, is a good scientist for which there is no substitute for any graduate student. Its significance is that we are able to study hormonally regulated PDE forms, but it represents only a small step in understanding hormone regulatory mechanisms and their pharmacological or genetic control. As for honors, what better for such an egotistical lot as we than to be recognized and mimicked by our colleagues?"

---

1. **Appleman M M & Kemp R G.** Puromycin: a potent metabolic effect independent of protein synthesis. *Biochem. Biophys. Res. Commun.* 24:564-8, 1966.
2. **Brooker G, Thomas L J, Jr. & Appleman M M.** The assay of adenosine 3',5'-cyclic monophosphate and guanosine 3',5'-cyclic monophosphate in biological materials by enzymatic radioisotopic displacement. *Biochemistry—USA* 7:4177-81, 1968.
3. **Thompson W J, Terasaki W L, Epstein P M & Strada S J.** Assay of cyclic nucleotide phosphodiesterase and resolution of multiple molecular forms of the enzyme. *Advan. Cyclic Nucl. Res.* 10:69-92, 1979.
4. **Garfield E.** *Citation Classics*—four years of the human side of science. *Current Contents* (22):5-16, 1 June 1981.

# This Week's Citation Classic

CC/NUMBER 17
APRIL 26, 1982

Salomon Y, Londos C & Rodbell M. A highly sensitive adenylate cyclase assay.
*Anal. Biochem.* 58:541-8, 1974.
[Section on Membrane Regulation, Lab. Nutrition and Endocrinology, Natl. Inst. Arthritis, Metabolism, and Digestive Diseases, NIH, Bethesda, MD]

This paper describes a technique which permits nearly complete separation (to the level of 2 ppm) of 3'5' cyclic [$^{32}$P] AMP from an [$\alpha$ $^{32}$P] ATP, by sequential chromatography on Dowex 50 and neutral alumina. Also described are the necessary procedures for long-term maintenance and reutilization of both types of columns. [The *SCI*® indicates that this paper has been cited over 1,105 times since 1974.]

Yoram Salomon
Department of Hormone Research
Weizmann Institute of Science
Rehovot 76100
Israel

December 23, 1981

"When I arrived as a postdoctoral fellow at Martin Rodbell's laboratory at the National Institutes of Health, adenylate cyclase was assayed by the method of Krishna et al.,[1] which separated 3'5' cyclic AMP from other phosphate-containing compounds by sequential chromatography on a cation exchange resin and nascent BaSO$_4$. Although this technique represented a major advance over earlier *in vitro* adenylate cyclase assays, it had several drawbacks. First, the background was too high for accurately measuring true initial reaction rates, a project that Michael Lin and I have begun. Second, differences among lots of [$\alpha$ $^{32}$P] ATP were reflected in highly variable background levels which led us occasionally to discard entire experiments. Finally, in order to minimize the problems with high $^{32}$P-backgrounds, disposable columns were prepared daily for both the ion exchange and BaSO$_4$ steps. This was a costly and time-consuming aspect of the assay.

"The above difficulties with the existing technique stimulated a search for a method that would permit better separation of 3'5' cyclic AMP from ATP and impurities. This I accomplished by combining the Dowex cation exchange step of the Krishna method[1] with the neutral alumina chromatographic procedure developed by White and Zenser[2] and Ramachandran.[3] The results obtained with the double chromatographic technique were even better than I had hoped for, since assay backgrounds were reduced to a level barely distinguishable from the machine background of the scintillation counter.

"Another postdoctoral fellow, Constantine Londos, and I then carried out a series of experiments in which we compared this new method with others in use at the time. The results prompted us to write the article which appeared in *Analytical Biochemistry*.

"Given the overwhelming acceptance of this method among investigators in the adenylate cyclase field, we are now pleased to have overcome several obstacles that nearly prevented its publication.

"First, as young postdoctoral fellows, we were unsure as to whether or not such a finding merited publication. Since Rodbell was abroad for several months, we consulted other colleagues who advised us not to waste time on a methods paper. Nevertheless, we could not avoid the feeling that many would welcome an efficient new method free of the problems mentioned above. Moreover, another aspect of the method proved to be a considerable relief. We found that both Dowex and alumina columns could be used repeatedly, without affecting the quality of the results. Additionally, columns once used could be set aside for weeks; the dried residue functions well upon renewal with the described technique. The notion that this labor-saving aspect, coupled with the extremely high sensitivity of the method, would be welcome led us after all to write the article. Finally, it is rather amusing that, in its present form, the paper was initially rejected for insufficient advancement. However, our persistence with the editor resulted in its acceptance.

"An up-to-date cookbook version of this method has recently appeared in *Advances in Cyclic Nucleotide Research*."[4]

1. **Krishna G, Weiss B & Brodie B B.** A simple, sensitive method for the assay of adenyl cyclase.
   *J. Pharmacol. Exp. Ther.* 163:379-85, 1968.
   [The *SCI* indicates that this paper has been cited over 1,215 times since 1968.]
2. **White A A & Zenser T V.** Separation of cyclic 3',5'-nucleoside monophosphates from other nucleotides on aluminum oxide columns: application to the assay of adenyl cyclase and guanyl cyclase.
   *Anal. Biochem.* 41:372-96, 1971. [The *SCI* indicates that this paper has been cited over 240 times since 1971.]
3. **Ramachandran J.** A new simple method for separation of adenosine 3',5'-cyclic monophosphate from other nucleotides and its use in the assay of adenyl cyclase. *Anal. Biochem.* 43:227-39, 1971.
   [The *SCI* indicates that this paper has been cited over 265 times since 1971.]
4. **Salomon Y.** Adenylate cyclase assay. *Advan. Cyclic Nucl. Res.* 10:35-55, 1979.

# This Week's Citation Classic

CC/NUMBER 29
JULY 16, 1984

Lindell T J, Weinberg F, Morris P W, Roeder R G & Rutter W J. Specific inhibition of nuclear RNA polymerase II by α-amanitin. *Science* 170:447-9, 1970.
[Dept. Biochemistry and Biophysics, Univ. California, San Francisco, CA]

α-Amanitin, the bicyclic octapeptide from *Amanita phalloides*, was found to be a potent and specific inhibitor of DNA-dependent RNA polymerase II from phylogenetically divergent organisms (rat and sea urchin). The utility of this inhibition was also demonstrated by differentially assaying RNA polymerase activities in isolated nuclei. [The *SCI*® indicates that this paper has been cited in over 450 publications since 1970.]

Thomas J. Lindell
Departments of Pharmacology
and
Molecular and Cellular Biology
University of Arizona
Tucson, AZ 85721

June 8, 1984

"I originally went to Bill Rutter's laboratory in fall 1968 on a postdoctoral fellowship to work on the comparative biochemistry of aldolases. Shortly after my arrival in Seattle, Bob Roeder began to obtain some striking results indicating that eukaryotes contained three separate RNA polymerases (I, II, and III).[1] Because of this excitement and the fact that I shared a laboratory with Bob, I chose to work on the RNA polymerase problem. The actual impetus for us to investigate the potential role of α-amanitin as an inhibitor of eukaryotic transcription was derived from a report by Stirpe and Fiume.[2]

"Rutter had obtained a sample of the toxin by telephoning T. Wieland[3] in Germany but Roeder did not have time to examine the effect of the compound on the individual RNA polymerases he had separated despite active discussions in the lab about the project. When Rutter moved from Seattle to San Francisco, I was the only person in the laboratory with experience who had actually separated and assayed the multiple RNA polymerases. The problem then became mine to determine whether α-amanitin inhibited any of these individual RNA polymerases.

"Fanyela Weinberg, who was a technician, and I did the original experiments with the sea urchin enzymes and Paul Morris repeated them on the separated enzymes from rat liver. As we were preparing a manuscript to submit to *Nature*, Pierre Chambon wrote to Rutter and enclosed a preprint on similar work describing an effect of α-amanitin on RNA polymerase B(II) which would be published shortly.[4] I was personally decimated by that communication, but because we had some different observations, we immediately submitted the paper to *Nature*, where it was rejected. Rutter was able to convince the editor of *Nature* to reconsider the paper but it was again rejected. Another paper describing the effect of α-amanitin on eukaryotic RNA polymerase by Jacob et al.[5] was published shortly thereafter in *Nature* and ours was finally published in *Science* in October 1970.

"The probable reason our paper was so highly cited was that ours was a more complete and comprehensive study and the results could be more easily extended to other work. Our paper also graphically demonstrated that RNA polymerase II from two highly divergent species was specifically inhibited after separation on DEAE Sephadex columns and that chain elongation was inhibited. Further, it demonstrated that a differential assay of the eukaryotic RNA polymerases could be obtained in isolated nuclei without resorting to their separation by column chromatography. Later, it was observed that RNA polymerase III was also inhibited by α-amanitin, but at higher concentrations.[6] Thus, using two different concentrations of α-amanitin, it is now possible to assay all three RNA polymerase activities in isolated nuclei."[7]

1. Roeder R G & Rutter W J. Multiple forms of DNA-dependent RNA polymerase in eukaryotic organisms. *Nature* 224:234-7, 1969. (Cited 670 times.)
2. Stirpe F & Fiume L. Studies on the pathogenesis of liver necrosis by α-amanitin. Effect of α-amanitin on ribonucleic acid synthesis and on ribonucleic acid polymerase in mouse liver nuclei. *Biochemical J.* 105:779-82, 1967. (Cited 160 times.)
3. Wieland T. Poisonous principles of mushrooms of the genus *Amanita*. *Science* 159:946-52, 1968. (Cited 230 times.)
4. Kedinger C, Gniazdowski M, Mandel J L, Jr., Gissinger F & Chambon P. α-Amanitin: a specific inhibitor of one of two DNA-dependent RNA polymerase activities from calf thymus. *Biochem. Biophys. Res. Commun.* 38:165-71, 1970. (Cited 370 times.)
5. Jacob S T, Sajdel E M & Munro H N. Specific action of α-amanitin on mammalian RNA polymerase protein. *Nature* 225:60-2, 1970. (Cited 95 times.)
6. Weinmann R & Roeder R G. Role of DNA-dependent RNA polymerase III in transcription of the tRNA and 5S RNA genes. *Proc. Nat. Acad. Sci. US* 71:1790-4, 1974. (Cited 345 times.)
7. Lindell T J & Duffy J J. Enhanced transcription by RNA polymerases II and III after inhibition of protein synthesis. *J. Biol. Chem.* 254:1454-6, 1979.

| Number 47 | **Citation Classics** | November 21, 1977 |

Temin H M. RNA-dependent DNA polymerase in virions of Rous sarcoma virus. *Nature* 226:1211-3, 1970.

The presence of a ribonuclease-sensitive endogenous DNA polymerase activity in particles of RNA tumor viruses was first reported in this paper and the accompanying paper by David Baltimore.[1] The enzyme responsible for this activity, now known as reverse transcriptase, is the key part of the mechanism for the transfer of information from RNA to DNA (the provirus) in the replication of RNA tumor viruses. Publication of these papers led to the immediate acceptance of the DNA provirus hypothesis, originally proposed in 1963 and 1964. [The *SCI*® indicates that this paper was cited a total of 664 times in the period 1961-1975.]

Dr. Howard M. Temin
McArdle Laboratory for Cancer Research
University of Wisconsin
Madison, Wisconsin 53706

October 4, 1977

"Since 1963-1964, I had been proposing that the replication of RNA tumor viruses involved a DNA intermediate. This hypothesis, known as the DNA provirus hypothesis, apparently contradicted the so-called 'central dogma' of molecular biology and met with a generally hostile reception. The hypothesis was not generally accepted until the publication of this and the accompanying paper by David Baltimore. These papers demonstrated the existence of an enzyme associated with an RNA template in particles of RNA tumor viruses that could carry out information transfer from RNA to DNA. This result might have been predicted from the previous discovery of polymerases in virus particles and the DNA provirus hypothesis. That the discovery took so many years might indicate the resistance to this hypothesis.

"The actual proof that information transfer from RNA to DNA occurred in the replication of RNA tumor viruses involved assay of infectious virus DNA and nucleic acid hybridization and was published in the next two or three years. However, discovery of an enzyme with the ability to copy RNA to give a DNA product in virions associated with a possible template RNA was already convincing to most scientists.

"These papers had their greatest and most permanent effect on the study of RNA tumor viruses. They also led to the recognition that there are other viruses related to the RNA tumor viruses that did not cause cancer. All of these viruses are now called retroviruses. The acceptance of the DNA provirus hypothesis provided a theoretical framework for the work with these viruses. The availability of specific DNA complementary to viral RNA and the specificity of the viral DNA polymerase provided sensitive assay techniques and markers for these viruses and their components.

"The later attributes were also important in the search for an RNA tumor virus involved in human cancer. Although there has been much excitement in this area, as yet there is no convincing proof of involvement of an infectious RNA tumor virus in the etiology of human cancer.

"Another consequence of the discovery was the use of reverse transcriptase, especially that purified from avian myeloblastosis virus, as a standard reagent in molecular biology to form complementary DNA to specific RNA. In this way, it has been used as the first step in forming recombinant DNA molecules starting with messenger RNA.

"Finally, these papers led to the proposal of the protovirus hypothesis and a hypothesis for the etiology of cancers in the absence of overt virus infection. Although there is some evidence consistent with these hypotheses, especially the mechanism of transformation by weakly transforming RNA tumor viruses involving genetic changes in the course of extended viral replication in an infected animal, these hypotheses are not yet established."

### REFERENCE

1. **Baltimore D.** Viral RNA-dependent DNA polymerase. *Nature* 226:1209-11, 1970.

# This Week's Citation Classic

Kacian D L, Watson K F, Burny A & Spiegelman S. Purification of the DNA polymerase of avian myeloblastosis virus.
*Biochim. Biophys. Acta* **246**:365-83, 1971.
[Institute of Cancer Research, College of Physicians & Surgeons, Columbia University, New York, NY]

Purification of reverse transcriptase from avian myeloblastosis virus is described. The enzyme is a complex of two subunits with molecular weights 110,000 and 69,000. The complex possesses the RNA-, DNA-, and RNA:DNA hybrid-directed DNA polymerase activities found in the virion. [The *SCI®* indicates that this paper has been cited in over 305 publications since 1971.]

---

Daniel L. Kacian
Clinical Microbiology Laboratory
Department of Pathology and
Laboratory Medicine
Hospital of the University
of Pennsylvania
Philadelphia, PA 19104

April 25, 1983

"When the discovery of RNA-dependent DNA-polymerase activity in retroviruses was announced, I was a graduate student in Sol Spiegelman's lab at Columbia University, one of the few in his group still working on Qβ replicase. The others were studying RNA tumor viruses. Spiegelman hoped these replicated via a template-specific polymerase and that variant viral RNAs could be made that would compete for the enzyme. If human cancers were caused by these viruses, the variant RNAs might be useful as agents for specifically destroying malignant cells. Only in infected cells containing the viral replicase would the variant RNAs multiply exponentially and so destroy them.

"The evidence presented by Baltimore[1] and Temin[2] that these viruses replicate via a DNA intermediate had profound implications for this approach. Spiegelman was at the meeting where the announcement was made, and he returned early the next day to redirect our efforts. I am told that the discoveries were received in other labs with equal excitement.

"Temin and Baltimore's work was confirmed and extended by several labs well before it appeared in print. DNA- and RNA: DNA hybrid-dependent DNA polymerase activities were found, indicating that within the virion were all the enzymes needed to make a double-stranded DNA copy of its genome to integrate into the host cell. But were one or several enzymes responsible for the nucleic acid polymerase activities observed and, more importantly, were they specific for their own template RNA? If specificity were found, the hope that variant RNAs could be useful therapeutic agents remained alive. If, however, the enzyme could copy any RNA, an important tool for molecular genetics would clearly be in hand.

"Working out the enzyme isolation, as is often the case, was essentially a lengthy, trial-and-error process. At the end, we had shown that a single enzyme complex with two subunits possessed all the nucleic acid polymerase activities previously found in the virion. More importantly, we found the enzyme was not template-specific and could be used to make DNA copies of a variety of RNAs.[3] It would be some years, however, before Jeanne Myers and I[4,5] were able to achieve another goal I'd set, the *in vitro* synthesis of full DNA copies of large, polycistronic RNAs.

"The usefulness of DNA transcripts of RNAs for hybridization probes and recombinant DNA studies undoubtedly accounts for the number of citations the paper has received. We deliberately tried to develop a procedure that was simple and gave high yields from the costly starting material. That researchers continue to find it useful is gratifying. The paper also provided information on the subunit structure and template specificities of the enzyme and is often cited in related studies.

"An excellent review of the early work on reverse transcriptase appeared in 1977."[6]

---

1. **Baltimore D.** RNA-dependent DNA polymerase in virions of RNA tumour viruses. *Nature* **226**:1209-11, 1970.
    [The *SCI* indicates that this paper has been cited in over 930 publications since 1970.]
2. **Temin H M & Mizutani S.** RNA-dependent DNA polymerase in virions of Rous sarcoma virus. *Nature* **226**:1211-13, 1970. [Citation Classic. *Current Contents* (47):14, 21 November 1977.]
3. **Spiegelman S, Watson K F & Kacian D L.** Synthesis of DNA complements of natural RNAs: a general approach. *Proc. Nat. Acad. Sci. US* **68**:2843-5, 1971.
4. **Kacian D L & Myers J C.** Synthesis of extensive, possibly complete, DNA copies of poliovirus RNA in high yields and at high specific activities. *Proc. Nat. Acad. Sci. US* **73**:2191-5, 1976.
5. **Myers J C, Spiegelman S & Kacian D L.** Synthesis of full-length DNA copies of avian myeloblastosis virus RNA in high yields. *Proc. Nat. Acad. Sci. US* **74**:2840-3, 1977.
6. **Verma I M.** The reverse transcriptase. *Biochim. Biophys. Acta* **473**:1-38, 1977.

# This Week's Citation Classic

CC/NUMBER 3
JANUARY 18, 1982

Sharp P A, Sugden B & Sambrook J. Detection of two restriction endonuclease activities in *Haemophilus parainfluenzae* using analytical agarose-ethidium bromide electrophoresis. Biochemistry 12:3055-63, 1973.
[Cold Spring Harbor Laboratory, Cold Spring Harbor, NY]

A rapid assay for restricting endonuclease was developed using electrophoresis of DNA fragments in agarose gels and detection of DNA bands by staining with the fluorescence dye ethidium bromide. Using this assay, two different restriction endonucleases were purified from extracts of Hemophilus parainfluenzae. [The SCI® indicates that this paper has been cited over 1,040 times since 1973.]

Phillip A. Sharp
Center for Cancer Research
Massachusetts Institute of Technology
Cambridge, MA 02139

October 6, 1981

"Advances in molecular biology are frequently the product of the development of new methodology. Perhaps the most striking recent example of this is the impact that DNA recombinant technology has had. Publications that make novel contributions to the development of new methodology are widely read and frequently referenced. Bill Sugden, Joe Sambrook, and I described this type of methodology at a time when the molecular biological community was discovering the multiple uses of restriction endonucleases.

"In 1971, I arrived at Cold Spring Harbor in New York to begin studying the molecular biology of DNA tumor viruses. James D. Watson had just become director of the laboratory and Sambrook was initiating a research effort to study the molecular biology of SV40. The main goal was to define the genetic structure of DNA tumor viruses in general. Fortunately for the whole field, Ham Smith at Johns Hopkins[1] had just discovered type II restriction endonucleases that cleaved DNA at defined nucleotide sequences. Dan Nathans, chairman of Smith's department, returned from a sabbatical with Ernest Winocour at the Weizmann Institute in Israel and used Smith's purified restriction endonuclease from *H. influenzae* to break down SV40 DNA into 11 pure fragments.[2] The ability to dissect viral genomes with restriction endonucleases obviously offered the possibility of defining the genetic structure of SV40 DNA. Not to be left in the dust at Cold Spring Harbor, we began an intense program to isolate and use restriction endonucleases.

"Smith and colleagues had used a decrease in viscosity of a DNA solution as an assay for endonuclease activity. Nathans introduced gel electrophoresis as a means of fractionating endonuclease cleaved $^{32}$P-labeled fragments; however, the distribution of fragments in the gel required radioautography, a time-consuming process. After having spent a postdoctoral fellowship at the California Institute of Technology and observing the enhanced fluorescence of the dye ethidium bromide upon binding to DNA, I decided to try to stain gels with ethidium bromide as a means of detecting the bands formed by different length fragments. The minimum concentration of ethidium bromide necessary for saturation of DNA was calculated and within three hours I had successfully detected DNA bands in gels with ethidium bromide.

"Staining of gels with ethidium bromide provided a rapid assay for detecting fragments produced by restriction endonucleases. Sugden was induced to help us purify the restriction endonucleases from *H. parainfluenzae* as this strain was thought to contain activities with different specificities than *H. influenzae*. We quickly discovered using the gel electrophoresis and ethidium bromide staining assay that there were two different endonucleases with different sequence specificities in these bacteria, Hpa I and Hpa II. Many other investigators have used these endonucleases or these methods in their experiments. In addition, a number of small companies have been established to supply such endonucleases to the molecular biological community."

1. Kelly T J, Jr. & Smith H O. A restriction enzyme from *Hemophilus influenzae*. II. Base sequence of the recognition site. *J. Mol. Biol.* 51:393-409, 1970.
2. Danna K & Nathans D. Specific cleavage of simian virus 40 DNA by restriction endonuclease of Hemophilus influenzae. *Proc. Nat. Acad. Sci. US* 68:2913-17, 1971.

# This Week's Citation Classic

**Kebabian J W, Petzold G L & Greengard P.** Dopamine-sensitive adenylate cyclase in caudate nucleus of rat brain, and its similarity to the "dopamine receptor."
*Proc. Nat. Acad. Sci. US* **69**:2145-9, 1972.
[Dept. Pharmacology, Yale Univ. Sch. Medicine, New Haven, CT]

Dopamine stimulates adenylate cyclase activity in a cell-free homogenate of the caudate nucleus of the rat brain. The ability of drugs to either mimic or block this effect of dopamine suggests this enzyme activity may be a receptor mechanism for dopamine. [The *SCI®* indicates that this paper has been cited in over 895 publications since 1972.]

John W. Kebabian
Biochemical Neuropharmacology Section
Experimental Therapeutics Branch
National Institute of Neurological &
Communicative Disorders & Stroke
National Institutes of Health
Bethesda, MD 20205

January 18, 1983

"In the early-1970s, Paul Greengard (PG) began his investigations of the role of cyclic AMP (cAMP) in simple, peripheral neural tissues, including the superior cervical ganglion. Working as PG's graduate student, I demonstrated that dopamine (as well as other catecholamines) increased the content of cAMP in the bovine superior cervical ganglion.[1] The choice of bovine tissue was fortuitous; ganglia from other species (e.g., rabbit or rat) are less responsive to dopamine. I came to use the bovine ganglion because of the classic 'fellow across the hall' who was using it in an unrelated project. An obvious extension of this project was to attempt to demonstrate a dopamine-stimulated accumulation of cAMP in striatal tissue. Because initial efforts using slices of bovine striatum were unsuccessful, PG decided that Gary Petzold (his 'supertech') and I should measure adenylate cyclase activity in homogenates of rat striatum. This latter approach was successful and was reported in the paper which ISI® has designated as a Citation Classic. My PhD thesis contained the data from the bovine ganglion and the rat striatum; it had only 48 pages of text and was, therefore, described as a 'master's thesis' by another graduate student.

"There are probably several reasons why this paper, demonstrating a dopamine-sensitive adenylate cyclase activity in a cell-free assay system, has been highly cited. First, it provided a cell-free model for a neurotransmitter receptor. Some of the biochemical phenomena demonstrated with this system apply to other neurotransmitter-sensitive adenylate cyclase systems. Second, the demonstration of this enzyme activity raised the possibility that some of the physiological effects of dopamine were initiated by a dopamine-stimulated accumulation of cAMP. Third, the enzyme system provided a rapid and quantitative procedure for screening dopaminergic agonists and antagonists. Two consequences of such screening were the following. The anticipated observation that many clinically active antipsychotic agents were potent competitive dopamine antagonists in the assay system led to speculation that a dopamine-sensitive adenylate cyclase might participate in the therapeutic effects of such drugs.[2] The unanticipated observation that lergotrile was a dopamine antagonist in the assay system[3] led Donald Calne and myself to propose the two dopamine receptor hypothesis.[4] In the past three years, my colleagues at the National Institutes of Health and I have attempted to verify this latter hypothesis."[5]

1. **Kebabian J W & Greengard P.** Dopamine-sensitive adenyl cyclase: possible role in synaptic transmission. *Science* **174**:1346-9, 1971.
2. **Clement-Cormier Y C, Kebabian J W, Petzold G L & Greengard P.** Dopamine-sensitive adenylate cyclase in mammalian brain: a possible site of action of antipsychotic drugs. *Proc. Nat. Acad. Sci. US* **71**:1113-17, 1974.
3. **Kebabian J W, Calne D B & Kebabian P R.** Lergotrile mesylate: an *in vivo* dopamine agonist which blocks dopamine receptors *in vitro*. *Commun. Psychopharmacol.* **1**:311-18, 1977.
4. **Kebabian J W & Calne D B.** Multiple receptors for dopamine. *Nature* **277**:93-6, 1979.
5. **Cote T E, Eskay R L, Frey E A, Grewe C W, Munemura M, Stoof J C, Tsuruta K & Kebabian J W.** Biochemical and physiological studies of the beta-adrenoceptor and the D-2 dopamine receptor in the intermediate lobe of the rat pituitary gland: a review. *Neuroendocrinology* **35**:217-24, 1982.

# This Week's Citation Classic

CC/NUMBER 29
JULY 20, 1981

Theorell H & Bonnichsen R. Studies on liver alcohol dehydrogenase. I. Equilibria and initial reaction velocities. *Acta Chem. Scand.* 5:1105-26, 1951.
[Medicinska Nobelinstitutets Biokemiska avdelning, Stockholm, Sweden]

This paper shows that liver alcohol dehydrogenase binds two molecules of NADH through sulfhydryl bonds with simultaneous shift of the absorption band at 340 to 325 nm. This discovery made possible closer studies of the mechanism of alcohol oxidation. NADH is bound more firmly than NAD and thus rate determining. [The *SCI®* indicates that this paper has been cited over 300 times since 1961.]

Hugo Theorell
Laboratory for Enzyme Research
Nobel Institute of Biochemistry
Karolinska Institute
Solnavägen 1
S-104 01 Stockholm 60
and
Roger Bonnichsen
Chemical Department
National Institute for Forensic Chemistry
Box 6209
S-104 01 Stockholm 60
Sweden

June 1, 1981

"Alcohol in its various aspects has interested mankind as long as it has been available. In our laboratory at the Nobel Institute of Biochemistry, Britton Chance had just found that a catalase-peroxide complex oxidizes alcohol. It was known that an oxidizing enzyme in liver converted alcohol to acetaldehyde and the question ensued: how is alcohol combusted in man? At that time Einar Lundsgaard from the Rockefeller Institute in Copenhagen was visiting us and during an afternoon discussion we decided to isolate and identify the dehydrogenase from liver. This was a long time ago and the isolating methods were rather primitive. The various chromatographic methods were not yet invented. For the kinetic studies we needed hundreds of mgs of enzyme and as most preparations of this kind sooner or later end up in the sink, preparations were a limiting factor. Subsequently, we found a better way to prepare the enzyme with higher yield. NAD, then DPN, was not commercially available and we had to prepare it by rather laborious methods until Neilands and Åkeson at the laboratory prepared NAD using an ion exchange column. We prepared the reduced form enzymatically and then extracted the coenzyme. We did not at that time realize that the enzyme preparations consisted of different variants of the enzyme. We usually got three peaks in the electrophoresis pattern, thought it to be impurities, and purified and recrystallized the preparation until we got one fraction by electrophoresis. Until the end of World War II the only spectrophotometer available was a Warburg-Negelein instrument and kinetic studies with this instrument were not possible. Britton Chance, however, brought with him the first Beckman spectrophotometer we had seen and we invested in a couple of these instruments.

"Misuse of alcohol is one of the greatest problems in our civilization and more than 100,000 papers have been written about alcohol and its effect on man. The knowledge of the ADH enzyme, as we coined it, seemed to us one of the keystones to solve or minimize the damage of alcohol. The discovery of the spectral shift of NADH-absorption from 340 to 325 nm following its binding with sulfhydryl groups made possible the kinetic studies in this and following works. We believe this to be the main reason why this work has been so much cited, as it started an extensive study of this enzyme, various isoenzymes with different properties and specificities, as well as studies of structure and reaction mechanism.[1] Accumulation of reduced NAD has been proved to be responsible for many of the toxic effects of alcohol. Still 30 years later there are many controversial theories concerning alcohol turnover, e.g., the role of microsomal tissue, catalase, peroxidase, and the small but very active, recently found variants of ADH, as well as its physiological substrate."

1. Brändén C I, Jörnvall H, Eklund H & Furugren B. Alcohol dehydrogenases. (Boyer P, ed.) *The enzymes.* New York: Academic Press, 1975. Vol. 11, pt. A. p. 103-90.

163

# This Week's Citation Classic
**NUMBER 17, APRIL 23, 1979**

Nachlas M M, Tsou K C, De Souza E, Cheng C S & Seligman A M. Cytochemical demonstration of succinic dehydrogenase by the use of a new *p*-nitrophenyl substituted ditetrazole. *J. Histochem. Cytochem.* 5:420-36, 1957.
[Sinai Hospital of Baltimore, Baltimore, MD, Johns Hopkins Univ. Sch. Med. Baltimore, MD, Dajac Labs., Chem. Div., Borden Co.]

This paper introduces the use of 2,2'-di(-p-nitrophenyl)-5,5'-diphenyl-3,3'-(3,3'-dimethoxy-4,4'-diphenylene)-di(tetrazolium chloride) (Nitro BT, or NBT) as the ideal substrate for the cytochemical demonstration of DPN and TPN diaphorase systems, as well as the succinic dehydrogenase system. The 'NBT' method has since become the most popular method for the demonstration of dehydrogenases in histochemistry. Diagnostic tests based on the use of NBT can be found today in most clinical laboratories. [The *SCI*® indicates that this paper has been cited over 1,150 times since 1961.]

K.C. Tsou
University of Pennsylvania
School of Medicine
Harrison Department of Surgical Research
Philadelphia, PA 19104

December 15, 1977

"The development of nitroblue tetrazolium, or NBT, for enzyme histochemistry, was indeed an important episode in our long-term effort to develop new and useful substrates for this still relatively young field of enzyme histo- and cytochemistry. In those days, as is even true today, many of us thought the route to studying respiratory enzymes should be through histochemistry. A. M. Rutenburg, R. Gofstein, and A. M. Seligman had earlier developed a method known as BT for succinic dehydrogenase, but it required the addition of potassium cyanide (KCN).[1] I asked Rutenburg how he could demonstrate oxidative phosphorylation in the presence of cyanide, but he was not pleased with my question. Seligman, on the other hand, encouraged me to look for better electron acceptors for succinic dehydrogenase. As I had already learned tetrazolium chemistry from earlier work at Harvard, I though it would be a cinch to find one with the right redox potential by putting electron withdrawing groups on the phenyl groups.

"In fact, it took a serious effort of almost three years to accomplish this phase. Today, I have a cryostat, a good microscope and other equipment to do better histochemistry. In those days, I used a very crude testing method of a chicken liver extract and looked for blue precipitate in the absence of KCN. Dr. C. S. Cheng, who was my postdoctoral associate then, and an extremely meticulous organic chemist, could never quite understand why I wasn't too happy with the beautifully crystallized compounds he prepared, until NBT came along.[2] I do not remember when, but I know it was some time in the fall of 1955, when I called Seligman one evening in Baltimore to ask if he was ready to test this preliminary sample which formed a beautiful blue precipitate with no trace of color in the supernatant on my bench. Mrs. Seligman told me that Arnold had gone out to pick the corn on his 'farm.' Nachlas called me back next morning and obtained this first sample the next day. A few days later, he called back to say that NBT worked! Our job, however, was not done, as we still had to improve the purification procedure so that he could do an extensive series of studies.

"This work stimulated my interest in the synthesis of substrates for electroncytochemistry and converted my career from organic to cytochemistry. The applications of NBT to isozyme research and medical diagnosis are still active even today, in spite of the fact that few users appreciate our difficulties in the development of this versatile reagent. The *SCI* citations remind me of many associates who contributed to the chemistry of this development and of my friendship with the late A. M. Seligman, whose name in histochemistry will be remembered by more than myself."

1. **Rutenburg A M, Gofstein R & Seligman A M.** Preparation of a new tetrazolium salt which yields a blue pigment on reproduction and its use in the demonstration of enzymes in normal and neoplastic tissue. *Cancer Res.* **10**:113-21, 1950.
2. **Tsou K C, Cheng C S, Nachlas M M & Seligman A M.** Synthesis of some p-nitrophenyl substituted tetrazolium salts as electron acceptors for the demonstration of dehydrogenases. *J. Amer. Chem. Soc.* **78**:6139-44, 1956.

## This Week's Citation Classic

CC/NUMBER 22
JUNE 1, 1981

Glenner G G, Burtner H J & Brown G W, Jr. The histochemical demonstration of monoamine oxidase activity by tetrazolium salts.
*J. Histochem. Cytochem.* 5:591-600, 1957.
[Lab. Pathol. and Histochem., National Institutes of Health, and Lab. Biochem. and Metab., NIAMD, National Institutes of Health, Bethesda, MD]

This paper describes a method for the localization of monoamine oxidase (MAO) enzymic activity in tissue sections based on the formation of a colored formazan by reduction of nitro-blue tetrazolium (NBT) using tryptamine as substrate. [The *SCI®* indicates that this paper has been cited over 480 times since 1961.]

---

George G. Glenner
Section on Molecular Pathology
National Institute
of Arthritis, Metabolism,
and Digestive Diseases
National Institutes of Health
Bethesda, MD 20014

April 17, 1981

"Much of the clinical excitement at the National Institutes of Health at this time came from the work in the Heart Institute on the carcinoid tumor syndrome and serotonin metabolism. The histochemical methods then available for localizing the tissue sites of the enzyme, MAO, implicated in the inactivation of vasoactive amines, e.g., serotonin, produced either poor or artifactual tissue localization or were unrelated to the activity of MAO on a strictly empirical basis. Dissatisfied with the results of this technique, I went through the list of available substrates with the urging of my colleague, James Longley, and found that tryptamine in the presence of the commercially available tetrazolium salts produced a reasonable cellular localization in frozen tissue sections. Concurrently, George Brown, a staff associate also in the Arthritis Institute, was engaged with me in synthesizing a tetrazolium salt (NBT) which could afford a more precise localization of flavoprotein-linked dehydrogenase activity. With Helen Burtner's technical assistance we applied NBT with tryptamine as substrate to localize MAO activity to cellular sites with astoundingly excellent results. We were pleased when the classic MAO inhibitors, Marsilid (iproniazid) and phenylhydrazine, inhibited the reaction. However, uncertainty set in when other carbonyl reagents such as hydrazine and cyanide ion also acted as inhibitors. This was at variance with known biochemical results. Herbert Weissbach, then a staff fellow in Sidney Udenfriend's group in the Heart Institute, was called in for consultation. Using a purified MAO preparation, he confirmed the histochemical results, but we remained unenlightened as to the anomalous inhibitor reactions, the mechanism producing NBT reduction, and the efficacy in our system of tryptamine, a less than optimal biochemical substrate for MAO. With Herb's biochemical confirmation we went to press with the answers to these questions to await further study. It was only later[1] that we discovered that the mechanism of reduction of the tetrazolium salt was due to instantaneous electron transfer by the labile aldehyde (indolyl-3-acetaldehyde) formed from tryptamine by MAO activity, explaining the inhibitor effect of the carbonyl reagents and the uniqueness of tryptamine as substrate. This was also the first application of a nonenzymatic sequence methodology for localizing enzymes in solid phase systems such as tissue sections and electrophoretic gels.

"This publication has been highly cited for several reasons. The use of newer aqueous-lipid insoluble formazan precursors has made the method applicable for electron microscopic purposes,[2] and its ready adaptability for localization of MAO species in electrophoretic gels have found extensive biochemical use (as have most specific histochemical methods). The accelerating interest in neuropathology, neuropharmacology, and neurophysiology, especially with the focus on MAO inhibitors, has undoubtedly produced a concomitant accelerating interest in this easily reproducible histochemical technique."

---

1. Glenner G G, Weissbach H & Redfield B G. The histochemical demonstration of enzymatic activity by a nonenzymatic redox reaction. Reduction of tetrazolium salts by indolyl-3-acetaldehyde.
   *J. Histochem. Cytochem.* 8:258-61, 1960.
2. Lojda Z, Grossrau R & Schiepler T H. *Enzyme histochemistry.* Berlin: Springer-Verlag. 1979. 339 p.

165

## This Week's Citation Classic

**Wurtman R J & Axelrod J.** A sensitive and specific assay for the estimation of monoamine oxidase. *Biochem. Pharmacol.* **12**:1439-41, 1963.
[Lab. Clinical Science, Natl. Inst. Mental Health, NIH, Bethesda, MD]

An assay for monoamine oxidase (MAO) is described using isotopically labeled tryptamine as substrate and a toluene:dilute HCl system for separating products from unreacted substrate. One worker could assay 60 or more samples in a morning. [The *SCI*® indicates that this paper has been cited in over 660 publications since 1963.]

---

Richard J. Wurtman
Laboratory of Neuroendocrine Regulation
Massachusetts Institute of Technology
Cambridge, MA 02139

February 17, 1982

"I studied philosophy in college, and decided to work on the 'mind-body problem.' Hence I enrolled in medical school and sought opportunities to do research on brain mechanisms underlying behavior. Having heard that infusions of epinephrine caused some people to become anxious, I approached Peter Dews, a Harvard pharmacology professor, about using his operant conditioning techniques to characterize epinephrine's behavioral effects: pigeons in Skinner boxes would be allowed access to food if they pecked at a key 15 or more minutes after their last feeding; they quickly developed a regular behavior pattern, wasting no energy on fruitless pecking for ten minutes, then pecking vigorously. Various drugs characteristically altered this pattern. With a classmate, Michael Frank, I found that epinephrine caused a dose-dependent change in the animals' behavior. We presented our findings at a FASEB meeting, and they were well received. Thirty minutes later, Julius Axelrod—then unknown to me—gave his classic paper[1] showing that exogenous catecholamines are unable to cross the blood-brain barrier (!). (To this day I have no idea how the epinephrine worked; perhaps it gave the birds a headache.) When I described what had transpired to Dews, he suggested that I learn neurochemistry under Axelrod. In 1962, I entered Axelrod's laboratory at the National Institute of Mental Health and began exploring catecholamine metabolism.

"Several years earlier Axelrod had shown that catecholamines are inactivated not primarily by an enzymatic mechanism but by re-uptake into nerve terminals.[2] Two enzymes also could initiate catecholamine metabolism: catechol-O-methyltransferase and monoamine oxidase (MAO); the latter apparently functioned to set catecholamine levels within nerve terminals. That MAO could be involved in human behavior had been suggested by the antidepressant activity of MAO inhibitors.

"Axelrod proposed that I examine the effects of hormones on the fate of circulating catecholamines in rats. For this I would need also to measure MAO. Existing assays, based on manometric or fluorimetric procedures, were cumbersome and insensitive, so Axelrod suggested that we develop our own, using an isotopically labeled substrate and an appropriate organic solvent to separate deaminated metabolites from the unutilized substrate. The substrate used, tryptamine, was highly charged at an acidic pH, and couldn't pass from the aqueous phase into toluene, the organic solvent. However, once deaminated it passed into the toluene. In a day or two we worked out an MAO assay based on 14C-tryptamine's deamination, and used it to characterize the effects of hormones on MAO activity[3] and the extent to which tissue MAO activity had to be inhibited before catecholamine metabolism actually was affected.[4] At first we planned only to describe our MAO assay in the methods sections of these reports. However, we subsequently decided that its speed and sensitivity might make it useful to other people, so we wrote the paper cited here. Apparently, this has been the case.

"The intellectual godfathers of our assay were Lyman Craig, who pioneered the use of liquid-liquid chromatographic systems, and Bernard Brodie, who, with Axelrod,[5] applied such systems for separating biologically active compounds (like amphetamine) from their metabolites. MAO is now known to be a family of enzymes, acting on different substrates.[6,7] Fortunately, MAO assays using the single substrate tryptamine often provide adequate information about the behavior of the family in general.

"The mind-body problem remains unsolved."

---

1. **Axelrod J, Weil-Malherbe H & Tomchick R.** Physiological disposition of $H^3$-adrenaline and its principal metabolite, metanephrine. *Fed. Proc.* **18**:364, 1959.
2. **Hertting G, Axelrod J, Kopin I J & Whitby L G.** Lack of uptake of catecholamines after chronic denervation of sympathetic nerves. *Nature* **189**:66, 1961.
3. **Wurtman R J, Kopin I J & Axelrod J.** Thyroid function and the cardiac disposition of catecholamines. *Endocrinology* **73**:63-6, 1963.
4. **Wurtman R J & Axelrod J.** Sex steroids, cardiac 3H-norepinephrine, and tissue monoamine oxidase levels in the rat. *Biochem. Pharmacol.* **12**:1417-19, 1963.
5. **Axelrod J.** Studies on sympathomimetic amines. II. The biotransformation and physiological disposition of d-amphetamine, d-p-hydroxyamphetamine and d-methamphetamine. *J. Pharmacol. Exp. Ther.* **110**:315-26, 1954.
6. **Haenick D, Boehme D H & Vogel W H.** Monoamine oxidase in four rat and human tissues. *Biochem. Med.* **26**:451-4, 1981.
7. **Giambalvo C T & Becker R E.** Modulators of monoamine oxidase in plasma. *Life Sci.* **29**:2017-24, 1981.

# This Week's Citation Classic

CC/NUMBER 5
JANUARY 31, 1983

Fraser D R & Kodicek E. Unique biosynthesis by kidney of a biologically active vitamin D metabolite. *Nature* **228**:764-6, 1970.
[Dunn Nutritional Laboratory, University of Cambridge, and Medical Research Council, Cambridge, England]

---

A vitamin D 1-oxygenating enzyme was discovered in chicken kidney mitochondria. This produces the metabolite now known to be 1,25-dihydroxyvitamin D (1,25(OH)$_2$D). In rats it was shown that the 1-hydroxylase was present only in kidney thus revealing a new endocrine function for this organ in controlling calcium metabolism. [The *SCI*® indicates that this paper has been cited in over 785 publications since 1970.]

---

David R. Fraser
Dunn Nutritional Laboratory
University of Cambridge
and
Medical Research Council
Cambridge CB4 1XJ
England

December 15, 1982

"In the 1960s, studies on the mechanism of action of vitamin D were testing the hypothesis that it functioned as a steroid hormone in cell nuclei and induced specific proteins involved in calcium transport. With the discovery by DeLuca's group in Wisconsin[1] of the biologically active metabolite, 25-hydroxyvitamin D, attempts were made to locate this substance in the nuclei of intestinal cells. Eric Lawson in Cambridge then found that there was a derivative of 25-hydroxyvitamin D in these nuclei: a metabolite characterised by its serendipitous loss of tritium from the 1-alpha position of tritium-labelled vitamin D.[2]

"As my own research was floundering at that time, the director of the Dunn, Egon Kodicek, suggested that I try to find an enzyme which produced this new metabolite. The enzyme could then be used to generate the metabolite *in vitro* in amounts required to complete its chemical identification. It was assumed first that the enzyme might be a steroid hydroxylating mixed-function oxidase and secondly that its location was probably at the site of highest metabolite concentration—the intestinal mucosa. The first assumption eventually proved to be correct but the second was dispiritingly wrong. After several futile months incubating labelled 25-hydroxyvitamin D with all manner of intestinal preparations, the search was diverted to other tissues. By a combination of homogenate incubations and surgical extirpations, a wide range of organs was found not to contain the enzyme. The most unpredicted and only remaining site, the kidney, was finally shown on October 14, 1969 to make the metabolite *in vitro*. Because this was the one positive finding, nephrectomized rats were then used to confirm that the enzyme was functioning only in the kidney *in vivo*.

"For the next 12 months, I had the delightful and luxurious privilege of studying a fascinating new renal enzyme system of which the world outside the Dunn was still unaware. This research was finally reported in *Nature* in November 1970. Subsequently, medical and endocrine use of 1,25(OH)$_2$D made the study of its formation in the kidney a very popular topic. For a recent review, see *Physiological Reviews*.[3] Despite its popularity, this *Nature* paper is not of special importance. It describes just one step in a sequence of research in several laboratories which revealed the functional metabolism of vitamin D. The discovery of the 1-hydroxylase was notable for a complete absence of intellectual wisdom—a search for a needle in a haystack by the tedious inspection of each straw. A number of other research groups would, with time, have stumbled on the same result. This particular search was successful because of a simple assay for a tritium-deficient product on thin-layer chromatography and a realisation that only μg amounts of 1,25(OH)$_2$D would be formed *in vitro*.

"This paper may be widely cited, partly because all the work was published only once and partly because there have now been large numbers of papers on vitamin D which all needed a convenient reference to the renal 1-hydroxylase. Regardless of any historical significance, this research was great fun to do, giving a lot of pleasure to myself and much satisfaction to Kodicek who sadly died in July 1982."

---

1. **Blunt J W, DeLuca H F & Schnoes H K.** 25-Hydroxycholecalciferol. A biologically active metabolite of vitamin D$_3$. *Biochemistry* **7**:3317-22; 1968.
2. **Lawson D E M, Wilson P W & Kodicek E.** New vitamin D metabolite localized in intestinal cell nuclei. *Nature* **222**:171-2, 1969.
3. **Fraser D R.** Regulation of the metabolism of vitamin D. *Physiol. Rev.* **60**:551-613, 1980.

## This Week's Citation Classic

NUMBER 16
APRIL 16, 1979

Dahlqvist A. Method for assay of intestinal disaccharidases.
*Anal. Biochem.* 7:18-25, 1964. (University of Lund, Lund, Sweden)

The article describes a technically very simple method for the assay of disaccharidase activity with a TRIS-buffered glucose oxidase reagent. The method is so sensitive that it can be used for assay of the enzyme activities in peroral biopsy specimens of the small-intestinal mucosa and is thus useful for clinical studies in human subjects. [The *SCI®* indicates that this paper has been cited over 540 times since 1964.]

Arne Dahlqvist
Department of Nutrition
University of Lund
Chemical Centre
Box 740
S-220 07 Lund 7, Sweden

April 10, 1978

"When this paper was written, I had studied the biochemistry of the small-intestinal disaccharidases for many years. Assay methods, based on the increase in reducing power during the hydrolysis of disaccharides, were used for these studies. With reducing disaccharides as substrates, e.g. maltose, isomaltose, and lactose, such methods have low sensitivity because a large fraction of the substrate has to be hydrolysed before the increase in reducing power can be measured with reasonable accuracy.

"At the time, recent findings of different groups of patients with impaired ability to digest disaccharides made it urgent to develop an accurate micromethod for disaccharidase activity which could be used for the analysis of such small amounts of tissue as would be obtained by peroral biopsy.

"It appeared obvious that replacement of reducing sugar methods with enzymatic assay of liberated glucose by glucose oxidase would yield considerable advantages. When different glucose oxidase preparations were tested, however, they were all found to contain strong contaminant disaccharidase activity, which prohibited their use.

"Considerable efforts were made to separate the glucose oxidase activity from these disaccharidase activities, but these attempts failed completely.

"The finding that the disaccharidase activities could be inhibited by TRIS, without any decrease in glucose oxidase activity, was made by pure chance. TRIS was not regarded as an enzyme inhibitor. In fact there was printed on the label of the TRIS bottle that it 'does not inhibit enzymes.' When I decided to look for an inhibitor for the disaccharidase activities which would not inhibit the glucose oxidase, I therefore tested numerous other substances than TRIS, without success. When for a certain experiment TRIS buffer was used instead of the usual buffer in order to avoid precipitation with other reagents, I found that TRIS had just the properties I had been looking for.

"The simple method for disaccharidase activity assay that was described in the paper has been used for numerous clinical investigations, and is now used in many hospitals as a routine method for clinical diagnosis in patients with intestinal disturbances.

"Few things are so good that they cannot be made better. It was later found that the method could be further simplified by using TRIS also to interrupt the first incubation step.[1] In this way one step of dilution, boiling, and pipettation could be omitted and at the same time a further tenfold increase in sensitivity was achieved."

### REFERENCE

1. **Dahlqvist A.** Assay of intestinal disaccharidases. *Enzymol. Biol. Clin.* 11:52-66, 1970.

# This Week's Citation Classic™

CC/NUMBER 52
DECEMBER 24-31, 1984

Osserman E F & Lawlor D P. Serum and urinary lysozyme (muramidase) in monocytic and monomyelocytic leukemia. *J. Exp. Med.* **124**:921-52, 1966.
[Department of Medicine, Columbia University College of Physicians & Surgeons, and Francis Delafield Hospital, New York, NY]

This paper describes the finding of markedly elevated concentrations of the enzyme lysozyme in the serum and urine of patients with monocytic and monomyelocytic leukemia. In addition to its clinical importance, this finding permitted extensive biochemical and physicochemical studies of human lysozyme [The *SCI*® indicates that this paper has been cited in over 825 publications since 1966.]

---

Elliott F. Osserman
Department of Medicine
Institute of Cancer Research
College of Physicians & Surgeons
Columbia University
New York, NY 10032

November 19, 1984

"For many years, we had investigated the immunoglobulin and Bence-Jones protein abnormalities in patients with multiple myeloma, a neoplastic disease of plasma cells. One of our patients had multiple myeloma and was excreting a kappa Bence-Jones protein in his urine. With chemotherapy, it was possible to suppress the abnormal plasma cells and virtually eliminate the production of Bence-Jones protein in the patient. Unfortunately, however, after 18 months of treatment, the patient suddenly developed monomyelocytic leukemia, which is a recognized complication of myeloma after long-term chemotherapy. At this point, we restudied his urine and confirmed that there was no Bence-Jones protein. However, electrophoresis of the urine proteins demonstrated a markedly basic, cationic protein (CP) in the far 'post-gamma' mobility range, an abnormality that neither we nor anyone else had previously seen.

"Within a few weeks, we had determined that virtually all patients with monocytic and monomyelocytic leukemia had this protein in their urine and it, therefore, was a specific biochemical marker for this group of leukemias. Over the next several months, we determined the molecular properties of the CP protein. Because of the circumstances of the first case, we were convinced that this was another 'piece' of immunoglobulin, and we attempted to prove this—obviously with thoroughly negative results.

"The 'breakthrough' came when I finally accepted the fact that CP was *not* an immunoglobulin fragment and cleared my mind of this wrong preconception. Having done this, I went back to the initial and obvious fact that monocytes are precursors of macrophages and produce a wide variety of enzymes, e.g., ribonuclease, cathepsin, proteases, glucuronidase, elastase, and also lysozyme. We made a list of these enzymes and prepared to assay our purified protein for these specific activities. Several of the assays were set up in our own laboratory and others were carried out by friends who were working on macrophage functions. Among these was Zanvil Cohn of Rockefeller University, whose lab was set up for doing lysozyme assays. A few days after he received our sample, he called and said, 'Elliott, I don't know what else you have in your sample, but there certainly is an enormous amount of lysozyme activity.' Indeed, since our protein had been purified to homogeneity, it was clear that this was the enzyme itself. That day was probably the happiest and most exhilarating in my research career because it was obvious that we had identified a phenomenon with extremely wide and potentially important implications.

"Over the next several months, we completed the basic biochemical, immunochemical, and clinical studies that were all assembled and published in this paper along with a description of a new assay method for quantifying lysozyme. This paper, which ran to over 30 pages in the *Journal of Experimental Medicine*, could easily have been divided into three or four separate publications, but it was my conviction that the information would be more readily accessible as a single publication. Happily, this proved to be true. Our finding made possible a host of subsequent studies of human lysozyme including the amino acid sequence determination that was carried out by my associate, Robert Canfield,[1] and crystallization, which we achieved.[2] The paper also permitted the crystallographic studies of lysozyme by Colin Blake and his associates at Oxford.[3] A very large number of other investigations have been and continue to be carried out on human lysozyme,[4] and this is the most gratifying and rewarding consequence of our work."

1. **Canfield R E, Kammerman S, Sobek J H & Morgan F J.** Primary structure of lysozymes from man and goose. *Nature* **232**:16-17, 1971. (Cited 115 times.)
2. **Osserman E F.** Crystallization of human lysozyme. *Science* **155**:1536-7, 1967. (Cited 20 times.)
3. **Blake C C F & Swan I D A.** X-ray analysis of structure of human lysozyme at 6 angstrom resolution. *Nature* **232**:12-15, 1971. (Cited 60 times.)
4. **Osserman E F, Canfield R E & Beychok S**, eds. *Lysozyme*. New York: Academic Press, 1974. 641 p.

## This Week's Citation Classic

Kato R & Gillette J R. Effect of starvation on NADPH-dependent enzymes in liver microsomes of male and female rats. *J. Pharmacol. Exp. Ther.* **150**:279-84, 1965.
[Lab. Chem. Pharmacol., Natl. Heart Inst., Natl. Insts. Health, Bethesda, MD]

This paper described the presence of androgen-dependent and -independent activities in drug-metabolizing enzymes of rat liver microsomes. The change in the activities by starvation was sex-related and the activities being stimulated by androgen were decreased. [The *SCI®* indicates that this paper has been cited in over 615 publications since 1965.]

---

Ryuichi Kato
Department of Pharmacology
School of Medicine
Keio University
Shinjuku-ku, Tokyo 160
Japan

May 29, 1984

"Before joining the National Institutes of Health (NIH), I worked in the Department of Pharmacology at the University of Milan, Italy, and I was interested in sex differences in the microsomal drug-metabolizing enzyme (DME). At NIH, I worked with J.R. Gillette in B.B. Brodie's lab. At that time, this lab was a mecca for the study of drug metabolism.

"The presence of sex differences in the activity of DME of rat liver had been shown by several investigators. We demonstrated that the magnitude of sex differences depends on the substrate. For example, there were marked sex differences in aminopyrine N-demethylation and hexobarbital hydroxylation, while only a little or no difference was detected in aniline and zoxazolamine hydroxylations. One of my major interests was to clarify factors that modify the activity of DME. However, at that time, most investigators used only male rats. Therefore, we first examined the effect of starvation on DME in male and female rats. We observed clear decreases in hexobarbital hydroxylation and aminopyrine N-demethylation in male rats, but we didn't observe any decrease in aniline and zoxazolamine hydroxylations. Indeed, the aniline hydroxylation in starved male rats was increased. To our surprise, the starvation caused an increase in the activity of DME in all female rats.

"This paper presented the fact that the magnitude of the sex differences in the activity of DME varied markedly with the substrate. Moreover, starvation of male rats impaired the activity of sex-dependent, but not sex-independent, enzymes in male rats and both enzymes in female rats. Castration and androgen-supplement experiments indicated that the impairment occurred only in the enzyme activities being stimulated by androgen, but this stimulation was impaired by starvation and caused marked sex differences in its effect. In the next paper,[1] we reported similar changes in the activity of DME in morphine- or thyroxine-treated, adrenalectomized male and female rats. Therefore, the high rate of citation of this paper may be due to its clear demonstration of the existence of sex-related and -unrelated activities in DME in rat liver and sex-related changes of DME under some abnormal physiological conditions. In addition, this paper was cited for its description of the method for aniline hydroxylation, until the method of Imai et al.[2] became more popular.

"All these results have indicated a possible presence of specific forms of cytochrome P-450 in liver microsomes of male rats. Recently, we have purified male- and female-specific cytochrome P-450 (P-450 male and P-450 female) from male and female rats, respectively.[3] The studies on the regulation mechanism of P-450 male will afford new insight to the sex difference of DME. Gillette is now the chief of the Laboratory of Chemical Pharmacology, National Heart, Lung, and Blood Institute, NIH, Bethesda, Maryland."

1. **Kato R & Gillette J R.** Sex differences in the effects of abnormal physiological states on the metabolism of drugs by rat liver microsomes. *J. Pharmacol. Exp. Ther.* **150**:285-91, 1965. (Cited 285 times.)
2. **Imai Y, Ito A & Sato R.** Evidence for biochemically different types of vesicles in the hepatic microsomal fraction. *J. Biochem. Tokyo* **60**:417-28, 1966. (Cited 400 times.)
3. **Kamataki T, Maeda K, Yamazoe Y, Nagai T & Kato R.** Sex difference of cytochrome P-450 in the rat. Purification, characterization, and quantitation of constitutive forms of cytochrome P-450 from liver microsomes of male and female rats. *Arch. Biochem. Biophys.* **225**:758-70, 1983.

# This Week's Citation Classic™

Schenkman J B, Remmer H & Estabrook R W. Spectral studies of drug interaction with hepatic microsomal cytochrome.
*Mol. Pharmacol.* 3:113-23, 1967.
[Dept. Biophys. and Phys. Biochem., Johnson Res. Foundation, Univ. Pennsylvania, Philadelphia, PA]

The addition of some two dozen substrates of the hepatic monooxygenase to liver microsomes was shown to evoke one of two types of spectral changes. The spectral changes were shown to be reversible and to be dependent upon both substrate and microsomal protein concentrations. The dissociation constants for substrates causing one type of spectral change (type I) were shown to be similar to the $K_m$ values for some of those substrates, and it was suggested that the spectral changes were indicative of enzyme substrate complexes. [The *SCI®* indicates that this paper has been cited in over 1,025 publications since 1967.]

---

John B. Schenkman
Department of Pharmacology
University of Connecticut Health Center
Farmington, CT 06032

December 27, 1983

"Shortly after I joined Ron Estabrook's lab at the Johnson Foundation as a postdoctoral fellow in 1964, Ron suggested I study the hepatic microsomal mixed function oxidase. It was a good suggestion. Tsuneo Omura from Ryo Sato's department had joined Ron and David Y. Cooper at the University of Pennsylvania in a study of the adrenal cortex steroid hydroxylase and was a ready source of knowledge of P-450. Herbert Remmer would soon be joining the lab from Germany and was a source of knowledge of drug metabolism. Shakunthala Narasimhulu, who worked with Dave, had noted steroids affected the absorption spectrum of adrenal cortex microsomes,[1] so Ron suggested I examine the liver microsomes spectrophotometrically. With great trepidation I approached the Yang machine, a Johnson Foundation creation that seemed, at my level of sophistication, to be a monochromator and photomultiplier caught in a spaghetti factory explosion. Work went well and by the spring of 1965 it appeared that we had a method conducive to the spectral examination of the monooxygenase. The excitement level ran high and Ron invited Jim Gillette to join the fun. Jim arrived with Henry Sasame. Dave joined us and soon the lab was full, with Herbert and me preparing microsomes, Henry and Jim making solutions, and Dave and Ron cranking the Yang machine. By 2 or 3 a.m. it was clear—the results were not artifact; substrates did perturb the absorption spectrum of the microsomes, probably by binding to cytochrome P-450.[2]

"I completed these studies over the next year with great enjoyment and we published our findings in the 1967 *Molecular Pharmacology* paper in which we described and named the interaction of cytochrome P-450 with substrates. We suggested that substrates affect the binding of a ligand to cytochrome P-450 and also showed substrate to perturb the EPR spectrum[3] of the cytochrome. Within a few years, it became clear that cytochrome P-450 undergoes substrate-dependent spin state changes.

"Perhaps the reason for the large number of citations to the 1967 paper lies in the ease with which this method measures the binding of compounds to the cytochrome P-450 monooxygenase, and in the recognition that most of the perturbants are also substrates. Many studies have since been reported on the spectral titration of the enzyme with substrates, and many investigators have examined the nature of the spectral changes in an attempt to understand the role of the spin shift in the mechanism of this enzyme action.[4] As we indicated in a recent report,[5] more and more studies will make use of physical/chemical approaches for analysis of the cytochrome P-450 system. To date, such studies have aided greatly in unraveling the mechanism of this enzyme system."

---

1. **Narasimhulu S, Cooper D Y & Rosenthal O.** Spectrophotometric properties of a triton clarified steroid 21-hydroxylase system of adrenocortical microsomes. *Life Sci.* 4:2101-7, 1965. (Cited 95 times.)
2. **Remmer H, Schenkman J, Estabrook R W, Sasame H, Gillette J, Narasimhulu S, Cooper D Y & Rosenthal O.** Drug interaction with hepatic microsomal cytochrome. *Mol. Pharmacol.* 2:187-90, 1966. (Cited 405 times.)
3. **Cammer W, Schenkman J B & Estabrook R W.** EPR measurements of substrate interaction with cytochrome P-450. *Biochem. Biophys. Res. Commun.* 23:264-8, 1966.
4. **Schenkman J B, Sligar S G & Cinti D L.** Substrate interaction with cytochrome P-450. *Pharmacol. Ther.* 2:43-71, 1981.
5. **Schenkman J B & Gibson G G.** Status of the cytochrome P-450 cycle. *Trends Pharmacol. Sci.* 2:150-2, 1981.

# This Week's Citation Classic

**Conney A H.** Pharmacological implications of microsomal enzyme induction. *Pharmacol. Rev.* **19**: 317-66, 1967.

This review discusses characteristics of the enzyme inducers, consequences of enzyme induction for the action of drugs, the presence of multiple monooxygenases in liver microsomes, the selective induction of microsomal monooxygenases, mechanisms of induction of microsomal enzymes, effects of drugs on electron transport systems in liver microsomes, enzyme induction in nonhepatic tissues, stimulatory effects of drugs on the metabolism of steroid hormones and other normal body constituents, enzyme induction in humans, and possible therapeutic applications of enzyme induction. [The *SCI*® indicates that this paper was cited 2019 times in the period 1967-1977.]

Allan H. Conney
Department of Biochemistry and Drug Metabolism
Hoffmann-La Roche Inc.
Nutley, New Jersey 07110

June 9, 1977

Studies by H L Richardson and his associates in 1952 demonstrated that 3-methylcholanthrene—a potent skin carcinogen—inhibited the hepatocarcinogenicity of 3'methyl-4-dimethylaminoazobenzene [1] Additional research by James and Elizabeth Miller and their associates at the University of Wisconsin demonstrated that several polycyclic aromatic hydrocarbons inhibited the carcinogenicity of 3'-methyl-4-dimethylaminoazobenzene and 2-acetylaminofluorene [2] While a graduate student with James and Elizabeth Miller from 1952-1956, I studied the stimulatory effect of polycyclic hydrocarbons on liver microsomal enzymes that metabolize aminoazo dyes to noncarcinogenic products These studies helped explain why polycyclic hydrocarbons inhibited the carcinogenic action of aminoazo dyes We also demonstrated that 3-methylcholanthrene and benzo(a)pyrene induced the synthesis of a liver microsomal enzyme system that hydroxylates benzo(a)pyrene.

"After leaving the University of Wisconsin in 1956, I entered the world of business for several months as a pharmacist in my father's drugstore. I wanted very much to return to research and to explore the possibility that microsomal enzyme induction was a broad phenomenon that could be applied to many drugs, pesticides, chemical carcinogens, and other environmental pollutants. I left the drugstore in 1957 and asked Bernard B. Brodie at the National Institutes of Health if I could study the induction of microsomal drug-metabolizing enzymes in his laboratory. Brodie tried to convince me that I should work with him on other exciting aspects of drug metabolism or on aspects of the biochemistry of neurotransmitters that were already in progress in his laboratory. When he realized that I really wanted to study the possibility of the induction of drug-metabolizing enzymes, he had me work with John J. Burns, who gave me the opportunity and encouragement to pursue studies on microsomal enzyme induction after the completion of my principal asignment, which was the identification of metabolites of two new drugs—zoxazolamine and chlorzoxazone.

"After demonstrating that the duration of action of zoxazolamine was 730 minutes in control rats and only 17 minutes in benzo(a)pyrene-pretreated rats, we found that benzo(a)pyrene and phenobarbital decreased the action of zoxazolamine and many other drugs by stimulating their metabolism. A series of investigations on the pharmacological implications of microsomal enzyme induction then led to my 1967 review article which has become a citation classic. I tried to bring together work from my own laboratory as well as research from other laboratories that pointed out the broad implications of microsomal enzyme induction.

"I believe that my article on the pharmacological implications of microsomal enzyme induction is a citation classic because it touches on many aspects of research in the biomedical sciences and because the article has broad implications for humans who are treated with drugs or who are exposed to environmental pollutants."

1. Richardson H L, Stier A R & Borsos-Nachtnebel E. Liver tumor inhibition and adrenal histologic responses in rats to which 3'-methyl-4-dimethylaminoazobenzene and 20-methylcholanthrene were simulaneously administered. *Cancer Res.* **12**:356-61, 1952.
2. Miller E C, Miller J A, Brown R R & MacDonald J C. On the protective action of certain polycyclic aromatic hydrocarbons against carcinogenesis by aminoazo dyes and 2-acetylaminofluorene. *Cancer Res.* **18**:469-77, 1958.

# This Week's Citation Classic™
CC/NUMBER 50
DECEMBER 10, 1984

Rubin E & Lieber C S. Hepatic microsomal enzymes in man and rat: induction and inhibition by ethanol. *Science* 162:690-1, 1968.
[Dept. Pathology, Mt. Sinai Sch. Medicine, City Univ. New York, and Sect. Liver Disease and Nutrition, Bronx Veterans Administration Hosp., NY]

In rat and in man, chronic alcohol feeding increased the activity of hepatic drug-metabolizing enzymes. *In vitro*, alcohol also inhibited the activities of these detoxifying systems. These observations provided evidence that, in addition to changes in the central nervous system, the resistance of chronic alcoholics to the action of many drugs is, in part, caused by metabolic adaptation in the liver, and that inhibition of drug-metabolizing enzymes in the liver contributes to the increased sensitivity of inebriated persons to sedatives. [The *SCI*® indicates that this paper has been cited in over 265 publications since 1968.]

Emanuel Rubin
Department of Pathology and
Laboratory Medicine
Hahnemann University School of Medicine
Philadelphia, PA 19102

November 13, 1984

"My interest in alcoholism and drug metabolism was first aroused some 14 years before this project was undertaken. When I was an intern at the Boston City Hospital, the indigenous population enjoyed the consumption of spirits to such an extent that we considered almost any patient an alcoholic until proved otherwise. When 'drying out,' these patients required remarkably large doses of sedatives, while during periods of inebriation they were unusually sensitive to such drugs. Some years later, the association between the induction of microsomal drug-metabolizing enzymes in the liver and hypertrophy of the smooth endoplasmic reticulum (SER) was demonstrated, as was the role of cytochrome P450.[1]

"I later reported that in rats and volunteers chronically given alcohol, hypertrophy of the SER occurred.[2,3] It seemed that alcohol might act as an inducer of the mixed-function oxidate system, thus leading to metabolic tolerance of drugs in the clinical situation to which I alluded. Accordingly, we fed rats alcohol for a few weeks and demonstrated in hepatic microsomes that a variety of drug-metabolizing enzymes were increased.

"Since the history of research in alcoholism was encumbered by numerous studies in the rat that had no relevance to man, I felt that these findings would be accepted as conclusive only if the induction of drug-metabolizing enzymes were demonstrated in man. I recruited three nonalcoholic volunteers and fed them alcohol for about two weeks. Homogenates of liver biopsy tissue displayed a two- to threefold increase in the capacity to metabolize pentobarbital *in vitro*, compared with specimens taken before the experiment.

"Since drugs oxidized by the hepatic mixed-function oxidase system, when administered simultaneously, were known to inhibit each other's metabolism, it seemed reasonable that if alcohol acted as a drug, its presence might inhibit the metabolism of other drugs. *In vitro*, alcohol indeed inhibited the microsomal metabolism of the same compounds whose oxidation had been induced in rat microsomes by chronic alcohol ingestion. These studies provided evidence that cross-tolerance to drugs in chronic alcoholics involved not only changes in the central nervous system, but metabolic adaptation in the liver as well. Moreover, apart from the additive depressant effects of alcohol and sedatives on the central nervous system, the inhibition of drug-metabolizing enzymes during acute alcohol consumption contributes to the dangers associated with the combined intake of sedatives and alcohol.

"I surmise that this paper is a *Citation Classic* because it explained well-known clinical phenomena on a molecular basis. The inhibition of drug-metabolizing enzymes was also one of the earliest observations of an effect of alcohol on the liver that is clearly not related to the effects of its metabolism—a concept that has received increased attention in recent years. It also provided strong evidence for the concept that chronic alcohol consumption leads both to injury and to adaptation, and that in some instances, these are closely related. I followed these observations with studies in man and rats showing that alcohol not only affects the activities of drug-metabolizing enzymes but, as predicted, that chronic alcohol ingestion accelerates the disappearance of drugs from the blood and that acute alcohol ingestion inhibits this clearance. These studies were confirmed by others in a variety of clinical and experimental situations,[4] and served as a basis for close attention to regulating the dosage of many types of drugs in alcoholics."

1. Conney A H. Pharmacological implications of microsomal enzyme induction. *Pharmacol. Rev.* 19:317-53, 1967.
   [See also: Conney A H. Citation Classic. *Current Contents/Life Sciences* 22(3):14, 15 January 1979.]
2. Rubin E & Lieber C S. Early fine structural changes in the human liver induced by alcohol. *Gastroenterology* 52:1-13, 1967. (Cited 140 times.)
3. Rubin E, Hutterer F & Lieber C S. Ethanol increases hepatic smooth endoplasmic reticulum and drug-metabolizing enzymes. *Science* 159:1469-70, 1968. (Cited 245 times.)
4. Deitrich R A & Petersen D R. Interaction of ethanol with other drugs. (Majchrowicz E & Noble E P, eds.) *Biochemistry and pharmacology of ethanol.* New York: Plenum Press, 1979. Vol. II. p. 283-302.

# This Week's Citation Classic

CC/NUMBER 40
OCTOBER 4, 1982

Kaplow L S. A histochemical procedure for localizing and evaluating leukocyte alkaline phosphatase activity in smears of blood and marrow.
*Blood* 10:1023-9, 1955.
[Labs. Mary Fletcher Hosp. and Dept. Pathology, Coll. Med., Univ. Vermont and State Agricultural Coll., Burlington, VT]

A cytochemical azo-dye coupling method is described for demonstrating and semiquantitating leukocyte alkaline phosphatase activity (LAPA) in circulating neutrophils. Cells are rated from 0 to 4+ based on subjective assessment of the amount of intracellular precipitated reaction product. Ratings multiplied by designated factors are summated to provide a 'score' for a given sample. [The *SCI®* indicates that this paper has been cited in over 470 publications since 1961.]

Leonard S. Kaplow
Laboratory Service
Veterans Administration Medical Center
West Haven, CT 06516
and
Department of Pathology
and Laboratory Medicine
Yale University School of Medicine
New Haven, CT 06501

May 18, 1982

"In mid-1950, the University of Vermont College of Medicine had an unusual array of established and budding talent in the field of histochemistry. Alex Novikoff was professor of experimental pathology, Bjarne Pierson was chairman, and Roy Korson, now professor, was then assistant professor of pathology. Vittorio Defendi, currently chairman of pathology at New York University, was a young Fulbright fellow recently arrived from Italy. This was prior to my medical training, while I was supervisory technologist at the Laboratories of Mary Fletcher Hospital and simultaneously pursuing a master's degree in pathology.

"Blood cells intrigued me then as now. Their ready availability, representing a biopsy of the circulation, their profound clinical importance, and the aesthetic pleasure in examining a well-stained blood smear explains this not uncommon affection and my interest in leukocyte alkaline phosphatase. During that period of histochemistry, the enzyme had been extensively studied in solid tissues. Paradoxically, blood, so easily obtainable, was greatly neglected. Although, Wachstein[1] in 1946 described a heavy metal method for demonstrating the enzyme in blood cells and reported a marked decrease in LAPA in chronic myelogenous leukemia.

"Stimulated especially by Novikoff and impressed by Dameshek's statement in 1947 that 'morphologic hematology is undoubtedly in for a renaissance...in which it is hoped that appropriate staining technics will point directly to chemical and physiopathologic alterations,'[2] I chose to study alkaline phosphatase in a cat model of leukemia. Despite intensive efforts, neutrophils would not stain. Numerous reports on an azo-dye coupling method suggested that this approach might be superior to the heavy metal method. I determined the optimum fixative and modified the staining method to yield maximum staining of neutrophils on human blood smears and devised a scoring method for assessing overall activity. It worked beautifully and resulted in this *Citation Classic*. Cat cells, however, still would not stain. Only years later did I fully appreciate the bizarre distribution of this enzyme in leukocytes of different species.[3,4]

"There are cogent reasons for the sustained interest in this paper. Methods with clinical applications will always be frequently cited if they are simple, easy to perform, and reliable. The technique described has stood the test of time. The manuscript was written in the early days of hematologic cytochemistry. It sparked an interest in such techniques which has steadily increased in intensity and is presently being actively extended to immunocytochemistry and automated hematology."

1. Wachstein M. Alkaline phosphatase activity in normal and abnormal human blood and bone marrow cells. *J. Lab. Clin. Med.* 31:1-17, 1946.
2. Dameshek W. Preface. (Dameshek W, ed.) *Morphologic hematology.* New York: Grune & Stratton, 1947. p. 1-2.
3. Kaplow L. S. Alkaline phosphatase in peripheral blood lymphocytes. *Arch. Pathol.* 88:69-72, 1969.
4. ––––––. Leukocyte alkaline phosphatase cytochemistry: applications and methods. *Ann. NY Acad. Sci.* 155:911-47, 1968.

Number 12                                                                                   March 21, 1977
# Citation Classics

De Duve C, Pressman B C, Gianetto R, Wattiaux R & Appelmans F. Tissue fractionation studies. 6. Intracellular distribution patterns of enzymes in rat-liver tissue. *Biochemical Journal* 60:604-17, 1955.

The finding that the acid phosphatase of rat liver is enclosed within a special type of cytoplasmic granule, with sedimentation properties intermediate between those of mitochondria and microsomes, has led to the development of a new scheme of fractionation, whereby enzymes attached to these granules can be readily identified. This paper describes the method for determining enzyme distribution patterns. The authors contend that specific enzymic species have single intracellular locations and that granules of a given class are enzymically homogeneous. [The *SCI*® indicates that this paper was cited 1,402 times in the period 1961-1975.]

---

Dr. Christian R. de Duve
Rockefeller University
1230 York Avenue
New York, New York 10021

December 3, 1976

"This paper is part of a series which eventually ran into 18 installments, published between the years 1951 and 1964. It is clearly not a 'classic' in the sense of the Watson and Crick or Jacob and Monod papers. But it is the most important paper of the series, and probably the most significant of my own publications. In this paper we were committing ourselves to a new approach and trusting it to the point of affirming the existence of a new cell particle on the basis of purely biochemical results.

"My collaborators were two postdoctoral fellows and two medical students. Berton Pressman had come from Henry Lardy's laboratory at the University of Wisconsin. He has since made a distinguished career for himself, and is best known for his work on ionophores and mitochondrial ion transport. He is now at the University of Miami. Robert Gianetto had come to us from the Universite de Montreal, where he is now a professor in the department of biochemistry and heads a clinical laboratory. Robert Wattiaux, who became professor of biochemistry in Namur after spending a number of years with us, has recently made important observations on the damage subcellular particles may sustain under the influence of the strong hydrostatic pressures generated during high-speed centrifugation. Francoise Appelmans left the laboratory after she graduated, to go into clinical practice....

"The experiments were largely completed by the spring of 1954. But it took me almost a year to write the paper. I remember mulling over the results and searching for various ways of presenting them, finally hitting upon the relative specific activity vs. protein histograms, now known in some laboratories as 'duvograms,' and in our own as 'submarines.' Altogether 13 distinct enzyme distribution patterns obtained with a new fractionation scheme were presented in this manner, together with a number of additional kinetic and latency results....

"In a classical fractionation experiment, the tissue, usually liver, was homogenized, and fractionated into 'nuclei,' 'mitochondria,' 'microsomes,' and 'supernatant.' The fractions were analyzed and the results reported as reflecting the properties of the corresponding cell components, with some distortion allowed for experimental artifacts. This manner of interpreting the results had two major weaknesses in my opinion: 1) it assumed that the fractions were pure, which they could not be with the crude separation procedures used, and obviously were not; 2) it assumed that the cytoplasm contained only two kinds of particular components, which was certainly not proven and even appeared unlikely. I therefore decided to use other assumptions, equally unproven, but at least more plausible, and, what was more important, capable of being experimentally disproved (or falsified, as Karl Popper would put it). These assumptions were that specific enzymic species have single intracellular locations and that granules of a given class are enzymically homogeneous. Within the limits of validity of these assumptions, which I later called the postulates of single location and of biochemical homogeneity, the enzymes themselves could be used as 'markers'--we said reference--for their host-particles. The 'submarines' provided a convenient immediate global view of the manner in which each enzyme, and therefore its host-particle, was distributed between the fractions (distribution pattern).

"Here, I believe, was the truly innovative aspect of this paper. In practice, of course, it led to the identification of lysosomes as a new group of cytoplasmic particles, and already hinted at the existence of yet another group of particles, now known as peroxisomes. I might add that the word 'lysosome' appears in print for the first time in this paper. This presumably explains why the paper has enjoyed such a high citation frequency."

# Citation Classics

**Barka T & Anderson P J.** Histochemical methods for acid phosphatase using hexazonium pararosanalin as coupler. *J. Histochem. Cytochem.* 10:741-53, 1962.

The authors analyze problems inherent in histochemical demonstration of acid phosphatase activity and describe methods which provide reliable, accurate localizations. [The *SCI*® indicates that this paper was cited 778 times in the period 1962-1976.]

Professor Tibor Barka
Anatomy Department
Mount Sinai School of Medicine
The City University of New York
New York, New York 10029

April 14, 1977

"Our paper is cited frequently, presumably because it describes a useful azo-dye technique for the histochemical localization of acid phosphatase activity. We felt the paper offered more. It reflected our long experience with the classical Gomori method; pointed out its pitfalls and the causes of artifacts, e.g., nuclear staining, obvious in many published reports; described a workable modification of Gomori's technique; and even alluded to some interesting substrate-specificity of phosphatases in mesangial cells of the kidney. Whatever the merits of the paper, the method described apparently fulfilled a need and became accepted for the simple reason: given the proper chemicals it always works.

"The method, and finally the publication, evolved almost spontaneously. The proper ingredients were, however, present: our long-standing interest in histochemical methods, and particularly phosphatases, going back to the late forties; a stimulating environment in close contact with Leonard Ornstein and B.J. Davis, who were in the exciting phase of developing the polyacrylamide gel electrophoresis and discovered that hexazotized p-rosanilin is an excellent coupler in azo-dye methods; and the great surge of interest in acid phosphatases due to the formulation of the lysosome concept by DeDuve and to the insight of Novikoff in identifying the lysosomes in hepatocytes by using electron microscopic and histochemical methods.

"At that time we were also engaged in studying lysosomes using Gomori's technique and were not spared occasional frustrations. We knew a great deal about the workings of Gomori's method and adapted an acceptable variation but, in the search for a more reliable method for acid phosphatases, we tried Davis and Ornstein's new diazonium compound with various naphthol substrates. The extraordinary capability of hexazonium pararosanalin as a capture reagent became apparent when I first attempted acid phosphatase localization using alpha-naphthyl phosphate as the substrate and briefly reported this method in 1960.[1]

"Minor technical problems (diffusion, etc.) prompted us to search for more substantive substrates and the substituted naphthol-AS compounds of Burstone appeared promising. Dr. Anderson synthesized a number of these substrates and reported improved localization of acid phosphatase in the nervous system using naphthol-AS-BI phosphate.[2] After further study, we settled on naphthyl-AS-TR phosphate as the most suitable substrate for this hexazonium coupler in the acid phosphatase reaction.

"Of course, we are gratified by the sustained interest in our paper, but do not fail to appreciate the originality of the contributions of Menten, Young and Green, Burstone and Ornstein and Davis. On a personal level, my memory goes back to many conversations with Paul Anderson, perhaps over a glass of Hungarian apricot brandy, when he reminisced about one of his most remarkable teachers, George Gomori, at the School of Medicine of the University of Chicago, who started all that, after all."

## REFERENCES

1. **Barka T.** A simple aso-dye method for histochemical demonstration of acid phosphatase. *Nature* 187:248-9, 1960.
2. **Anderson P J & Song S K.** Acid phosphatase in the nervous system. *J. Neuropathol. Exp. Neurol.* 21:263-74, 1962.

# This Week's Citation Classic

CC/NUMBER 14
APRIL 4, 1983

Smith R E & Farquhar M G. Lysosome function in the regulation of the secretory process in cells of the anterior pituitary gland. *J. Cell Biol.* 31:319-47, 1966.
[Dept. Pathology, Univ. California Sch. Med., San Francisco, CA]

By electron microscopy and staining for acid phosphatase we investigated the importance of enzyme alterations relevant to the role of the Golgi apparatus and lysosomes in the secretory process of mammotrophic hormone-producing cells (MT) in pituitary glands from lactating rats and in involving MT cells induced by cessation of lactation. [The *SCI*® indicates that this paper has been cited in over 795 publications since 1966.]

---

Robert E. Smith
Lawrence Livermore National Laboratory
University of California
Livermore, CA 94550

January 31, 1983

"My fascination with the anterior pituitary gland first began in undergraduate study at Earlham College and was further kindled while a graduate student at Indiana University by Fernandus Payne. While a medical student there, I was asked to take over, as a guardian of sorts, an RCA-II electron microscope (EM) which provided me an opportunity to undertake EM studies of the mouse anterior pituitary. As I recall, a paramount question was: what are these round structures 500 mµ in diameter in the gonadotrophic cells without cristae, not mitochondria? With a few 'good' micrographs (1959 vintage) and many questions, I wrote to Marilyn Farquhar, which culminated with an invitation for me to work in her laboratory. In February 1962, as a student at the University of California, San Francisco, and working with Farquhar, we were the first to distill the magic fixative, glutaraldehyde, and stained for acid phosphate which resulted not only in lead phosphate deposition in the 500 mµ granules but also around some secretion granules and Golgi cisternae of the various cells. It was evident that there was a critical need for better morphology which precipitated development of the Smith-Farquhar Tissue Sectioner. In the spring of 1962, R. Marion Hicks convinced us that we were staining lysosomes, and a letter from Alex Novikoff expressed elation over our micrograph showing acid phosphatase staining of Golgi cisternae and vesicles in rat anterior pituitary cells.

"But what were lysosomes doing in cells with little phagocytic capability? Based on my light microscopic trichrome studies with Payne, I posed the question: what do pituitary cells do with secretion granules that cannot be released? Stimulation of mammotrophic cells by suckling and abrupt removal of nursing young was normal but produced an exaggerated physiologic condition ideal to study the role of the Golgi apparatus and lysosomes in the secretory process. It had been shown that when suckling young were removed, prolactin levels in the serum of the mother rat dropped quickly; however, assays of the glands revealed more hormone, which disappeared slowly in two to three days. Two years later, the theory of granulophagy (subsequently defined by de Duve as crinophagy) could be considered an alternative to secretion granule release whereby cells disposed of excess or outdated secretory material by an internal mechanism. The mechanism has been observed and studied morphologically, cytochemically, and biochemically in various types of secretory cells, while the cardinal events remain essentially the same; the principal reason I believe the paper has become a *Citation Classic*. However, the paper has also been referenced frequently because of its contribution to understanding the polarity and compartmentalization of the Golgi cisternae and their relation to the formation of Golgi vesicles, lysosomes, and the condensation of secretory material into granules.

"I recognized we were only staining for a marker enzyme, and in reality proteases were the enzymes essentially controlling crinophagy. I later worked with J. Ken McDonald to develop synthetic peptide substrates for anterior pituitary proteases. Then, with a move from Stanford VA Hospital to Eli Lilly and Company, my interest shifted to fluorogenic protease substrates designed to study the conversion of proinsulin to insulin and for the assay of thrombin and plasminogen activator.[1] In 1975, at Lawrence Livermore National Laboratory, I approached the study of proteases in the pituitary by flow cytometry[2] and then diverted my interests in the direction of clinical application of proteases and assisted in the development of the American-Dade Protopath and other clinical systems.[3,4] I am now back into the basics of cell biology, studying glandular kallikrein in the cat submandibular gland and saliva with John R. Garrett,[5] and defining isoenzyme patterns of cathepsin B relevant to limited proteolysis in the cells of the good old rat anterior pituitary gland."

---

1. **Smith R E & Van Frank R M.** The use of amino acid derivatives of 4-methoxy-β-naphthylamine for the assay and subcellular localization of tissue proteinases. (Dingle J T & Dean R T, eds.) *Lysosomes in biology and pathology.* Amsterdam: North-Holland, 1975. Vol. 4. p. 193-249.
2. **Smith R E & Dean P N.** A study of acid phosphatase and dipeptidyl aminopeptidase II in monodispersed anterior pituitary cells using flow cytometry and electron microscopy. *J. Histochem. Cytochem.* 27:1499-504, 1979.
3. **Huseby R M & Smith R E.** Synthetic oligopeptide substrates—their diagnostic application in blood coagulation, fibrinolysis, and other pathological states. (Mammen E F, ed.) *Seminars in thrombosis and hemostasis.* New York: Thieme-Stratton, 1980. p. 173-314.
4. **Pearson K W, Smith R E, Mitchell A R & Bissel E R.** Automated enzyme assays by use of a centrifugal analyzer with fluorescence detection. *Clin. Chem.* 27:256-62, 1981.
5. **Garrett J R, Smith R E, Kidd A, Kylacou K & Grabske R J.** Kallikrein-like activity in salivary glands using a new tripeptide substrate, including preliminary secretory studies and observations on mast cells. *J. Histochemistry* 14:967-79, 1982.

# This Week's Citation Classic

CC/NUMBER 11
MARCH 12, 1984

Koelle G B & Friedenwald J S. A histochemical method for localizing cholinesterase activity. *Proc. Soc. Exp. Biol. Med.* **70**:617-22, 1949.
[Wilmer Ophthalmological Institute, Johns Hopkins University and Hospital, Baltimore, MD]

A histochemical method is presented for localizing cholinesterase activity by incubating tissue sections in a medium containing acetylthiocholine, copper glycinate, and copper thiocholine. Results obtained with several tissues containing specific cholinesterase are described and illustrated. [The *SCI®* indicates that this paper has been cited in over 840 publications since 1955.]

---

George B. Koelle
Department of Pharmacology
School of Medicine
University of Pennsylvania
Philadelphia, PA 19104

January 10, 1984

"Following my discharge from the Army in 1946, I joined Jonas S. Friedenwald as half-time Chalfant Fellow in Ophthalmology (for the specific purpose of elucidating the role of adrenochrome in secretion of the aqueous humor), while attending medical school at Johns Hopkins University. The cited publication was a natural outcome of this collaboration. During the war, I had worked on the anticholinesterase nerve gases; Friedenwald, while primarily an ophthalmologist, was also one of the pioneers in enzymatic histochemistry in addition to his many other accomplishments (he was in fact the only genius I have ever known). At a point when the adrenochrome work appeared to have reached an impasse, I suggested to him that we might develop a histochemical method for the localization of cholinesterases. The copper-thiocholine procedure was the result. It has been cited recently (by an old friend) as 'the first long step towards bringing neurohistology into the twentieth century.'[1] The year following its publication, at the instigation of Julius H. Comroe, I received the Abel prize for a modification which permitted the separate localizations of acetylcholinesterase (AChE) and butyrylcholinesterase (BuChE).[2]

"The method and its several modifications (the most important of which is that of Karnovsky and Roots[3]) have been used extensively in our laboratory and in many others.[4,5] However, for a number of reasons it is not satisfactory for electron microscopy. For this purpose we developed the bis-(thioacetoxy) aurate method, which affords an extremely high level of resolution but is less specific; accordingly, it requires the use of rather elaborate controls for the accurate localizations of AChE and BuChE.[6] Comparison of results obtained by the two methods, by light and electron microscopy, in the normal and denervated superior cervical ganglion of the cat[7,8] led to the conclusion that a neurotrophic factor is essential for the maintenance of AChE and BuChE at postsynaptic sites in noncholinergic neurons.[9] We have recently demonstrated the presence of such a factor in extracts of the central nervous system,[10] and are now involved in its characterization and determination of its mechanism of action.

"Prior to the publication of our original findings, our colleague at the Wilmer Institute, Stephen Kuffler, asked Friedenwald and me if he might show one of our slides at a symposium he was attending in Spain. We readily agreed. Afterward, Steve told us that the participants were impressed by the picture but baffled by the unknown names of its producers. He enlightened them by identifying us as a medical student and an ophthalmologist. Another ophthalmologist, Bernie Becker, predicted, 'You can work on this for the next 20 years!' After nearly twice that long, I am still at it."

1. Garrett J R. Adventures with autonomic nerves. *Proc. Roy. Microscopical Soc.* **17**:242-53, 1982.
2. Koelle G B. The histochemical differentiation of types of cholinesterases and their localizations in tissues of the cat. *J. Pharmacol. Exp. Ther.* **100**:158-79, 1950. (Cited 220 times since 1955.)
3. Karnovsky M J & Roots L. A "direct-coloring" thiocholine method for cholinesterases. *J. Histochem. Cytochem.* **12**:219-21, 1964. (Cited 1,210 times.)
4. Koelle G B. Cytological distributions and physiological functions of cholinesterases. (Koelle G B, ed.) *Cholinesterases and anticholinesterase agents.* Berlin: Springer-Verlag, 1963. p. 187-298.
5. Silver A. *The biology of cholinesterases.* New York: American Elsevier Publishing Co., 1974. 596 p.
6. Koelle G B, Davis R, Smyrl E G & Fine A V. Refinement of the bis-(thioacetoxy) aurate (I) method for the electron microscopic localization of acetylcholinesterase and nonspecific cholinesterase. *J. Histochem. Cytochem.* **22**:252-9, 1974.
7. Davis R & Koelle G B. Electron microscope localization of acetylcholinesterase and butyrylcholinesterase in the superior cervical ganglion of the cat. I. Normal ganglion. *J. Cell Biol.* **78**:785-809, 1978.
8. ——————. Electron microscope localization of acetylcholinesterase and butyrylcholinesterase in the superior cervical ganglion of the cat. II. Preganglionically denervated ganglion. *J. Cell Biol.* **88**:581-90, 1981.
9. Koelle G B. Unsolved questions on the localization of cholinesterases. (Lee C Y, ed.) *Advances in neuropharmacology.* Taipei: Academia Sinica, 1982. p. 37-43.
10. Koelle G B & Ruch G A. Demonstration of a neurotrophic factor for the maintenance of acetylcholinesterase and butyrylcholinesterase in the preganglionically denervated superior cervical ganglion of the cat. *Proc. Nat. Acad. Sci. US—Biol. Sci.* **80**:3106-10, 1983.

# Citation Classics

Number 22      May 30, 1977

Ellman G L, Courtney K D, Andres V & Featherstone R M. A new and rapid colorimetric determination of acetylcholinesterase activity. *Biochemical Pharmacology* 7:88-95, 1961.

A spectrophotometric method for determining acetylcholinesterase activity of tissue extracts, homogenates, cell suspensions, etc., has been described. The activity is measured by following the increase of yellow color produced when the thio anion produced by the enzymatic hydrolysis of the substrate (acetylthiocholine) reacts with DTNB. The method was used to study the activity of human erythrocytes and homogenates of rat brain, kidney, lungs, liver, and muscle tissue. [The *SCI®* indicates that this paper was cited 995 times in the period 1961-1975.]

George Ellman, Ph.D.
University of California
Langley Porter Neuropsychiatric Institute
San Francisco, California 94143

April 15, 1977

"The germ of the idea for using the sulfur analogue of acetyl choline (first used by George Koelle in his famed histochemical procedure) and the sulfhydryl reagent which I invented when I was working at Dow Chemical Company actually occurred many months before I tried it. It was late one Friday afternoon when everyone had gone home and I was alone in the lab that I tried it out. I added the necessary substrate, enzyme, and DTNB to the spectrophotometer while the recorder was going, and, lo and behold, the color began to appear in a fantastically linear way that is the delight of every biochemist's heart. Additions of enzyme and substrate seemed to follow the usual laws of enzymology; all of this I managed to do in about an hour and a half.

"At this point, I went into Robert Featherstone's office with the recordings and said, 'Look here, I can measure cholinesterase.' I was, of course, somewhat dumbfounded to note that his expression was not one of particular interest, but more along the lines of 'Well, that's nice. So, how do you know it is cholinesterase?' He said, 'Is it inhibited by physostigmine?' 'Well I suppose it would be,' I said. 'Have you got any around?' We went to the stock room and got some and went back to the lab and set up the reaction again and put in the physostigmine in the middle of the run and, sure enough, the reaction stopped in about twenty seconds; then Robert Featherstone got all excited. I guess that just proves that the things which make a biochemist happy are a little different than those which make a pharmacologist happy.

"The subsequent work was put together shortly and was sent off to a journal. Some months passed and, much to our chagrin, we received a rejection slip. The referees had many questions which we had not answered in our paper. Then began the serious consideration of whether we should bother with all of those puzzles, when, in fact, we knew it worked.

"Bob Featherstone felt there were things we could do that would make the paper publishable and pressured the people in my lab and myself to go ahead with it. And so we did. Thanks to Diane Courtney and Val Andres, the necessary details of getting a paper published were accomplished. This included such things as studying the activity of various tissues, determining kinetic constants for the red cell enzyme, substrate inhibition, curves, and data on the inhibitory constants from a variety of known cholinesterase inhibitors.

"It is, as Dr. Lowry has pointed out in his Citation Classic commentary of January 3, 1977, not always a great contribution to produce a method, and yet it is, of course, the essence of science upon which all good things are based. So one need not feel too chagrined that a method is produced which is adaptable and has been used by many people. In fact, one of the nicest things is that my technician comes in and says, 'It is still working, just like it was 15 years ago.' It gives one a nice feeling, frankly."

# This Week's Citation Classic

Korn E D. Clearing factor, a heparin-activated lipoprotein lipase. I. Isolation and characterization of the enzyme from normal rat heart.
*J. Biol. Chem.* **215**:1-14, 1955.
[Lab. Cellular Physiol., Natl. Heart Inst., NIH, US Public Health Serv., Bethesda, MD]

This paper showed that the 'clearing factor' that appears in blood plasma following the injection of heparin is a lipoprotein-specific lipase normally present in heart and other tissues where it functions in the transport of fats from the blood into cells. [The *SCI®* indicates that this paper has been cited over 565 times since 1961.]

---

Edward D. Korn
Laboratory of Cell Biology
National Heart, Lung, and Blood Institute
National Institutes of Health
Bethesda, MD 20014

May 27, 1981

"A few months before I was to leave Jack Buchanan's laboratory at the University of Pennsylvania for postdoctoral training with H.A. Barker at Berkeley, it became necessary to make other plans. Arthur Kornberg kindly invited me to join him at Washington University but, wishing to gain experience with problems unrelated to my doctoral research on purine biosynthesis, I declined and thus made my contribution to a Nobel prize. Parenthetically, 16 years later, when Kornberg and I met and shared space in Alec Bangham's laboratory in Babraham, England, he remembered the incident. Paul Berg was to come to his laboratory that year and Kornberg had anticipated a publication coauthored by Korn, Berg, and Kornberg (an aborted *Citation Classic?*).

"Instead, I went to the National Heart Institute. Chris Anfinsen, a graduate student with Buchanan under Baird Hastings at Harvard, had established a group of young investigators to study the structure and metabolism of plasma lipoproteins, then recently implicated by John Gofman and others in the etiology of coronary artery disease.[1] One of the clinical associates was Don Fredrickson, now director of the National Institutes of Health.

"Ten years previously, P.F. Hahn had observed that fat that appears in blood following a meal is removed more rapidly when the anticoagulant heparin is injected intravenously.[2] Moreover, when post-heparin lipemic plasma is removed from the animal, the opalescent plasma continues to 'clear' in the test tube. This phenomenon had aroused considerable interest among clinical investigators and physiologists but was unknown to biochemists, most of whom, like me, had never heard of chylomicrons and low density lipoproteins; we were still eating eggs!

"To an enzymologist, it seemed obvious that, since the turbidity of lipemic plasma was caused by an increased level of triglycerides, the 'clearing factor' must be a lipase that catalyzed their hydrolysis to less turbid molecules. It was a simple matter to demonstrate an increase in fatty acids and glycerol accompanying 'clearing' *in vitro* (glycerol is water-soluble; fatty acids become so by binding to serum albumin). Moreover, since addition of heparin to lipemic plasma *in vitro* had no effect, injected heparin must cause the release of the lipase from tissues. It was routine enzymology to demonstrate that lipoprotein lipase is present normally in heart and other tissues.

"I like to think my first independent publication, one year out of graduate school, became a *Citation Classic* because it explained an important physiological and medical problem in molecular terms, the only way in which such problems can be understood. More likely, the many later workers who discovered most of what is now known about the biochemistry of plasma triglycerides just felt obligated to refer to the paper in which lipoprotein lipase was christened. A review of this field was recently published in the *Annual Review of Biochemistry.*"[3]

1. **Gofman J W.** Biophysical approaches to atherosclerosis. *Advan. Biol. Med. Phys.* **2**:269-80, 1951.
2. **Hahn P F.** Abolishment of alimentary lipemia following injection of heparin. *Science* **98**:19-20, 1943.
3. **Nilsson-Ehle P, Garfinkel A S & Schotz M C.** Lipolytic enzymes and plasma lipoprotein metabolism. *Annu. Rev. Biochem.* **49**:667-93, 1980.

# This Week's Citation Classic

CC/NUMBER 16
APRIL 18, 1983

Rosalki S B. An improved procedure for serum creatine phosphokinase determination. *J. Lab. Clin. Med.* **69**:696-705, 1967.
[St. Mary's Hospital, London, England]

The creatine kinase activity of serum is determined by a procedure in which adenosine triphosphate, liberated by the action of the enzyme on creatine phosphate and adenosine diphosphate, is linked to the reduction of nicotinamide-adenine dinucleotide phosphate with glucose, hexokinase, and glucose-6-phosphate dehydrogenase, and the reaction followed spectrophotometrically at 340 nm. All reagents are combined in a single stable lyophilisate requiring only reconstitution with water. [The *SCI®* indicates that this paper has been cited in over 825 publications since 1967.]

Sidney B. Rosalki
Department of Chemical Pathology
Royal Free Hospital and School of Medicine
London NW3 2QG
England

February 22, 1983

"I had previously described a test for myocardial infarction ('α-hydroxybutyrate dehydrogenase').[1] I found that an American company was marketing the procedure with altered determination conditions so that its diagnostic performance would be impaired, and wrote complaining to the company president. On a visit to the United Kingdom in November 1964, he invited my wife and me to join him for dinner. During the course of the meal, he suggested that instead of complaining about an existing product, I might propose something novel. At that time, creatine kinase methodology was abysmal. The only convenient test[2] frequently gave negative values, and other methods were so prolonged and labour-intensive that few laboratories carried out determinations. On the back of the menu card, I sketched out my idea for modification of the Kornberg adenosine triphosphate (ATP) assay[3] to creatine kinase measurement by the addition of creatine phosphate and adenosine diphosphate (ADP), and the combination of all reagents in a single lyophilisate which would require only aqueous reconstitution and sample addition. In correspondence, I outlined more fully the required reagent composition, which was prepared and presented in individual gelatin capsules.

"Method optimisation and evaluation were carried out under very adverse conditions. At the time, I was a consultant in clinical pathology and responsible for all the clinical chemistry, haematology, microbiology, and even histopathology for two acute children's hospitals—all on a part-time basis! I had received negligible support from my hospital for my research work; I had no research assistant and no suitable apparatus.

"I persuaded the American company to donate to me a simple fixed-wavelength (340 nm) single cuvette spectrophotometer, cost $250. To measure enzyme reaction rates on this instrument, a needle was held at null point by rotating a knob connected to a numbered dial. Each Sunday, my wife would accompany me to the hospital. I would set up the reaction and read out the figures on the dial at minute intervals with a stopwatch. My wife would record the readings. Subsequently, we would calculate and plot the change in absorbance per minute. The reaction had a six-minute lag phase and the linear phase required a further five-minute monitoring period. Each single enzyme determination required 15 minutes instrument time. Each reaction mixture contained ten constituents, each of which had to be individually varied during optimisation studies. The labour involved was considerable, and can scarcely be imagined in these days of multi-sample, microprocessor-controlled automated enzyme analysers.

"Despite all the difficulties, method and clinical studies were completed in 1965. Details were submitted and immediately accepted for publication in 1966, and appeared in 1967.

"This paper has been highly cited because the hitherto complex creatine kinase determination was now so simple, requiring only sample addition to a single pre-prepared substrate mixture, and so sensitive that the procedure was adopted worldwide for creatine kinase determination in clinical biochemistry laboratories. This facilitated wider recognition of the outstanding value of creatine kinase determination in the investigation of heart and muscle disease, and in turn prompted increased use of the method."

1. **Rosalki S B & Wilkinson J H.** Reduction of α-ketobutyrate by human serum. *Nature* **188**:1110-11, 1960.
2. **Tanzer M L & Gilvarg C.** Creatine and creatine kinase measurement. *J. Biol. Chem.* **234**:3201-4, 1959.
3. **Kornberg A.** Reversible enzymatic synthesis of diphosphopyridine nucleotide and inorganic pyrophosphate. *J. Biol. Chem.* **182**:779-93, 1950.

## This Week's Citation Classic

Nebert D W & Gelboin H V. Substrate-inducible microsomal aryl hydroxylase in mammalian cell culture. I. Assay and properties of induced enzyme.
*J. Biol. Chem.* **243**:6242-9, 1968.
[Chemistry Branch, National Cancer Institute, National Institutes of Health, Bethesda, MD]

The aryl hydrocarbon (benzo[a]pyrene) hydroxylase (EC 1.14.14.1) assay is detailed. This rapid and sensitive method offers great promise for studying metabolism of the environmental carcinogen not only in laboratory animal liver, but also in extrahepatic tissues and in cell culture systems. [The *SCI®* indicates that this paper has been cited in over 890 publications since 1968.]

---

Daniel W. Nebert
Laboratory of Developmental Pharmacology
National Institute of Child Health
& Human Development
National Institutes of Health
Bethesda, MD 20205

October 28, 1983

"As a graduate student with Howard S. Mason (University of Oregon Medical School, Portland), I had already developed a fascination with carcinogenesis and drug metabolism.[1,2] When a foreign chemical enters the body, is it broken down by constitutive enzymes or are new enzymes mobilized to challenge this foreign substance? Is the resulting metabolism of carcinogens a cause, or a prevention, of tumorigenesis? These same questions are being asked today, although much more is understood.

"During my postdoctoral fellowship at the National Cancer Institute, NIH, several important events converged. I had gained experience in Mason's lab with the drug-metabolizing enzyme 'cytochrome P-450' (previously also called 'microsomal $Fe_x$'). Polycyclic hydrocarbon 'transformation' experiments in the tissue culture laboratory of Larry J. Alfred and Joseph A. Dipaolo,[3] across the hall from my mentor, Harry V. Gelboin, could best be explained on the basis of induction of polycyclic hydrocarbon metabolism in fetal hamster secondary cultures. This idea was heresy at the time. Further, there was a published spectrophotofluorometric method[4] for detecting the conversion of benzo[a]pyrene to hydroxylated products. It thus seemed ideal for me to develop a rapid and sensitive assay for this enzyme of carcinogen metabolism and to study its induction properties in a well-controlled cell culture system instead of the intact animal.

"Our first two publications[5] were accepted with relative ease and enthusiasm. I believe the first paper has been so highly cited because it described all the enzyme assay conditions. The second paper dealt with the hydroxylase induction kinetics and demonstrated, for the first time, inducible P-450 in cultured cells. I feel the second paper was more exciting than the first. A third paper[6] demonstrated the sensitivity of this induction process to actinomycin D and cycloheximide in cell culture.

"The two 1968 papers represented a 'shot heard 'round the world' in the combined fields of cell culture and pharmacology. Dozens of laboratories began studying, and presently continue to study, the induction of drug-metabolizing enzymes in cell culture. Hundreds of laboratories are characterizing the induction of P-450-mediated enzyme activities in the intact animal. Hardly an issue of any pharmacology or toxicology journal now can be found without the term 'aryl hydrocarbon hydroxylase' or the magic abbreviation 'AHH' appearing somewhere.

"The paper I regard as my major breakthrough, however, is the one which reported our discovery of Mendelian inheritance among inbred strains of mice lacking, or not lacking, the normal AHH induction response by polycyclic hydrocarbons.[7] This key finding led to the thorough characterization of the *Ah* receptor and more recently the cloning of the entire mouse $P_1$-450 gene.[8] Dozens of other labs are now in the process of cloning various rat and rabbit P-450 genes. Not necessarily from the 'most-cited' 1968 paper, but rather from my persistence in this line of work during the past two decades, I have received several recognitions, including Annual Pfizer Lectureship Awards, a scholarship from the Japanese Society for the Promotion of Science, and the US Public Health Service Meritorious Service Medal.

"The next ten years of research in this field should tell us: (i) the evolution of this P-450 superfamily (from *Pseudomonads*, yeast, and plants to the human); (ii) the mechanism of control of P-450 gene expression by one or more unique receptors that bind avidly to foreign chemicals; and (iii) perhaps development of an assay to determine within the human population individuals at increased risk for certain types of environmentally caused cancers and drug toxicities."

---

1. Nebert D W & Mason H S. An electron spin resonance study of neoplasms. *Cancer Res.* **23**:833-40, 1963.
2. ———————. A microsomal difference between normal liver and "minimal deviation" hepatoma 5123 detectable by electron spin resonance. *Biochim. Biophys. Acta* **86**:415-17, 1964.
3. Dipaolo J A & Donovan P J. Properties of Syrian hamster cells transformed in the presence of carcinogenic hydrocarbons. *Exp. Cell Res.* **48**:361-77, 1967. (Cited 115 times.)
4. Wattenberg L W, Leong J L & Strand P J. Benzpyrene hydroxylase activity in the gastrointestinal tract. *Cancer Res.* **22**:1120-5, 1962. (Cited 345 times.)
5. Nebert D W & Gelboin H V. Substrate-inducible microsomal aryl hydroxylase in mammalian cell culture. II. Cellular responses during enzyme induction. *J. Biol. Chem.* **243**:6250-61, 1968. (Cited 195 times.)
6. ———————. The role of ribonucleic acid and protein synthesis in microsomal aryl hydrocarbon hydroxylase induction in cell culture: the independence of transcription and translation. *J. Biol. Chem.* **245**:160-8, 1970. (Cited 80 times.)
7. Gielen J E, Goujon F M & Nebert D W. Genetic regulation of aryl hydrocarbon hydroxylase induction. II. Simple Mendelian expression in mouse tissues *in vivo*. *J. Biol. Chem.* **247**:1125-37, 1972. (Cited 225 times.)
8. Nakamura M, Negishi M, Altieri M, Chen Y T, Ikeda T, Tukey R H & Nebert D W. Structure of the mouse cytochrome $P_1$-450 genomic gene. *Eur. J. Biochem.* **134**:19-25, 1983.

# This Week's Citation Classic

Bruchovsky N & Wilson J D. The conversion of testosterone to 5α-androstan-17β-ol-3-one by rat prostate in vivo and in vitro. J. Biol. Chem. 243:2012-21, 1968.
[Department of Internal Medicine, University of Texas Southwestern Medical School, Dallas, TX]

Nuclear association of testosterone 5α-reductase, intranuclear abundance of dihydrotestosterone, and retention of dihydrotestosterone exclusively by target tissues for testosterone were reported here. On this evidence it was suggested that dihydrotestosterone is the active form of testosterone in peripheral tissues. [The *SCI®* indicates that this paper has been cited over 640 times since 1968.]

---

Nicholas Bruchovsky
Department of Cancer Endocrinology
Cancer Control Agency
of British Columbia
Vancouver V5Z 3J3
Canada

July 7, 1980

"Recalling this past experience is doubly pleasurable since the effort not only brings to mind a stimulating and productive time in the United States, but also affords an occasion to read 56 letters written to my then future wife. My surprise upon her welcome revelation that the letters were still in existence was possibly exceeded by the satisfaction of being able to sharpen my memory concerning the following events.

"After completing MD and PhD programs at the University of Toronto in 1966, I had hoped to continue with residency training in internal medicine at the Southwestern Medical School in Dallas. Since all positions were filled, I accepted an alternative appointment of postdoctoral fellow and non-remunerative resident. The latter assured me of free meals in the Parkland Hospital cafeteria, conceivably an important benefit, in return for modest clinical duties.

"The objective of the research project, started with J.D. Wilson in October, was to determine whether a testosterone binding protein could be isolated from prostatic nuclei. After an abortive attempt to purify nuclear phosphoprotein, I tried a less specific approach. Animals were injected with tritiated testosterone and binding of radioactivity to nuclear components was then successfully shown by the relatively new procedure of gel-exclusion chromatography. The need to ascertain the identity of the nuclear radioactivity was obvious, but unexpectedly little of this material was recoverable in the form of testosterone when analysed by thin-layer chromatography.

"In February and March of 1967 the careful examination of chromatograms in discrete sections revealed that the majority of radioactivity co-migrated with a potent metabolite of testosterone, dihydrotestosterone. At first this observation was viewed with skepticism, and Wilson and I found ourselves debating the merits of different follow-up experiments. On one occasion the discussion became so lively that I thought my postdoctoral position would soon have to be vacated. For a few evenings I attended to my anxiety by the frequent playing of Chopin's Prelude No. 20 for piano, which I discovered by chance to be a good musical sedative.

"Doubts about dihydrotestosterone were dispelled in the next two months as it became clear that the prostate contained enzymes very active in converting testosterone to dihydrotestosterone, and the latter to androstanediol. With help and much work, enough data were available by August for us to consider writing a report. The third draft of a paper was finished by the end of October and distributed locally for comment. On the suggestion of M.D. Siperstein, more emphasis was given in the final version to the possible role of dihydrotestosterone as the active androgen. Only minor changes were necessitated by subsequent editorial review."

"This paper is cited because it was the first to attach biological significance to the formation of dihydrotestosterone within target cells for testosterone."

## This Week's Citation Classic

CC/NUMBER 38
SEPTEMBER 19, 1983

Lieberman J. Heterozygous and homozygous alpha$_1$-antitrypsin deficiency in patients with pulmonary emphysema. N. Engl. J. Med. 281:279-84, 1969.
[Dept. Respiratory Diseases, City of Hope Med. Ctr., Duarte; Dept. Medicine, Veterans Admin. Hosp., Long Beach; and UCLA Sch. Med., Los Angeles, CA]

Measurement of the serum trypsin inhibitory capacity (STIC) in 39 relatives of a homozygous proband for alpha$_1$-antitrypsin ($\alpha_1$AT) deficiency and in 66 patients with pulmonary emphysema at a Veterans Administration hospital revealed that there is an increased prevalence of both homozygous and heterozygous $\alpha_1$AT deficiency in patients with pulmonary emphysema. [The *SCI*® indicates that this paper has been cited in over 225 publications since 1969.]

---

J. Lieberman
Veterans Administration Medical Center
16111 Plummer Street
Sepulveda, CA 91343

June 23, 1983

"The studies reported in this paper were performed while I was a clinical investigator and section chief at the Long Beach, California, Veterans Administration (VA) Hospital. My research had involved studies of blood and tissue proteolytic enzymes, making me especially interested in the reports by Laurell and Eriksson[1,2] of an association between severe alpha$_1$-antitrypsin ($\alpha_1$AT) deficiency and pulmonary emphysema. A phone call informed me of a 50-year-old patient with far advanced pulmonary emphysema and a strong family history of this disease. I rapidly set up an assay to measure the serum trypsin inhibitory capacity (STIC) in this patient, and, as expected, found a severe deficiency of $\alpha_1$AT. To my delight, 39 relatives were also available for a pedigree study revealing three with documented pulmonary emphysema and STIC levels in the *intermediate* deficiency range. At that time, only a severe deficiency of $\alpha_1$AT was thought to predispose to pulmonary emphysema. However, if an intermediate deficiency also predisposed to the development of emphysema, the number of susceptible individuals in the population would increase from 0.04 to five percent. I therefore undertook an investigation of the STIC levels in patients at the Long Beach VA Hospital who were coded as having pulmonary emphysema.

"The manner in which I undertook this study was fortuitous; I obtained names and phone numbers from the charts, then called the patients to come to my laboratory from their homes to provide blood samples. By so doing, I inadvertently avoided acutely ill hospitalized patients with severe infection. $\alpha_1$AT is an acute phase reactant protein whose serum level fluctuates in response to bodily stresses such as acute infection. Had I utilized acutely ill patients, I probably would not have detected the 15.2 percent with intermediate $\alpha_1$AT deficiency.

"Some investigators who initially attempted to confirm my report failed because they studied hospitalized patients in whom a rise of STIC to low normal had occurred, or they studied patients with emphysema in old age homes. We had found that $\alpha_1$AT deficiency is seen mostly in younger patients with emphysema so that a study of patients over 60 years of age would discover few with the deficiency.

"The reasons for this article becoming a *Citation Classic* are: 1) it renewed interest in $\alpha_1$AT deficiency as a significant predisposing factor to pulmonary emphysema rather than a mere medical curiosity; 2) it initiated an ongoing controversy as to whether the heterozygote, intermediate deficiency state of $\alpha_1$AT actually predisposed to pulmonary emphysema. Current work indicates that an *acquired* relative deficiency of $\alpha_1$AT can also develop in heavy cigarette smokers (increases neutrophilic elastase and decreases elastase-inhibitory activity) and contribute to the development of pulmonary emphysema.[3] If so, the lower baseline levels of $\alpha_1$AT found in heterozygotes would make them even more prone than others to develop a protease-inhibitor imbalance, so that any argument regarding whether or not an intermediate deficiency of $\alpha_1$AT may predispose to emphysema in smokers is unwarranted. Thus, I believe that this paper still relays an important and practical message: SMOKING IS BAD FOR YOUR LUNGS, ESPECIALLY IF YOU INHERIT AN *INTERMEDIATE OR SEVERE* DEGREE OF $\alpha_1$AT DEFICIENCY."[4,5]

---

1. Laurell C B & Eriksson S. Electrophoretic $\alpha_1$-globulin pattern of serum $\alpha_1$-antitrypsin deficiency. Scand. J. Clin. Invest. 15:132-40, 1963.
2. ─────. Serum $\alpha_1$-antitrypsin in families with hypo-$\alpha_1$-antitrypsinemia. Clin. Chim. Acta 11:395-8, 1965.
3. Janoff A & Dearing R. Alpha$_1$-proteinase inhibitor is more sensitive to inactivation by cigarette smoke than is leukocyte elastase. Amer. Rev. Resp. Dis. 126:691-4, 1982.
4. Gelb A F, Klein E & Lieberman J. Pulmonary function in nonsmoking subjects with alpha$_1$-antitrypsin deficiency (MZ phenotype). Amer. J. Med. 62:93-8, 1977.
5. Lieberman J, Gaidulis L & Roberts L. Racial distribution of $\alpha_1$-antitrypsin variants among junior high school students. Amer. Rev. Resp. Dis. 114:1194-8, 1976.

# This Week's Citation Classic

CC/NUMBER 48
NOVEMBER 26, 1984

Hamberg M & Samuelsson B. Prostaglandin endoperoxides. Novel transformations of arachidonic acid in human platelets. *Proc. Nat. Acad. Sci. US* 71:3400-4, 1974.
[Department of Chemistry, Karolinska Institutet, Stockholm, Sweden]

Arachidonic acid added to human blood platelets is transformed by two dioxygenase pathways, i.e., one initiated by fatty acid cyclo-oxygenase and another initiated by arachidonic-acid-12-lipoxygenase. The structures of three major end-products are described. [The *SCI®* indicates that this paper has been cited in over 900 publications since 1974.]

---

Mats Hamberg and Bengt Samuelsson
Department of Chemistry
Karolinska Institutet
S-104 01 Stockholm 60
Sweden

September 17, 1984

"In 1972-1973, we isolated and characterized two unstable endoperoxide intermediates in prostaglandin biosynthesis, i.e., prostaglandins $G_2$ and $H_2$.[1,2] The endoperoxides proved to be potent stimulators of vascular and respiratory smooth muscle.[1-3] We were also interested in examining the possible effects of endoperoxides on blood platelets. Platelets have a contractile system and in many respects may be looked upon as a smooth-muscle tissue. At that time, the role of prostaglandins in platelet function was poorly understood. In fact, a paradoxical situation existed since a) prostaglandins had been found to be mainly anti-aggregative agents and b) aspirin and other nonsteroidal anti-inflammatory drugs, inhibitors of prostaglandin biosynthesis,[4] were known to inhibit platelet aggregation and to increase the bleeding time. Thus, if prostaglandins played a role in platelet aggregation, it would be logical to expect them to have a pro-aggregative effect rather than anti-aggregative. We will never forget the day when we prepared a suspension of human platelets, added pure prostaglandin $H_2$, and found that this endoperoxide indeed caused platelet aggregation. The same was true for prostaglandin $G_2$, although this endoperoxide was somewhat more potent. The finding that the cyclo-oxygenase system of platelets has a pro-aggregative role initiated further work on the mechanism of action, analogs, inhibitors, and so on. It seemed important next to investigate the capacity of human platelets to synthesize endoperoxides and to study the further metabolism of endoperoxides in platelets. We chose $^{14}C$-arachidonic acid for this task since this fatty acid is the predominant cyclo-oxygenase substrate present in platelet lipids.

"Three major stable end-products formed from arachidonic acid by two pathways were isolated and characterized. Two compounds, i.e., thromboxane $B_2$ (provisionally called 'PHD') and 12L-hydroxy-5,8,10-heptadecatrienoic acid (12-HHT) were formed by reactions initiated by fatty acid cyclo-oxygenase. As expected, formation of these products was inhibited by aspirin and related drugs. The third compound, 12L-hydroxy-5,8,10,14-eicosatetraenoic acid (12-HETE), was formed by a pathway initiated by arachidonic-acid-12-lipoxygenase. This enzyme was not inhibited by aspirin but by several antioxidant drugs and by 5,8,11,14-eicosatetraynoic acid. An interaction between the lipoxygenase and cyclo-oxygenase pathways in platelets has been found; however, the exact biological role of 12-HETE and its hydroperoxide precursor remains unclear.

"We think the reasons for the high citation rate of this study are a) the discovery of a true lipoxygenase in mammalian tissue, inspiring further studies on transformations of polyunsaturated fatty acids by mammalian lipoxygenases[5] and b) the isolation of thromboxane $B_2$, which subsequently led to the detection of thromboxane $A_2$, an unstable, extremely potent aggregating agent in platelets."[6]

1. Hamberg M & Samuelsson B. Detection and isolation of an endoperoxide intermediate in prostaglandin biosynthesis. *Proc. Nat. Acad. Sci. US* 70:899-903, 1973. (Cited 540 times.)
2. Hamberg M, Svensson J, Wakabayashi T & Samuelsson B. Isolation and structure of two prostaglandin endoperoxides that cause platelet aggregation. *Proc. Nat. Acad. Sci. US* 71:345-9, 1974. (Cited 855 times.)
3. Hamberg M, Hedqvist P, Strandberg K, Svensson J & Samuelsson B. Prostaglandin endoperoxides IV. Effects on smooth muscle. *Life Sci.* 16:451-62, 1975. (Cited 195 times.)
4. Vane J R. Inhibition of prostaglandin synthesis as a mechanism of action for aspirin-like drugs. *Nature New Biol.* 231:232-5, 1971. [See also: Vane J R. Citation Classic. *Current Contents/Life Sciences* 23(42):12, 20 October 1980.]
5. Samuelsson B. Leukotrienes: mediators of immediate hypersensitivity reactions and inflammation. *Science* 220:568-75, 1983.
6. Hamberg M, Svensson J & Samuelsson B. Thromboxanes: a new group of biologically active compounds derived from prostaglandin endoperoxides. *Proc. Nat. Acad. Sci. US* 72:2994-8, 1975. [See also: Hamberg M. Citation Classic. *Current Contents/Life Sciences* 26(2):19, 10 January 1983.]

# This Week's Citation Classic

CC/NUMBER 18
APRIL 30, 1984

Moncada S, Gryglewski R, Bunting S & Vane J R. An enzyme isolated from arteries transforms prostaglandin endoperoxides to an unstable substance that inhibits platelet aggregation. *Nature* **263**:663-5, 1976.
[Dept. Prostaglandin Res., Wellcome Res. Labs., Langley Court, Beckenham, Kent, England]

This work describes the discovery of PGX, a powerful vasodilator and inhibitor of platelet aggregation, generated by vascular tissue. The potential biological importance of this compound is related to the homeostatic regulation of platelet vessel wall interactions. [The *SCI*® indicates that this paper has been cited in over 1,325 publications since 1976.]

Salvador Moncada
Department of Prostaglandin Research
Wellcome Research Laboratories
Langley Court
Beckenham, Kent BR3 3BS
England

March 7, 1984

"The period comprising the summer, autumn, and winter of 1975/1976 was one of the most exciting times in the life of our laboratory in Beckenham. We had been chiefly dedicated to work on inflammation for several of the preceding years, but in July 1976 at the International Congress on Prostaglandins in Florence, Bengt Samuelsson announced the structure of the elusive rabbit aorta contracting substance (RCS) described by Piper and John Vane in 1969.[1] That compound, which was released from lungs and from platelets, was renamed thromboxane $A_2$ and its most important biological activities were its vasoconstrictor and proaggregating actions. We decided to look for the enzyme responsible for the transformation of the prostaglandin endoperoxides into thromboxane $A_2$. This led, in a short period, not only to the isolation and partial characterization of an enzyme which we named 'thromboxane synthetase,' but also to the development of bioassay tissues capable of detecting unstable substances, which were to be so important in the discovery of prostacyclin. In addition to the rabbit aortic strip which had been used in cascade superfusion by Piper and Vane for the discovery of RCS, we started to prepare, with Stuart Bunting, strips of the coeliac and mesenteric arteries of the rabbit, which differentiated between the prostaglandin endoperoxides and thromboxane $A_2$. An important addition to this dynamic bioassay system was the inclusion of studies of platelet aggregation. During that time, we also found the first specific inhibitors of thromboxane synthetase.

"Our intention was to continue this work and, as Vane put it, to 'map the body' for sites of thromboxane $A_2$ synthesis. I was especially interested in reexamining the metabolism of arachidonic acid in the vessel wall. The reason for this was twofold. First, I had recently been impressed by an article by Morrison and Baldini[2] showing that platelets and the vessel wall share some common proteins, and, second, recent results in our laboratory measuring cutaneous bleeding time in the rat had shown very erratic results which needed explanation. My hypothesis was simple, namely, that thromboxane $A_2$ generated by the vessel wall might synergize with thromboxane $A_2$ synthesized by the platelets, for the formation of the haemostatic plug. In autumn 1975, Richard Gryglewski joined us for a six-month sabbatical from Krakow University, and we searched for generation of thromboxane in many tissues, especially the vessel wall.

"Our results were disappointing. We could not find other tissues like the platelets which were able to produce large quantities of bioassayable thromboxane $A_2$. For the vessel wall, too, an enzyme preparation made from pig aortae obtained from a local abattoir failed to generate thromboxane $A_2$. When a second experiment was done, after another disappointing day, we noticed that something unexpected was happening: although thromboxane $A_2$ was not formed, the precursor used in the reaction, the prostaglandin endoperoxide, seemed to disappear. After looking at the results of the experiment, I suggested two further experiments to distinguish between simple inactivation of the endoperoxides and the possibility that a new active substance was being formed which our bioassay tissues did not detect. These experiments showed that indeed we had discovered a new metabolic pathway which produced an unstable substance which relaxed the strips of mesenteric and coeliac arteries. We called this substance PGX.

"The next important step came several weeks later when it occurred to me that since we had been looking for thromboxane $A_2$, which was a vasoconstrictor and aggregator of platelets, and we had found instead a vasodilator, the equation would be completed if the newly found substance also inhibited platelet aggregation. On that day, which was probably the most exciting for me in the whole process of discovery because it gave a clear idea of the biological implications of this compound, we found that PGX was a strong inhibitor of platelet aggregation. Later developments have demonstrated that prostacyclin has not only physiological importance, but also that its absence might underlie some pathological conditions such as atherosclerosis. It is also likely that prostacyclin will provide the basis for the development of potent antithrombotic compounds. The discovery of prostacyclin has increased our understanding of those aspects of the cardiovascular system concerning the interaction between platelets and the vessel wall. For a recent review, see reference 3."

1. **Piper P J & Vane J R.** Release of additional factors in anaphylaxis and its antagonism by anti-inflammatory drugs. *Nature* (London) 223:29-35, 1969. (Cited 640 times.)
2. **Morrison F S & Baldini M G.** Antigenic relationship between blood platelets and vascular endothelium. *Blood* 33:46-57, 1969. (Cited 35 times.)
3. **Moncada S,** ed. Prostacyclin, thromboxane and leukotrienes. (Whole issue.) *Brit. Med. Bull.* 39(3), 1983. 91 p.

# This Week's Citation Classic

Brady R O, Gal A E, Bradley R M, Martensson E, Warshaw A L & Laster L.
Enzymatic defect in Fabry's disease: ceramidetrihexosidase deficiency.
N. Engl. J. Med. 276:1163-7, 1967.
[Lab. Neurochem., Natl. Inst. Neurological Dis. and Blindness, and Sect. Gastroenterol., Metabolic Dis. Branch, Natl. Inst. Arthritis and Metabolic Dis., NIH, Bethesda, MD]

Fabry's disease is an inherited metabolic disorder in which a lipid called ceramidetrihexoside accumulates throughout the body. The condition was shown to be due to a deficiency of the enzyme that catalyzes the hydrolytic cleavage of the terminal molecule of galactose of ceramidetrihexoside. [The *SCI®* indicates that this paper has been cited in over 290 publications since 1967.]

---

Roscoe O. Brady
Developmental and Metabolic
Neurology Branch
National Institute of Neurological
and Communicative Disorders and Stroke
National Institutes of Health
Bethesda, MD 20205

July 5, 1984

"Fabry's disease is transmitted as an X-chromosome-linked recessive disorder. Hemizygous males frequently have a reddish-purple maculopapular rash on their skin. They experience acroparesthesias in the hands and feet that get worse with exercise and hot weather. The clinical picture is further characterized by corneal opacities, tortuosity of retinal vessels, generalized atherosclerosis, propensity to premature myocardial infarction and stroke, and eventual renal shutdown. Heterozygotes usually have much milder manifestations although severe signs may be apparent in occasional individuals.

"Original descriptions of this condition were published by dermatologists W. Anderson[1] and J. Fabry[2] in 1898. Eventually, it became apparent that there was a generalized accumulation of lipid in the tissue of these patients. In 1963, Charles C. Sweeley and Bernard Klionsky reported that ceramidetrihexoside [galactosylgalactosylglucosylceramide, (CTH)] was the major accumulating material in Fabry's disease.[3] Much of this lipid appears to be derived from glycolipids in the stroma of senescent erythrocytes.

"In 1965 and 1966, David Shapiro, Julian N. Kanfer, and I synthesized several sphingolipids labeled with $^{14}C$ with which the metabolic defects in Gaucher's disease and Niemann-Pick disease were established. Based on these findings, I anticipated that Fabry's disease was caused by a deficiency of an enzyme that cleaves the terminal molecule of galactose from CTH.[4] However, in 1966, it was not possible to synthesize CTH chemically with a radioactive tracer in the critical terminal molecule of galactose. Andrew E. Gal joined my group and succeeded in labeling ceramidetrihexoside throughout the molecule by exposing it to radioactive hydrogen gas in a sealed vessel (the Wilzbach procedure). Since the terminal galactose contained radioactive 3H, we were able to trace the fate of this moiety. We discovered that mammalian tissues contain an enzyme that catalyzes the hydrolytic cleavage of this galactose and that intestinal mucosa had the highest activity in this regard.[5] Optimal conditions for measuring the activity of this enzyme were determined and lactosylceramide was identified as the product of the reaction. When this information became available, Andrew L. Warshaw and Leonard Laster obtained biopsy specimens of human small intestinal mucosa from 12 controls, from men with Fabry's disease, and from the mother of one of the patients. Ceramidetrihexosidase activity was readily demonstrated in the specimens from the controls, whereas no activity was detected in the biopsies from the men with Fabry's disease. The activity of the enzyme in the sample from the female carrier was 25 percent of the mean of the controls, a value considerably less than might have been expected for a heterozygote, but compatible with the Lyon hypothesis for X-chromosome inactivation. Furthermore, this level of enzyme activity was consistent with her clinical presentation.

"I believe the paper is frequently cited because Fabry's disease is one of the more common sphingolipid storage disorders. Numerous investigations on the pathogenesis of the clinical manifestations in this condition and approaches to the control[6] and therapy[7] of this disorder have been reported. A review of historical aspects and summary of recent studies on Fabry's disease is available."[8]

---

1. **Anderson W.** A case of angiokeratoma. *Brit. J. Dermatol.* 10:113-17, 1898. (Cited 70 times since 1955.)
2. **Fabry J.** Ein Beitrag zur Kenntnis der Purpura haemorrhagica nodularis (Purpura papulosa haemorrhagica Hebrae). *Arch. Dermatol. Syphilis* 43:187-200, 1898. (Cited 105 times since 1955.)
3. **Sweeley C C & Klionsky B.** Fabry's disease: classification as a sphingolipidosis and partial characterization of a novel glycolipid. *J. Biol. Chem.* 238:PC3148-50, 1963. (Cited 230 times.)
4. **Brady R O.** Sphingolipidoses. *N. Engl. J. Med.* 275:312-18, 1966. (Cited 110 times.)
5. **Brady R O, Gal A E, Bradley R M & Martensson E.** The metabolism of ceramide trihexosides. I. Purification and properties of an enzyme that cleaves the terminal galactose molecule of galactosylgalactosylglucosylceramide. *J. Biol. Chem.* 242:1021-6, 1967. (Cited 75 times.)
6. **Brady R O, Uhlendorf B W & Jacobson C B.** Fabry's disease: antenatal diagnosis. *Science* 172:174-5, 1971. (Cited 90 times.)
7. **Brady R O, Tallman J F, Johnson W G, Gal A E, Leahy W E, Quirk J M & Dekaban A S.** Replacement therapy for inherited enzyme deficiency: use of purified ceramidetrihexosidase in Fabry's disease. *N. Engl. J. Med.* 289:9-14, 1973. (Cited 120 times.)
8. **Brady R O.** Fabry's disease. (Dyck P J, Thomas P K, Lambert E H & Bunge R, eds.) *Peripheral neuropathy.* Philadelphia: Saunders, 1984. p. 1717-27.

# This Week's Citation Classic

Okada S & O'Brien J S. Tay-Sachs disease: generalized absence of a beta-D-N-acetylhexosaminidase component. Science 165:698-700, 1969.
[Dept. Neurosciences, Sch. Medicine, Univ. California, La Jolla, CA]

The primary defect in Tay-Sachs disease is the absence of the lysosomal enzyme, hexosaminidase A. Deficiency of this enzyme leads to the storage of ganglioside $GM_2$ in neurons and resultant cerebral degeneration. Carrier detection and prenatal diagnosis by enzyme assay have led to a large reduction in the incidence of Tay-Sachs disease in North America in the past decade. [The SCI® indicates that this paper has been cited over 475 times since 1969.]

John S. O'Brien
Department of Neurosciences
School of Medicine
University of California
La Jolla, CA 92093

September 11, 1981

"Shintaro Okada and I were pleased to learn that this publication has become a Citation Classic. Okada joined my laboratory in 1967 and we began to work on the enzyme defects in hereditary ganglioside storage diseases. In 1968 we discovered the primary defect in $GM_1$ gangliosidosis, a deficiency of β-galactosidase.[1] We then turned to the problem of Tay-Sachs disease, a $GM_2$ gangliosidosis.

"At that time little was known about enzymes which degraded gangliosides. I strongly suspected a defect of ganglioside degradation in Tay-Sachs disease, a hexosaminidase deficiency being most likely. We began to work on human hexosaminidases using synthetic substrates for assay. I became aware of the work of Don Robinson and John Stirling[2] at Queen Elizabeth's College in London, who demonstrated two different electrophoretic forms of hexosaminidase in human tissues, an acidic form, hexosaminidase A, and a basic form, hexosaminidase B. Okada carried out starch gel electrophoresis of hexosaminidase A and B and found that hex A was absent in Tay-Sachs disease tissues. We next demonstrated the defect in freshly prepared leukocytes from patients and found a partial deficiency of hex A in serum from carriers, confirming the hereditary transmission of the defect.

"We were then at University of California, San Diego, working in temporary quonset huts situated on the cliffs overlooking Black's beach in La Jolla. Okada and I celebrated our discovery with some Old Bushmills. As the sun set on the blue Pacific, we contemplated upon a Japanese and an Irishman cracking a Jewish disease.

"We sent our manuscript to Science and I presented our findings at the Gordon Conference on Lysosomes in June 1969. There I learned that Konrad Sandhoff had found the same deficiency but was not sure it was the primary defect. After hearing our work, he published his findings.[3]

"We then perfected the carrier test using serum[4] and established a reliable prenatal test for Tay-Sachs disease using amniotic cells.[5]

"I believe this article has been widely cited because it led to the first prospective prevention program for a human genetic disease by carrier screening of an at-risk population. Michael Kaback organized and led the first screening program for Tay-Sachs carriers among the Jewish citizens of Baltimore and Washington in 1971. In ten years, more than 312,000 individuals have been screened worldwide in 73 cities from 13 countries, using an improved automated version of our serum test.[4] This has led to the identification of 12,763 carriers and 268 at-risk couples, none of whom had a previous family history of Tay-Sachs disease. Tay-Sachs disease has been diagnosed before birth in 175 of 814 pregnancies; 639 unaffected babies have been born. A recent calculation[6] indicates that the program has resulted in a 65-85 percent reduction of Tay-Sachs disease in North America within the past decade.

"Okada is now associate professor of pediatrics at Osaka University and is still active in research in human biochemical genetics. Both of us are extremely gratified that our work has helped to diminish the suffering caused by this fatal disease."

1. Okada S & O'Brien J S. Generalized gangliosidosis: beta-galactosidase deficiency. Science 160:1002-4, 1968.
2. Robinson D & Stirling J L. N-acetyl-β-glucosaminidases in human spleen. Biochemical J. 107:321-7, 1968.
3. Sandhoff K. Variation of β-N-acetyl-hexosaminidase pattern in Tay-Sachs disease. FEBS Lett. 4:351-4, 1969.
4. O'Brien J S, Okada S, Chen A & Fillerup D L. Tay-Sachs disease: detection of heterozygotes and homozygotes by serum hexosaminidase assay. N. Engl. J. Med. 283:15-20, 1970.
5. O'Brien J S, Okada S, Fillerup D L, Veath M L, Adornato B & Brenner P H. Tay-Sachs disease: prenatal diagnosis. Science 172:61-4, 1971.
6. Kaback M M. Unpublished data. Los Angeles, CA: University of California, 1980.

# This Week's Citation Classic

CC/NUMBER 17
APRIL 27, 1981

McCord J M & Fridovich I. Superoxide dismutase: an enzymic function for erythrocuprein (hemocuprein). *J. Biol. Chem.* 244:6049-55, 1969.
[Dept. Biochemistry, Duke Univ. Medical Center, Durham, NC]

The paper characterizes a new enzymic activity, superoxide dismutase, and ascribes this activity to a copper-containing protein of heretofore unknown function. The purification of the enzyme from bovine erythrocytes is described, as is a 'standard assay' for superoxide dismutase activity. [The *SCI®* indicates that this paper has been cited over 1,200 times since 1969.]

Joe M. McCord
Department of Biochemistry
College of Medicine
University of South Alabama
Mobile, AL 36688

March 24, 1981

"In June 1967, I entered Irwin Fridovich's lab as a graduate student at Duke. For a decade Fridovich and Handler had studied xanthine oxidase, observing that its action on hypoxanthine initiated the free radical chain oxidation of sulfite, suggesting an intermediate in the enzymic reaction was a free radical. The enzyme reduced cytochrome c only in the presence of oxygen or an electron carrier such as methylene blue. Furthermore, preparations of myoglobin or carbonic anhydrase competitively inhibited the aerobic reduction of cytochrome c. The hypothesis was that xanthine oxidase passed electrons to cytochrome c only if $O_2$ were bound at the active site as an electron-conducting 'bridge,' with the free radical $O_2^-$ an enzyme-bound intermediate. This site could be covered by the binding of myoglobin or carbonic anhydrase to xanthine oxidase.

"My project was to demonstrate by physical methods the binding of carbonic anhydrase to xanthine oxidase. Ten months were devoted to ever more elaborate experiments designed to show that which did not exist. As frustration mounted, I began to rethink the problem, finally suspecting that the negative data were telling me something.

"The breakthrough came on April 2, 1968. I was studying the methylene blue mediated reduction of cytochrome c by xanthine oxidase—a process understood to involve reduced methylene blue in free solution. The process showed saturation by cytochrome, but kinetic constants differed. At low cytochrome concentrations the reduced dye autoxidized; at higher cytochrome concentrations saturation occurred. That evening it occurred to me that oxygen could carry electrons in an analogous manner, through free solution. Saturation would be seen because at low cytochrome concentration the superoxide radicals could dismute: $O_2^- + O_2^- + 2H^+ \rightarrow H_2O_2 + O_2$. If this were true, more cytochrome should be required for saturation at higher rates of $O_2^-$ production. Because dismutation is second order in $O_2^-$, the $K_m$ for cytochrome should be a power function of xanthine oxidase concentration! Kinetic experiments the next day confirmed the prediction. Oxygen carried electrons through solution as $O_2^-$, a free radical. How, then, did the inhibitory proteins work? From stoichiometric considerations, they clearly had to remove $O_2^-$ from solution *catalytically*: they were *superoxide dismutases*. It quickly became clear that myoglobin and carbonic anhydrase were not themselves the culprits—both preps contained superoxide dismutase as a minor impurity.

"The first publication,[1] rarely cited, is the important one. The present publication was useful because it defined the new enzymic activity and described its isolation and assay, rendering it available as a highly specific tool for those investigating oxygen metabolism and toxicity. Two international conferences have yielded proceedings of broad scope."[2-4]

1. McCord J M & Fridovich I. The reduction of cytochrome *c* by milk xanthine oxidase. *J. Biol. Chem.* 243:5753-60, 1968.
   [The *SCI®* indicates that this paper has been cited over 180 times since 1968.]
2. Michelson A M, McCord J M & Fridovich I, eds. *Superoxide and superoxide dismutases.*
   New York: Academic Press, 1977. 568 p.
3. Bannister J V & Hill H A O, eds. *Chemical and biochemical aspects of superoxide and superoxide dismutase.*
   New York: Elsevier/North Holland. In press, 1981.
4. Bannister W H & Bannister J V, eds. *Biological and clinical aspects of superoxide and superoxide dismutase.*
   New York: Elsevier/North Holland. In press, 1981.

# Citation Classics

Number 37 — September 11, 1978

Loomis W D & Battaile J. Plant phenolic compounds and the isolation of plant enzymes. *Phytochemistry* 5:423-38, 1966.

Inability to isolate active enzymes from certain plants was found to be due to plant phenolic compounds. These bind to proteins more strongly than biochemists had realized. Polymeric adsorbents, adapted from brewing chemistry, released the plant enzymes from these complexes. [The *SCI®* indicates that this paper was cited 304 times in the period 1966-1977.]

---

W.D. Loomis
Department of Biochemistry and Biophysics
Oregon State University
Corvallis, OR 97331

January 19, 1978

"This paper was born of frustration. We had set out to study the biosynthesis of monoterpenes in plants, and peppermint, which is an important crop in Oregon, looked like an ideal subject. Eventually we realized that the lack of publications in this area was not due to lack of effort, but to lack of results. Labeled precursors weren't incorporated effectively, and the only enzyme we could demonstrate was polyphenoloxidase, i.e., extracts turned brown.

"We tried every conventional technique of enzyme extraction, and several unconventional ones. We could prevent the browning, but we still couldn't detect other enzymes. We think the idea of adding insoluble protein to the homogenates came from a plant pathology paper describing the use of hide powder to adsorb tannins in isolating a virus from cocoa leaves. We didn't have any hide powder, but a neighboring laboratory had some beef heart muscle. We tried it, and it seemed to work. We then tested hide powder, collagen, and cottage cheese. They all gave us colorless extracts, with active enzymes, but we wanted something non-biological so we could be sure that we weren't adding contaminating enzymes. Discussing this with a colleague who does research on hops, I learned that the brewing chemists had recently solved the chill-haze problem in beer by adding synthetic polymers to adsorb plant phenolics. I went to the brewing literature and found that the brewing chemists had already done the basic plant biochemistry for us. Their best phenol adsorbent was insoluble PVP (Polyclar AT), and it was the key to isolating active enzymes from peppermint.

"The solution to our problem had come from unexpected sources. The background literature to explain the results came from even more diverse areas, including insect physiology, leather chemistry, and food technology. More recently we have improved our methods further by using adsorbent polystyrene beads that were originally developed for industrial waste treatment. We have become convinced that scientific advances often result from combining information from seemingly unrelated fields.

"We have also become convinced that there is much useful information and valuable insight in the older scientific literature. The older work should not be forgotten, but should be regarded as a starting point for current research. This is especially true in plant biochemistry, because plants were very popular subjects with the early biochemists. Their popularity waned because those biochemists had the same problem we did: it was hard to get active enzymes. The interest in our paper indicated that we were not the only ones still frustrated by plant enzymes."

# This Week's Citation Classic

Brewbaker J L, Upadhya M D, Mäkinen Y & Macdonald T. Isoenzyme polymorphism in flowering plants. III. Gel electrophoretic methods and applications. *Physiologia Plantarum* 21:930-40, 1968.
[Dept. Horticulture, Univ. Hawaii, Honolulu, HI]

The article summarized inexpensive, do-it-yourself hardware and techniques for separating plant enzymes on horizontal gels. Applications were cited for several widely distributed enzymes of high stability and ease of handling in genetic and other biological studies. [The *SCI®* indicates that this paper has been cited in over 130 publications since 1968.]

James L. Brewbaker
Department of Horticulture
College of Tropical Agriculture
and Human Resources
University of Hawaii
Honolulu, HI 96822

October 18, 1983

"The discovery of gel electrophoresis was like the discovery of microcomputers to my team of plant breeders and geneticists, a looking glass through which we could step magically into a new world. Our research in tropical crop improvement often focused on plants that had not been studied genetically, plants with long life cycles. The ability to identify isoenzymes easily on gels provided a powerful genetic tool for trees, wild plant species, and other living organisms in which genetic study had previously been tedious or impossible.

"Our paper reflected the motivation of the College of Tropical Agriculture to provide techniques for colleagues abroad that could be widely and simply adapted. The coauthors were from India, Finland, and New England, each bringing his sense of minimal cost and maximal utility. We assembled our electrophoresis power sources from do-it-yourself kits, and bought most of our hardware in local grocery stores. We even tried replacing commercial (and expensive) starch for gels with supermarket cornstarch, since much of our work was on corn, but no matter how well purified, it failed! Our students, L. Espiritu, E. Hamill Johnson, J. Scandalios, and a visiting Swedish geneticist, Lars Beckman, contributed importantly to the refinement of these methodologies.

"Among the authors, we had prior experience applying gel electrophoresis in a surprising diversity of plants and plant tissues (reflecting our cosmopolitan backgrounds). We knew that simple techniques were needed for the large-scale screening of hundreds of genetic progenies. Enzymes of relative stability and easily sampled plant tissues were favored in these studies. We promoted isoenzymes as 'tools' for the plant breeder, a view that has been widely verified.

"Looking back, it seems to me we were overly modest in our assessment of the value of a paper such as this on techniques. It was widely cited primarily because it described simple and inexpensive methods adaptable to any research laboratory. So often, contributions of this type are 'put on the back burner' as we press to get out our hot new findings from application of the techniques. We published this as the third in a series of eight papers entitled 'Isoenzyme polymorphism in flowering plants,' and more than 30 papers have followed on uses and roles of plant isoenzymes. Among later papers reviewing this work is that entitled 'Polymorphisms of the major peroxidases of maize.'[1]

"The preparation of a techniques paper is reserved in many instances until authors feel their techniques error-proof. Such is rarely the case, however, as modifications inevitably occur, often involving newly available equipment. Indeed, this paper would not have been published so soon had my conservative professorial view prevailed. Colleagues M.D. Upadhya and Scandalios sensed much more the urgency of publication. As it was, the publication included two errors that have annoyed me for years, simple modifications that should have been refined before going to press.

"It seems in retrospect a classic enigma in my publishing experience, whether to wait until all the loose ends are tied up or publish with the view that other scientists may benefit immediately from your experience. This is a particular challenge for scientists in developing countries who cannot participate in international scientific workshops and annual meetings. In the future, we are challenged to develop an 'editable' publication format, with use of microprocessors to make minor corrections in manuscripts on techniques as the years add refinements. How delightful if computer-updated versions could be available annually as reprints!"

1. Brewbaker J L & Hasegawa Y. Polymorphisms of the major peroxidases of maize. *Isozymes—Curr. Top. Biol. Med. R.* 3:659-73, 1975.

# This Week's Citation Classic

CC/NUMBER 52
DECEMBER 24-31, 1979

Lardy H A, Johnson D & McMurray W C. Antibiotics as tools for metabolic studies. I. A survey of toxic antibiotics in respiratory, phosphorylative and glycolytic systems. *Arch. Biochem. Biophys.* 78:587-97, 1958.
[Institute for Enzyme Res., Univ. Wisconsin, Madison, WI]

Antibiotics toxic to animals were found to have specific inhibitory effects on mitochondrial metabolism—some as inhibitors of respiratory enzymes, others as inhibitors of phosphorylation (oligomycin), some as uncouplers or inhibitors of respiration specific for certain substrates. Subsequent work in many laboratories confirmed the prediction that 'toxic antibiotics might prove to be generally useful tools for investigating metabolic systems.' [The *SCI*® indicates that this paper has been cited over 560 times since 1961.]

Henry A. Lardy
Institute for Enzyme Research
University of Wisconsin
Madison, WI 53706

April 12, 1978

"In the 1950s the field of oxidative phosphorylation was so nebulous and confused that all possible concepts, facilities, and tools were needed to advance knowledge of the subject. When Hotchkiss found that gramicidin blocked phosphate uptake by *Staphylococci*, we had postulated that this antibiotic uncoupled phosphorylation from oxidation—a concept we had first proposed for the metabolic effect of 2,4,-dinitrophenol (DNP).[1,2] From then on, we collected antibiotics from many sources hoping that some would be useful inhibitors of cellular respiration and phosphorylation. The first of these found to be effective was usnic acid[3] which, in micromolar concentrations, uncoupled phosphorylation in washed liver particles. We also discovered that antimycin A blocked the respiratory system of bakers yeast and enhanced aerobic fermentation to the anaerobic rate. This information was passed on to Ahmad and Strong in the next laboratory, who were working intensively on antimycin.

"By 1958 we had tested more than 60 antibiotics and had found about one of 10 to have interesting effects on mitochondrial metabolism and function. When we later confined our screening to antifungals that were non-toxic to anaerobic bacterial growth, one of every three or four new antibiotics was a 'keeper.'

"The 1958 paper reported the inhibition of mitochondrial respiration by oligomycin and its reversal by DNP. It inhibited ATP hydrolysis induced by DNP, $Ca^{2+}$, deoxycholate, and triiodothyroacetic acid. We concluded that oligomycin acts on an enzyme involved in phosphate fixation or in phosphate transfer rather than on enzymes in electron transport. Regardless of which theory of oxidative phosphorylation an investigator espoused, he was soon finding oligomycin useful in his work. We kept no record of the number of requests we received for this antibiotic but I know we disposed of a gross of small vials before we began folding samples into glazed paper and sending them by letter mail.

"Valacidin (identical with Pfizer's streptonigrin) was found to reverse respiratory inhibition by oligomycin or antimycin 'indicating that it acts as an electron carrier from DPNH to cytochromes.'

"The paper also reported for the first time the effects of nigericin and dianemycin on mitochondrial respiration, phosphorylation, and ATP hydrolysis.

"Dianemycin was named for Diane Johnson, the second author of the paper. She is a 'classic' in her own right. After 10 years as my technician Johnson decided to take a doctorate in history of science. Although her undergraduate academic record was brilliant, as a Ph.D. candidate she suffered the disgrace of her only B grade—in a one-credit seminar. (The professor who awarded it has since left this University and I have wondered whether he was encouraged to do so by the assistant dean of his college—Diane Johnson.) She is also chairman of the Drug Quality Council for the State of Wisconsin—their task, to determine whether generic drug products are equivalent to brand name products. Believe it or not, she has just been appointed assistant director of athletics at the University of Wisconsin!

"Bill McMurray was a postdoctorate fellow from the University of Western Ontario, where he is now professor of biochemistry and where he later discovered the effects of valinomycin on mitochondria.

"I have always considered this one of the trivial papers from our laboratory but am pleased that others have found it useful enough to cite it."

1. **Lardy H A & Elvehjem C A.** Biological oxidations and reductions. *Ann. Rev. Biochem.* 14:1-30, 1945.
2. **Lardy H A & Phillips P H.** The effect of thyroxine and dinitrophenol on sperm metabolism. *J. Biol. Chem.* 149:177-82, 1943.
3. **Johnson R B, Feldott G & Lardy H A.** The mode of action of the antibiotic, usnic acid. *Arch. Biochem.* 28:317-23, 1950.

# Citation Classics

**Number 16** — April 17, 1978

Mitchell P. Chemiosmotic coupling in oxidative and photosynthetic phosphorylation. *Biol. Rev. Cambridge Phil Soc.* 41:445-502, 1966.

This paper reviews and discusses the chemiosmotic hypothesis of coupling between oxidoreduction and phosphorylation in oxidative and photosynthetic phosphorylation systems. [The *SCI®* indicates that this paper was cited 605 times in the period 1966-1976.]

---

Peter Mitchell
Glynn Research Laboratories
Bodmin, Cornwall PL 30 4AU
England

November 25, 1977

"The basic concept discussed in this paper was that coupling in oxidative and photosynthetic phosphorylation systems could be achieved by the circulation of protons or their equivalent through the proton-conducting aqueous media on either side of a topologically-closed insulating membrane of low proton conductance. The respiratory chain or photoredox chain was conceived as being a generator of proticity plugged through the membrane, while the ATP synthase was conceived as being a biochemically separate reversible protonmotive ATPase, also plugged through the membrane, that would phosphorylate ADP when reversed by the flow of proticity from the respiratory or photoredox chain. Proticity is the protonic analogue of electricity.

"To facilitate testing, the chemiosmotic hypothesis was based on four main postulates: (1) The ATP synthase is a reversible protonmotive ATPase of characteristic $\rightarrow H^+/P$ stoichiometry; (2) Respiratory and photoredox chains are protonmotive systems of characteristic $\rightarrow H^+/2e^-$ stoichiometry, and appropriate polarity; (3) There are proton-linked solute porter systems for osmotic stabilisation and metabolite transport; (4) Systems 1 to 3 are plugged through a topologically-closed membrane of low permeability to solutes in general and to $H^+$ and $OH^-$ ions in particular. These fundamental postulates were conceived as the basis of a network of functional interrelationships, constituting the rationale of chemiosmotic coupling.

"I suppose that my 1966 review has been frequently cited because it was the first paper in which the chemiosmotic rationale was discussed in depth, and examined stoichiometrically, thermodynamically, kinetically, and mechanistically in the light of available experimental data on the systems described by the four postulates.

"In 1961, the four postulates were almost entirely hypothetical. After sixteen years of experimental scrutiny, they are now generally accepted as experimentally tested facts. This procedure, from detailed conjecture, through rigorous experimental testing, to acceptable rationale or theory, is more characteristic of physics than of biochemistry. That may partly explain the frequency of citation of my 1966 review, which has acted as a source of the chemiosmotic conjecture that numerous experimentalists rightly sought to test to destruction. The survival of this conjecture has led to the general acceptance of the chemiosmotic principles in membrane bioenergetics. My 1966 review has therefore been cited by those who have sought to rationalise and explain their experimental observations in terms of the chemiosmotic theory, and also by those (including me) who have sought to develop the theory further. Proticity is now recognised as a major means of power distribution in biology.

"The fate of the chemiosmotic hypothesis was, and still is, in the lap of the gods of natural phenomena. It is they, not I, who have begun to make such a success of it!"

## This Week's Citation Classic

CC/NUMBER 34
AUGUST 24, 1981

Dalziel K. Initial steady state velocities in the evaluation of enzyme-coenzyme-substrate reaction mechanisms. *Acta Chem. Scand.* 11:1706-23, 1957.
[Dept. Biochemistry, Nobel Medical Inst., Stockholm, Sweden]

Initial rate equations derived for several plausible mechanisms are particular cases of a common equation involving four kinetic coefficients, which can be estimated experimentally. Mechanisms are more easily distinguished by relations between these coefficients than by Haldane relations. Complex mechanisms that account for nonlinear Lineweaver-Burk plots and substrate inhibition and activation are considered. [The *SCI®* indicates that this paper has been cited over 315 times since 1961.]

Keith Dalziel
Department of Biochemistry
University of Oxford
Oxford OX1 3QU
England

June 17, 1981

"I went to the Nobel Institute on a Rockefeller Fellowship in 1955, to work for a year with Hugo Theorell. My visit was remarkably well-timed. Two-substrate enzyme kinetics was in its infancy—and Theorell received the Nobel prize that year.

"I had already spent a few weeks at the Nobel Institute in the previous year, making some magnetic susceptibility measurements with their uniquely sensitive balance (with Anders Ehrenberg) on a supposed haemoglobin-$H_2O_2$ complex I had come across in kinetic studies of oxyhaemoglobin dissociation. I enjoyed that brief visit enormously, and was delighted when K.G. Paul suggested that I should try to go again for a longer period.

"Haemoglobin kinetics and the development of a constant-flow rapid reaction apparatus had been my life till then. That was changed when I got to Stockholm in 1955. Theo suggested that I might spend 'a few weeks' looking at the effects of pH change and NaCl in the kinetics of liver alcohol dehydrogenase, before continuing with my haemoglobin project. I had the advantage of complete ignorance of enzyme kinetics, and of the few good papers on two-substrate kinetics that had appeared at that time.[1] So I could look at the design and interpretation of our experiments without preconceived notions about the fundamental significance of a Michaelis constant, or that two-substrate kinetics are a simple extension of one-substrate kinetics. As it happened, I spent a lot of time, guided by Åke Åkeson, on further purification of alcohol dehydrogenase and separating a minor active component that turned out later to be the sterol isoenzyme. Then I ruined my first kinetic studies by forgetting about the effects of 0.5 M NaCl on the pH of phosphate buffer.

"This paper was written as the basis for interpreting our kinetic data, mostly obtained during a five-month extension of the original year. This enabled me to meet Robert Alberty, whose 1953 paper[2] was the starting point of mine. I suppose that my paper happened to appear at just the right time, and offered a simple, direct method for evaluating kinetic coefficients($\Phi$'s) that have the dimensions of reciprocal rate constants, are independent of enzyme concentration, and describe the initial rate behaviour of two-substrate reactions in the simplest way. Compared with $K_m$'s, the $\Phi$'s for most mechanisms are less complex functions of the rate constants, which simplifies the representation and manipulation of rate equations, especially for three-substrate reactions.[3] They also vary independently of one another with inhibitor concentration and pH. The representation of $k_m/k_{cat}$ for a single substrate reaction as $\Phi$ never caught on, however, despite the fundamental significance of this ratio. I suspect that the frequency of citation of the paper simply reflects the need for authors to refer to some source for their rate equations and their method for estimating parameters—and the benefit of advertisement of the paper in the 1964 edition of Dixon and Webb![4]

"Were there any obstacles to the research or its publication? No. Under Theorell, the Nobel Institute was the ideal place for work and fun."

1. Schwert G W & Hakala M T. Lactic dehydrogenase. I. Kinetics. *Arch. Biochem. Biophys.* 38:55-65, 1952.
2. Alberty R A. The relationship between Michaelis constants, maximum velocities and the equilibrium constant for an enzyme-catalyzed reaction. *J. Amer. Chem. Soc.* 75:1928-32, 1953.
3. Dalziel K. The interpretation of kinetic data for enzyme-catalysed reactions involving three substrates. *Biochemical J.* 114:547-56, 1969.
4. Dixon M & Webb E C. *Enzymes.* London: Longmans Green, 1964. 950 p.

# Citation Classics

Number 28 — July 11, 1977

Cleland, William W. The kinetics of enzyme-catalyzed reactions with two or more substrates or products. I. Nomenclature and rate equations.
*Biochimica et Biophysica Acta* 67:104-37, 1963.

A nomenclature is proposed to facilitate discussion of possible mechanisms for enzyme-catalyzed reactions with more than one substrate or product. A general method for expressing the full steady-state rate equations for these mechanisms in terms of measurable kinetic constants is explained, and the resulting rate equations are given for a number of mechanisms with two or three substrates or products. [The *SCI*® indicates that this paper was cited 819 times in the period 1961-1975.]

William W. Cleland
Department of Biochemistry
University of Wisconsin
Madison, Wisconsin 53706

January 31, 1977

"Enzyme kinetics has clearly come of age when a theoretical paper published in 1963 makes the 'most cited' list. The article is probably cited as much for its definition of nomenclature as for the theory and equations themselves, and it is gratifying that both the nomenclature and equations seem to have stood the test of time and are used as widely as they are today. I am pleased that there is so little of what I said in 1963 that I would change today. We certainly understand the fundamental reasons behind the observed kinetic patterns today, while in 1963 we could only predict them by empirical rules. Many new types of kinetic experiments have been devised, but all of our new knowledge supplements, rather than replaces, what was said then.

"The story of how this work came to be published in *BBA* is amusing in retrospect, although it was traumatic at the time. When the work was in preprint form as three long articles, I sent it to *Biochemistry* early in 1962 and asked whether they would consider publishing such material. Dr. Neurath replied that with their bimonthly publication schedule, they really could not publish material of such length. I then wrote to three other journals, enclosing preprints and asking them whether they'd be interested. 'If so,' I wrote, 'I will put them in the proper format, make some small revisions, and submit them formally.' Both the *Journal of Theoretical Biology* and *Journal of Biological Chemistry* replied favorably to this letter within two weeks, but I did not hear from *BBA*. On July 5, therefore, I formally submitted the papers to the *Journal of Biological Chemistry*. I then received a letter from *BBA* dated July 12, saying, 'I take pleasure to inform you that we have accepted your papers.... Please send the corrections you wrote about in your letter of May 18.' This letter caused real panic, but I decided to play dead and imagine I was on vacation and not receiving my mail. It was not until a month later that I finally received a very detailed 5-page letter from John Edsall expressing interest, but requesting considerable revision and changes in nomenclature. It closed with, 'We recognize that the above rather drastic revisions may be disheartening after your efforts in writing three articles. We have taken this trouble, however, because we believe you have developed a system of substantial value which is well worthy of publication in the *Journal of Biological Chemistry*.' I felt unhappy that Dr. Edsall had spent so much time on the manuscript, but with an ace in the hole I sat down and wrote him that I wished to withdraw the papers, rather than revise them as drastically as he suggested, and I then sent the final corrected copies to *BBA*.

"It is interesting to speculate what might have happened if I had not also sent the preprints originally to *BBA* or they had not misunderstood my letter. Such now familiar terms as 'ping pong' might never have seen the light of day! It is gratifying that Dr. Edsall recognized the value of the theory, but in retrospect I think I was right and he was wrong about the nomenclature. This should perhaps be a lesson to all of us; brash youth is probably a better judge of nomenclature in a new field than established authority!"

# Chapter 6
# Physical Analysis and Instrumentation

This chapter is divisible into two distinct subsections by either of two separate plans. On the basis of time, one group of Citation Classics became such in 1983 or 1984, and a second group achieved Classic status in 1977 and 1978, the first two years that Classics were published in *Current Contents*. The first group of ten Classics relate largely to instrumental methods designed for specific purposes, whereas the second group unites Classics more generally related to physical analysis. Accordingly, in the first group are articles on flow colorimetry, the protein sequenator, and volumetric gas analysis, and the second group has papers on X-ray reflection, nuclear magnetic resonance, mass spectrometry, and other "heavy instrumentations."

Many of these Classics became tremendously popular and as a group probably have a higher citation record than any other. Of the twenty-six articles listed here, eleven have been cited more than 1,000 times each and one has been cited 945 times. Since many of these calculations date back to 1977 and 1978, the current figures would be much larger. Among the high scorers are Bray's scintillation cocktail (9,305 citations), Laskey and Mills' sensitive fluorographic method for detecting $\beta$ emitters on X-ray film (2,110 citations), Stewart et al. on X-ray scattering by bonded hydrogen atoms (1,879), and Edman and Begg's protein sequenator (1,700).

The invention of liquid scintillation counters enabled a far more sensitive means of detecting atomic decompositions than the Geiger counter, but this instrument was not immediately applicable to counting in self-quenching aqueous solutions. At least this was the case until Jeffay read in his Sunday supplement how $CO_2$ was removed from the air in submarines by absorption into ethanolamine. It was a start, and his idea to oxidize $^{14}C$ compounds to produce $^{14}CO_2$ would have had more use but for one thing. Only a few months earlier Bray had published the recipe for his scintillation cocktail, and this soon eclipsed Jeffay's oxidative procedure as the "method of choice." Skeggs' development of the single-channel, automatic colorimeter for the quantitation of urea nitrogen became the forerunner of the twenty-channel machine as it is now produced. When one contemplates the number of these instruments in use worldwide, it is difficult to believe that the original was hard to market. These instruments are criticized for providing more data than a physician ordinarily needs to establish a diagnosis, but the economy of determining the concentration of twenty separate blood components by one automated operation must be balanced against the operator labor and machine cost when only selected determinations are chosen.

Probably very few are aware that the use of discontinuous buffer systems for the electrophoretic separation of proteins was based on the unintended omission of the same buffer from the electrode vessel that was present in the gel (p. 202). How many readers knew that the protein sequenator was invented in Australia (p. 203)? It is interesting too that a pKa guessing game that Good played with a colleague eventually led to an entirely new concept in the construction of organic buffers. Hepes, Tricine, and others of their ilk are the preferred buffers for many systems.

The success that Laskey and Mills had in increasing the fluorographic sensitivity of X-ray for the detection of $^3$H and $^{14}$C by flash preexposure provides an interesting waffle story. First the film manufacturers stated it won't work, then later said it might be worth trying.

The second half of this chapter will be a little heavier reading for biologists who have forgotten their physics. Among these Classics, however, are an equal number of highly cited articles as were present in the first grouping.

## This Week's Citation Classic

CC/NUMBER 10
MARCH 5, 1984

Jeffay H & Alvarez J. Liquid scintillation counting of carbon-14. Use of ethanolamine-ethylene glycol monomethyl ether-toluene.
*Anal. Chem.* 33:612-15, 1961.
[Department of Biochemistry, College of Medicine, University of Illinois, Chicago, IL]

A method is described in which $^{14}C$ in a biological sample is oxidized to $^{14}CO_2$, trapped in ethanolamine, and counted in a liquid scintillation counter as a toluene solution of ethylene glycol monomethyl ether. [The *SCI®* indicates that this paper has been cited in over 450 publications since 1961.]

---

Henry Jeffay
Department of Biological Chemistry
College of Medicine
University of Illinois
Chicago, IL 60612

January 5, 1984

"In the 1950s, we were limited in doing radioactive tracer studies with $^{14}C$ by the problem of the self-absorption of weak $\beta$ emissions in the Geiger-Müller counters. The invention of liquid scintillation counters for $^{14}C$ offered great hope for improving our ability to move quickly and to more accurately compare samples with very different specific activities. In those days, the biochemists followed the advice of the manufacturers and stayed away from water-containing samples in order to reduce quenching and efficiency. Unfortunately, most biochemicals are not directly soluble in solvents suitable for liquid scintillation counting.

"We needed a method to do balance sheet studies on *in vivo* protein catabolism. We wanted to measure the total activity in large and small samples with low or high specific activities. We needed a method that would be linear (directly proportional to the true radioactive concentration) and independent of the amount or volume of unlabeled material in the sample. Other investigators tried preparing gels or suspensions. Those interested in counting proteins or tissues also tried dissolving them in an alkali compatible with toluene or dioxane solutions of scintillators. However, most biological samples had a very limited solubility in Hyamine (quaternary base), typical of the bases used. Oxidation methods were used, trapping the $CO_2$ in alkali. But often the carbonate precipitated out of the scintillation media when present in modest amounts. Also, most organic bases had limited capacity to dissolve large amounts of $CO_2$.

"Our own intuition led us to believe an oxidation would be a universal and useful method if a $CO_2$ trapping agent with a large capacity and solubility in toluene could be found. After many failures we tried using a polar organic solvent to increase the solubility of the base in toluene. For more than a year we tried, but finally gave up.

"One Sunday morning I was reading a feature story about life in a nuclear submarine, and the problems of staying submerged for long periods. They removed the respiratory $CO_2$ with ethanolamine because it was nonvolatile and stable, with a high capacity and instantaneous complete absorption of $CO_2$. (Furthermore, the $CO_2$ could be removed and pumped out of the submarine.) Immediately a light bulb flashed in my mind and I called my lab technician, Julian Alvarez. On Monday we found a sample of ethanolamine, tried it as a $CO_2$ trap with our oxidation system of a $^{14}C$ sample, and counted it with toluene and enough ethylene glycol monomethyl ether to obtain one phase. It worked. Working almost without stopping, both Julian and I obtained all the data we needed to prove the method was perfectly linear over a wide range of radioactive concentrations, and totally independent of the specific activity. The entire project was completed in a few days. Immediately we redesigned our protein catabolism experiments to incorporate this new measurement of $^{14}C$.

"Several months later, I became convinced that we had a universal method for the liquid scintillation counting of $^{14}C$. So I decided to publish it. Of course I was wrong. Bray's solution became popular.[1,2] Oxidation was not convenient for a large number of samples. Now we have other means of doing the job. I would guess that this paper has been highly cited because it was published at the beginning of the era when we changed from Geiger-Müller counting of radioactivity in tracer studies to the use of the now universally used technique of liquid scintillation counting. This change was made possible by the development of methods of the types reported in this paper."

---

1. Bray G A. A simple efficient liquid scintillator for counting aqueous solutions in a liquid scintillation counter. *Anal. Biochem.* 1:279-85, 1960. (Cited 9,425 times since 1960.)
2. ............. Citation Classic. Commentary on *Anal. Biochem.* 1:279-85, 1960. *Current Contents* (2):16, 10 January 1977.

# Citation Classics

Number 2 — January 10, 1977

Bray G A. A simple efficient liquid scintillator for counting aqueous solutions in a liquid scintillation counter. Anal. Biochem. 1:279-285, 1960.

...The author asserts that a "modification of the naphthalene-dioxane system" for "counting radioactive compounds in aqueous solutions using a liquid scintillation counter" has a relatively high efficiency "for both carbon-14 and tritium." The modification of the "naphthalene-dioxane" procedure has "been used to count glycolytic intermediates on paper chromatograms. It has also been used to determine radioactive glucose incorporation into glycogen by counting the acid hydrolyzate of the isolated glycogen."

Dr. George A. Bray
University of California, Los Angeles

"The march of science is paced by the appearance of new ideas. Broadly speaking, these ideas can be divided into two groups: those which provide new methods and those which provide a broad conceptual framework for thought. Among the former are the printing press, the microscope, the stethoscope, X-rays and liquid scintillation counting. In the latter category are the heliocentric theory of the planetary system, the circulation of the blood, the theory of evolution and the microbial theory of disease.

The paper which is described above contributes to one aspect of a major methodologic advance. The use of liquid scintillation techniques for assaying the radioactivity of beta-emitting isotopes has been a significant and important advance in technology. The ability to measure beta-emitting isotopes in aqueous solutions provided one corner of this methodology.

In looking back at the development of the liquid scintillation cocktail cited above, it seems appropriate to note the contribution of two individuals without whose ideas this method would not have been developed. In 1958, liquid scintillation counting was in its infancy. When radioactive compounds containing carbon-14 or tritium were obtained from biological fluids or in aqueous solution, they usually had to be extracted or otherwise treated in order to determine the amount of radioactivity. As a renal physiologist, I was confronted with the problem of counting tritium-labelled compounds in the urine from dogs. I consulted with Dr. Daniel Steinberg about my problem and he suggested the use of anthracene and triton, a method which he had just published.[1] Unfortunately, this technique had an efficiency of only 1% for tritiated compounds. Dr. Steinberg then suggested that I try a mixture of dioxane and napthalene but the samples froze when put in the freezers used to house counting vials. The obvious solution was to add antifreeze. A 20% mixture of ethylene glycol in dioxane and naphthalene worked very well and for over a year the cocktail had this composition (dioxane, ethylene glycol, naphthalene and appropriate scintillation compounds).

It was at this time that Dr. Jean Wilson entered the scene. He too was faced with counting aqueous solutions of radioactive compounds. We shared an office which also housed an analytical balance for his laboratory. One day when he was weighing out 3.0 gm aliquots of anthracene, I suggested he might like to try the cocktail that I had been using for the past year. A few weeks later, Jean asked where I had published this method. Since there was no manuscript, he encouraged my literary efforts. Before publication, however, the efficacy of this 'homemade' cocktail had to be tested to see whether it might not be modified to make it more useful. After trying a variety of antifreeze solutions and other ingredients, the final composition was established and in due course published."[2]

1. Steinberg D. A new approach to radioassay of aqueous solutions in the liquid scintillation spectrometer. **Anal. Biochem.** 1:23-39, 1960.
2. Bray G A. Personal communication, November 15, 1976.

# Citation Classics

Number 36 — September 4, 1978

Skeggs L T Jr. An automatic method for colorimetric analysis.
*Amer. J. Clin. Pathol.* 28:311-22, 1957.

**This paper described the first completely automatic method for colorimetric analysis. It employed a new analytical technique that was performed in a continuously flowing stream. The determination of urea nitrogen in whole blood was described as one application of a generally applicable method. [The *SCI*® indicates that this paper was cited 356 times in the period 1961-1977.]**

---

Leonard T. Skeggs
Veterans Administration Hospital
10701 East Boulevard
Cleveland, OH 44106

January 17, 1978

"The method described in this paper was conceived as a result of a serious need. I started my career as a supervisor of a clinical chemistry laboratory. We were understaffed, with far too much work to do. As a consequence, the quality of the work suffered. It seemed inexcusable to me that some of the results we were reporting might be inaccurate and misleading. I longed for a method that would perform analyses quickly and precisely without the possibility of human error.

"Suddenly one day in the laboratory, it occurred to me that one might do analyses continuously in a flowing stream rather than in batches of test tubes. In the course of the next two years I constructed several models of such a machine in my basement shop at home, using such equipment as I could make or adapt for the purpose. These first machines were successful and gave satisfactory analytical results for blood urea or glucose at the rate of up to 40 samples per hour.

"In general, samples were introduced in succession into a flowing stream of reagents. Air bubbles were introduced into the stream in order to maintain separation between samples. The stream might then be dialyzed, heated or otherwise processed and finally be passed through a flowcell in a colorimeter equipped for recording.

"The equipment that was needed to carry out the method included a turntable for introduction of the samples, a multiple channel peristaltic pump, a dialyzer and a heating bath capable of processing flowing streams, and a recording flowcell colorimeter. Since none of these was available at the time, it seemed necessary for a manufacturer to make them if my method was to be generally useful.

"There followed a long and discouraging period during which I attempted to interest several different manufacturers in my method. Finally, the Technicon company undertook the project. Three years later, they brought my method to the market in the form of the Autoanalyzer. This first automatic single-channel analyzer was very successful and led directly to the development of today's large, multiple-channel, continuous-flow analyzers that can perform 20 different analyses on a blood sample every 20 seconds.

"I had great difficulty in getting my paper accepted by the *American Journal of Clinical Pathology*. When it did appear, it attracted very little attention. However, the situation changed rapidly after the autoanalyzer was introduced. People began to adapt the equipment to their own problems and soon there were publications citing my paper and describing the determination of many substances that could be performed by the continuous flow method of analysis."

# This Week's Citation Classic

CC/NUMBER 25
JUNE 20, 1983

van Holde K E & Baldwin R L. Rapid attainment of sedimentation equilibrium.
*J. Phys. Chem.* **62**:734-43, 1958.
[Depts. Biochemistry, Chemistry, and Dairy and Food Industries,
Univ. Wisconsin, Milwaukee, WI]

Sedimentation of macromolecules can be attained rapidly by the use of short solution columns. The approach to equilibrium can be quantitatively predicted. Experiments with sucrose and ribonuclease demonstrate that very accurate molecular weight values can be obtained by this method. [The *SCI*® indicates that this paper has been cited in over 590 publications since 1961.]

---

K.E. van Holde
Department of Biochemistry and Biophysics
Oregon State University
Corvallis, OR 97331

April 14, 1983

"My graduate studies at the University of Wisconsin were with J.W. Williams, who had, in turn, worked with T. Svedberg. In using the sedimentation equilibrium method to study synthetic polymers, I was impressed with its potential accuracy, but dismayed by the time required—often a week or more—for each experiment. Aside from the inconvenience, this made the method impractical for sensitive biological materials.

"Upon joining the research laboratories of E.I. du Pont de Nemours, I began some studies to seek ways to shorten this time. In 1955, I left industrial research, and returned to Wisconsin as a postdoctoral fellow. Shortly thereafter, I began a theoretical analysis of the problem, which showed that the time to reach equilibrium depended on the *square* of the solution column length. Clearly, using very short columns was the answer, but an experimental demonstration was essential. However, I had no idea of how to do an experiment that would convince biochemists of the method's utility.

"At this point, I had the good luck to renew an acquaintance with R.L. Baldwin, who had just joined the faculty at Wisconsin. We found that we had been working along very similar lines, so we decided to join forces. Baldwin had a new Spinco ultracentrifuge and, more importantly, a thorough understanding of contemporary problems in biochemistry. He suggested that we employ ribonuclease as a test substance, for sequencing of that protein had proceeded to the point where the molecular weight was exactly known. My own knowledge of biochemistry was so meager that I was quite unclear as to the difference between ribonuclease and ribonucleic acid! Baldwin was, in addition, an expert on ultracentrifuge theory, and together we were able to work out new methods for data analysis.

"With Baldwin's expertise, and his superbly tuned centrifuge, the experiments went splendidly. Equilibrium could be attained in hours, rather than days, and the theory for the approach to equilibrium was quantitatively confirmed. The known molecular weights of sucrose and ribonuclease were reproduced almost exactly.

"In a sense, the paper made practicable the application of sedimentation equilibrium to biochemical problems. Extensions of the method soon followed. Especially notable were the use of the interferometric optical system[1,2] and the development by D.A. Yphantis of the technically simpler, yet accurate, 'meniscus depletion method.'[3] Later, the extension of the technique to reversibly associating macromolecules[4,5] opened a whole new area of research. The extensive use of sedimentation equilibrium during the following years doubtless explains the numerous citations. For a recent review, see 'Sedimentation analysis of proteins.'[6]

"In retrospect, it seems to me that the work was possible only because of the standards of excellence that had been set by Williams and his protégés. Perhaps the most important of these to the younger scientists was Louis Gosting, who inspired each of us to do every experiment with uncompromising attention to detail."

---

1. **Richards E G & Schachman H K.** Ultracentrifuge studies with Rayleigh interference optics. I. General applications. *J. Phys. Chem.* **63**:1578-91, 1959.
2. **LaBar F E & Baldwin R L.** A study by interference optics of sedimentation in short columns. *J. Phys. Chem.* **66**:1952-9, 1962.
3. **Yphantis D A.** Equilibrium ultracentrifugation of dilute solutions. *Biochemistry* **3**:297-317, 1964.
4. **Adams E T, Jr. & Fujita H.** Sedimentation equilibrium in reacting systems. (Williams J W, ed.) *Ultracentrifugal analysis in theory and experiment.* New York: Academic Press, 1963. p. 119-29.
5. **Adams E T, Jr. & Williams J W.** Sedimentation equilibrium in reacting systems. II. Extension of the theory to several types of association phenomena. *J. Amer. Chem. Soc.* **86**:3454-61, 1964.
6. **van Holde K E.** Sedimentation analysis of proteins. (Neurath H & Hill R L, eds.) *The proteins.* New York: Academic Press, 1975. Vol. I. p. 225-91.

# This Week's Citation Classic™
CC/NUMBER 15
APRIL 9, 1984

Poulik M D. Starch gel electrophoresis in a discontinuous system of buffers.
*Nature* **180**:1477-9, 1957.
[Dept. Public Health, Univ. Toronto, Canada]

---

The introduction of starch gel electrophoresis in a discontinuous system of buffers was instrumental in the discovery of many polymorphic systems especially of serum proteins and enzymes. The underlying principle (Kohlrausch regulating function) accounts for the superior resolving power of similar electrophoretic methods, e.g., disc polyacrylamide electrophoresis. [The *SCI*® indicates that this paper has been cited in over 1,560 publications since 1957.]

---

Miroslav Dave Poulik
Division of Immunopathology
Department of Clinical Pathology
William Beaumont Hospital
Royal Oak, MI 48072

March 9, 1984

"In 1948, I escaped from my native country, Czechoslovakia, and in 1953, I entered the University of Toronto Medical School to complete my medical education. To support myself, I worked in the department of hygiene and preventive medicine. My work was directed by F.T. Frazer, a prominent immunologist at that time. My project was to enhance the purity of diphtheria toxin that was treated with formalin to produce the diphtheria toxoid for mass immunization.

"I had previously developed a prototypic immunolectrophoresis method[1] that I now applied to the problem of identifying the multitude of chromogens and proteins present in the crude diphtheria toxin. However, this original method was not suited to produce a toxin more pure than that already available. Fortunately, O. Smithies was working in a nearby laboratory and he was developing the methodology for starch gel electrophoresis: the most powerful electrophoretic method known at the time. I adapted Smithies's techniques in my project. However, each experimental separation of the impurities in the toxin required eight hours of electrophoresis. This consumption of time was not conducive to a smooth course in either my medical school work or my marriage, even though my wife, Emily, was working with me as a research technician. Consequently, we both tried to devise a less time-consuming procedure.

"By chance, I had noticed an article describing the use of tris (hydroxymethyl) aminomethane for acidimetric work.[2] I thought it would be interesting to try this compound in electrophoresis, and suggested that Emily prepare starch gel in 0.076 molar tris and adjust the pH to 8.65 with 0.005 molar citric acid. She did so and applied our standard 'dirty' toxin to the slab for separation at about 9 a.m. one day. When I arrived at the laboratory after my morning classes to have lunch with Emily, to my amazement the 'run' was nearly completed. All of the chromogenic bands known to be in the toxin were visibly separated and a strange 'brown-line' was present at the position of the fastest migrating protein. This proved later to be impurities of the buffer compounds stacked at the highest voltage gradient along the gel. The experiment proved to be reproducible and separation markedly improved.

"Further experimentation showed that the reason for the superior resolving power of this new system was the serendipitous omission by Emily of not placing the tris-citrate buffer in the electrode vessels as well as in the gel slab. The new 'discontinuous' system (tris-citrate buffer in the gel and borate buffer in the electrode vessels) worked in accordance with the Kohlrausch regulating function described in 1897 for separation of ionic species.[3] A thorough study of the phenomenon by L. Ornstein[4] explained the 'steady-state stackup' and led to the development of the disc polyacrylamide electrophoresis technique that eventually replaced starch gel electrophoresis in many laboratories. In the years from 1957 to 1964, starch gel electrophoresis in a discontinuous system of buffers became a method of choice, and was used by a great number of investigators and thus led to the discovery of a multitude of polymorphic systems of proteins. See reference 5 for a report of my most recent work."

---

1. **Poulik M D.** Filter paper electrophoresis of purified diphtheria toxoid. *Can. J. Med. Sci.* **30**:417-19, 1952.
2. **Fossum J H, Markunas P C & Riddick J A.** Tris (hydroxymethyl) aminomethane as an acidimetric standard. *Anal. Chem.* **23**:491-3, 1951. (Cited 55 times.)
3. **Kohlrausch F.** Über Konzentrations Verschiebungen durch Electrolyse im Innern von Lösungen und Lösungsgemischen. *Ann. Phys.* **62**:209-11, 1897. (Cited 110 times since 1955.)
4. **Ornstein L.** Disc electrophoresis. I. Background and theory. *Ann. NY Acad. Sci.* **121**:321-49, 1964. (Cited 3,520 times.)
5. **Lilleho J E & Poulik M D.** $\beta_2$-microglobulin and membrane proteins. (Ioachim H L, ed.) *Pathobiology annual 1979.* New York: Raven Press, 1979. Vol. 9. p. 49-80.

# This Week's Citation Classic™
CC/NUMBER 9
FEBRUARY 27, 1984

Edman P & Begg G. A protein sequenator. *Eur. J. Biochem.* **1**:80-91, 1967.
[St. Vincent's School of Medical Research, Melbourne, Australia]

The protein sequenator is an instrument for the automatic determination of amino acid sequences in proteins and peptides. The degradation proceeds at a rate of 15.4 cycles in 24 hours with a yield in the individual cycle in excess of 98 percent requiring approximately 0.25 μmoles of protein. [The *SCI*® indicates that this paper has been cited in over 1,700 publications since 1967.]

Geoffrey Begg
St. Vincent's School of
Medical Research
Fitzroy, Victoria 3065
Australia

November 11, 1983

"In 1958, I joined St. Vincent's School of Medical Research as a junior laboratory assistant to Pehr Edman, who was developing his isothiocyanate method for protein sequencing.

"After observing the sequencing of proteins, I soon realised how repetitive it was. At a morning tea break one day, I suggested that a machine could be made to do this work, but because of my junior status, and the fact that the Edman degradation was a highly skilled technique, my comments met much derision from the other staff. About a week later, Edman (who, unknown to me, was also considering such a machine) took up the suggestion and after a day of intense discussion we had a clear idea of how a protein sequencing machine would work. We started with a simple glass chromatographic column, tilted, so that the protein would not simply fall out. Only a small part of the lower inclined surface would support the protein which certainly could be washed out. We realised that if the tilted column were rotated about its axis more of the inside surface would be used, but the speed of rotation would have to be high so as not to form a simple spiral. At high speeds the protein would be held on the wall by centrifugal force but liquids entering at the top would not just pass down and out the tip but would rise up again when they reached the bottom. A vertical rotating column with a base would allow extracting liquids introduced to the bottom to rise up over protein on the wall and be scooped off at the top. Placed into a chamber that could be evacuated, we had the 'machine.'

"We first experimented with an old electric motor and a simple blown glass cup attached with sealing wax to the shaft. After a few hours we had the machine running and were satisfied that our initial ideas were feasible in theory and practice. Within 12 months (after building almost every piece three times), we had a prototype machine sequencing 34 amino acids in 53 hours.[1] Continually referring to the 'machine' soon became tiring so we affectionately named it 'Matilda' (Waltzing Matilda), but when it came time to publish, we renamed it 'sequenator.'

"We knew the sequenator had a much greater potential so we delayed publication of the complete paper for another three years until the performance was improved.

"The paper is highly cited because, until the invention of the sequenator, no single technique for sequencing was universally accepted. The sequenator immediately proved the isothiocyanate method as a procedure *par excellence* for sequencing proteins and peptides. We did not seek to patent the sequenator and the technology is freely available to all.

"Since publication, two significant developments[2,3] have improved the performance and the introduction of the gas-phase sequenator[4] could further advance protein sequencing. For a good overall review, see *Protein Sequence Determination*."[5]

1. **Edman P.** Determination of amino acid sequences in proteins. *Thromb. Diath. Haemorrhag.* (Suppl. 13):17-20, 1964.
2. **Zimmerman C L, Appella E & Pisano J J.** Rapid analysis of amino acid phenylthiohydantoins by high-performance liquid chromatography. *Anal. Biochem.* **77**:569-73, 1977. (Cited 290 times.)
3. **Tarr G E, Beecher J F, Bell M & McKean D J.** Polyquarternary amines prevent peptide loss from sequenators. *Anal. Biochem.* **84**:622-7, 1978. (Cited 240 times.)
4. **Hewick R M, Hunkapiller M W, Hood L E & Dreyer W J.** A gas-liquid solid phase peptide and protein sequenator. *J. Biol. Chem.* **256**:7990-7, 1981.
5. **Edman P & Henschen A.** Sequence determination. (Needleman S B, ed.) *Protein sequence determination: a sourcebook of methods and techniques.* New York: Springer-Verlag, 1975. p. 211-79. (Cited 330 times.)

## This Week's Citation Classic™
CC/NUMBER 49
DECEMBER 3, 1984

Scholander P F. Analyzer for accurate estimation of respiratory gases in one-half cubic centimeter samples. *J. Biol. Chem.* **167**:235-50, 1947.
[Edward Martin Biological Laboratory, Swarthmore College, PA]

A volumetric gas analyzer is described that will determine carbon dioxide, oxygen, and nitrogen in samples of ≤ 0.5 cc with an accuracy of ±0.015 volume percent. It directly handles samples containing from 0 to over 99 percent absorbable gases. [The *SCI*® indicates that this paper has been cited in over 1,130 publications since 1955.]

---

Mrs. Per F. Scholander
8374 Paseo del Ocaso
La Jolla, CA 92037

June 28, 1984

"My husband, Per Scholander, developed this technique while a research associate in physiology at Swarthmore College. He was very surprised that so many requests were made for this paper; apparently the method was of some use. The reason for developing the technique was that the standard, an excellent Haldane apparatus with similar accuracy, needed a large sample and could not analyze high oxygen contents. His comparative studies called for small samples and very high oxygen content. The main trick in this minianalyzer was the use of a simple volumetric micrometer burette, based on displacement of mercury from a standard micrometer plunger enclosed in a glass tube.

"The principle of the ½ cc analyzer is as follows: a gas sample is introduced into a reaction chamber connected to a micrometer burette and is balanced by means of an indicator drop in a capillary against a compensating chamber. Absorbing fluids for carbon dioxide can be tilted into the reaction chamber without causing any change in the total liquid content of the system. During absorption, mercury is delivered into the vibrating reaction chamber from the micrometer burette, while maintaining the balance of the sample against the compensating chamber. Volumes are read in terms of micrometer divisions. Rinsing fluid and absorbents are accurately adjusted to have the same vapor tension.

"The analyzer permits the determination of $CO_2$, $O_2$, and $N_2$ in samples of respiratory gases ≤ 0.5 cc with an accuracy of ±0.015 percent. It will directly handle samples containing up to 99 percent absorbable gases. An analysis requires from 6 to 10 minutes.

"An essential trick is to isolate the surface tension in the reaction chamber from that in the readout capillary. This requires application of a ring of a hydrophobic coating (rosin) about once in every 150-200 analyses. Without that coating, the accuracy is gone. Per had seen this important point disregarded so many times that whenever he saw one of these analyzers in a lab he closed his eyes and hurried past it.

"His lab has used this analyzer on expeditions from the Tropics to the Arctic, in tents and on ships. It was even taken on a Mt. Everest expedition. Per had not used it for years, however, and a number of electronic gadgets have now taken over. The original technique continues its usefulness at least as a standard for calibration of commercially available analyzers."[1,2]

[Per F. Scholander, emeritus professor of physiology, Scripps Institution of Oceanography, died in June 1980.]

---

1. **Davis R W.** Lactate and glucose metabolism in the resting and diving harbor seal (*Phoca vitulina*). *J. Comp. Physiol.* **153**:275-88, 1983.
2. **Douglas N J, White D P, Pickett C K, Weil J V & Zwillich C W.** Respiration during sleep in normal man. *Thorax* **37**:840-4, 1982.

## This Week's Citation Classic

CC/NUMBER 40
OCTOBER 3, 1983

Good N E, Winget G D, Winter W, Connolly T N, Izawa S & Singh R M M.
Hydrogen ion buffers for biological research. *Biochemistry—USA* 5:467-77, 1966.
[Department of Botany and Plant Pathology, Michigan State University, East Lansing, MI]

To the dismay of experimental biologists who did not want to spend time reinvestigating phenomena already worked over, 12 radically different hydrogen ion buffers were introduced. These were consciously designed to have desirable physical, chemical, and biological properties. They were also designed to have buffering (pKa's) in the biologically important range of 6.0 to 8.5. [The *SCI*® indicates that this paper has been cited in over 1,025 publications since 1966.]

Norman E. Good
Department of Botany
and Plant Pathology
Michigan State University
East Lansing, MI 48824

July 5, 1983

"The idea of tailoring molecules for specific buffering purposes came from a game I used to play with one of my colleagues, Richard O'Brien, at the University of Western Ontario. He was interested in the effects of ionization on the activity of nerve poisons and related drugs. We developed between us a game of guessing pKa's on the basis of the oftentimes rather complex structures of the drugs he was studying. Meanwhile I had been investigating electron transport and ATP synthesis in chloroplast lamellar preparations, and I had discovered that various anions commonly used in buffers could uncouple electron transport from phosphorylation. How then to avoid the anion uncoupling and still have essential control of hydrogen ion concentration? Since the inner salt glycine had absolutely no uncoupling effect, it occurred to me that I might modify glycine in such a manner that its new pKa would bring it into an appropriate buffering range (glycine itself buffers at far too high a pH for chloroplast research). It seemed obvious that replacing the amino group of glycine with tris(hydroxymethyl)-aminomethane (Tris) would produce an amino acid with a pKa rather similar to that of Tris itself. Thus, Tricine was born.

"Subsequently, it also occurred to me that I could use sulfonic acids instead of carboxylic acids and thereby avoid much of the binding of polyvalent metal ions. Consequently, when I moved to Michigan State University, I ordered all of the simple primary and secondary amines commercially available and started making the appropriate N-substituted taurines and glycines, drawing on the intuition developed in the above-mentioned pKa guessing game. Hence the origin of most of the other buffers listed in the paper.

"Almost all experimental biologists and not a few analytical chemists were interested in the new buffers and consequently the paper received instant attention. The fact that the buffers often proved superior to anything before available made their wide use inevitable. It is a measure of their acceptance that the biological and chemical journals now allow the use of the trivial names of several—e.g., Hepes, Tricine—without definition and without reference.

"In a desultory way, I have continued to make and introduce more buffers along the same lines and I have occasionally enlisted the assistance of real chemists. A series of N-substituted 3-aminopropanesulfonic acids was prepared using propane sultone and more recently, W.J. Ferguson and I[1] have introduced a series of N-substituted 3-amino-2-hydroxypropanesulfonic acids, all excellent buffers.

"A summary of available biological buffers was compiled in 1972 by me and my colleague, S. Izawa."[2]

1. Ferguson W J, Braunschweiger K I, Braunschweiger W R, Smith J R, McCormick J J, Wasmann C C, Jarvis N P, Bell D H & Good N E. Hydrogen ion buffers for biological research. *Anal. Biochem.* 104:300-10, 1980.
2. Good N E & Izawa S. Hydrogen ion buffers. *Meth. Enzymology* 24:53-68, 1972.

# This Week's Citation Classic

CC/NUMBER 13
MARCH 28, 1983

Laskey R A & Mills A D. Quantitative film detection of $^3$H and $^{14}$C in polyacrylamide gels by fluorography. *Eur. J. Biochem.* 56:335-41, 1975.
[Medical Research Council Lab. Molecular Biology, Univ. Postgrad. Med. Sch., Cambridge, England]

Pre-exposure of X-ray film to a flash of light was found to increase fluorographic sensitivity of X-ray film, improving isotope detection efficiency. It also allowed quantitative interpretation of film images. These findings subsequently allowed us to apply intensifying screens to detection of $^{32}$P and $^{125}$I.[1] [The *SCI®* indicates that this paper has been cited in over 2,110 publications since 1975.]

---

R.A. Laskey
Laboratory of Molecular Biology
Medical Research Council Centre
University Medical School
Cambridge CB2 2QH
England

December 21, 1982

"In 1974, William Bonner and I showed that tritium could be detected in polyacrylamide gels using a solution of the scintillator PPO in dimethyl sulphoxide.[2] The scintillator converts β particles from the sample to blue light which is recorded by X-ray film at -70°C. In order to use this method quantitatively, Tony Mills and I tested the relationships between image absorbance and amount of radioactivity or exposure time. Neither was linear; small amounts of radioactivity were seriously underrepresented. While preparing a statement to warn other users of the method, we sought a procedure for overcoming the nonlinearity. An anxious reading of photographic physics immediately identified the problem as 'low intensity reciprocity failure,' arising from subdivision of each β particle to multiple smaller quanta of light. The single silver atom produced by a single photon is unstable and reverts in approximately 1 s at ambient temperature. However, once two silver atoms are formed in a grain they remain stable and all subsequent photons have an equally high chance of contributing to the image. This explained why exposure at -70°C is necessary, because it would increase the half-life of the first atom, increasing the chance of forming a stable pair. In addition, it should then be possible to bypass the troublesome reversible phase of image formation by pre-exposing the film to an instantaneous flash of light so that each silver halide grain acquires a stable pair of silver atoms and can then accumulate further atoms linearly in response to each new photon from the sample.

"We consulted a major film manufacturer who assured us that this would not work. Nevertheless, we tried pre-exposure and found that it did overcome the nonlinearity of the film response and thereby it greatly increased sensitivity for small amounts of radioactivity. Two weeks later, the film manufacturer phoned to say that it might be worth trying. We assured them it was. Ironically, we found later that a similar technique of pre-exposure had been used to sensitize film in the cinema industry and astronomy for many years.

"We also considered another possible method of pre-exposure, namely, exposing films to tritiated luminous paint at -70°C. However, the small container of paint we received contained 0.25 curies suggesting that this might never become a highly cited method. The reasons why our published procedure has been cited frequently appear to be that it provides a simple method for extending the limits of sensitivity of radioisotope detection and it allows accurate quantitation of the isotope distribution.

"A by-product of this work was the method of enhancing detection of $^{32}$P and $^{125}$I using intensifying screens at -70°C since it was clear that low temperature and pre-exposure should also allow the blue light produced by intensifying screens to be recorded efficiently at low intensities.[1]

"Mills has since left his technician post to become a schoolteacher."

1. Laskey R A & Mills A D. Enhanced autoradiographic detection of $^{32}$P and $^{125}$I using intensifying screens and hypersensitized film. *FEBS Lett.* 82:314-16, 1977.
2. Bonner W M & Laskey R A. A film detection method for tritium-labelled proteins and nucleic acids in polyacrylamide gels. *Eur. J. Biochem.* 46:83-8, 1974.
   [Citation Classic. *Current Contents/Life Sciences* 26(1):16, 3 January 1983.]

# This Week's Citation Classic

NUMBER 44
OCTOBER 29, 1979

Björndal H, Lindberg B & Svensson S. Mass spectrometry of partially methylated alditol acetates. *Carbohyd. Res.* 5:433-40, 1967.
[Inst. Organic Chem., Stockholm Univ., Stockholm, Sweden]

The mass spectrometry of partially methylated alditol acetates was investigated and it was demonstrated that these substances are readily identified from their mass spectra. Analysis of mixtures of such derivatives by gas chromatography-mass spectrometry has become a standard method in structural studies of complex carbohydrates. [The *SCI®* indicates that this paper has been cited over 205 times since 1967.]

---

Bengt Lindberg
Department of Organic Chemistry
Arrhenius Laboratory
University of Stockholm
S-106 91 Stockholm
Sweden

July 23, 1979

"Methylation analysis, introduced by Haworth and coworkers in the thirties, is the most important method in structural studies of complex carbohydrates such as polysaccharides, glycoproteins, and glycolipids. It involves methylation of all free hydroxyl groups, followed by acid hydrolysis. During this hydrolysis the glycosidic linkages are cleaved but the methyl ether linkages are stable. The product is a mixture of partially methylated sugars, and the free hydroxyls in these mark the positions to which the sugars are linked in the starting material. A qualitative and quantitative analysis of this mixture therefore gives considerable structural information. Methylation analysis was, however, time-consuming as each partially methylated sugar in the generally complex mixture had to be isolated and characterized, and this also required fairly large amounts of material. Furthermore, the method was not very accurate, and minor components, which might be structurally significant, were often overlooked.

"We were studying structures of biologically important polysaccharides from bacteria. It was, however, quite clear that new and better methods had to be developed if we were ever to cover even a small part of this vast field. Gas chromatography would be the best method for separation and quantitative analysis of the mixture of methylated sugars obtained on methylation analysis. However, these would first have to be transformed into stable, volatile materials. The alditol acetates were thought suitable, and we had found that they were readily separated by gas chromatography on some newly developed column materials.

"The components had, of course, also to be identified. With the aid of combined gas chromatography-mass spectrometry, for which instruments had recently become commercially available, it should have been possible to solve this problem. The only instrument at the university belonged to the department of analytical chemistry and it was already extensively used. My research students, Bjorndal and Svensson, however, did not mind working after hours, and the mass spectrometry was done at night, when we could use this instrument.

"The mass spectra of the partially methylated alditols were simple and it was easy to identify these substances. We had therefore reached our object, a fast and accurate method for methylation analysis, which required only small amounts of material. This method, in conjunction with a methylation technique developed by Hakomori, has been used in numerous structural studies of complex carbohydrates, and almost totally replaced other methods.[1] With the current interest in glycoproteins and glycolipids on the surfaces of cells, it is necessary to go from the milligram scale, which we use, to the nanogram scale, and there is consequently a need for further improvements."

1. Hakomori S. A rapid permethylation of glycolipid and polysaccharide catalysed by methylsulfinyl carbanion. *J. Biochem.* 55:205-8, 1964.

# Citation Classics

**Bloembergen N, Purcell E M & Pound R V. Relaxation effects in nuclear magnetic resonance absorption.** *Physical Review* **73**:679-712, 1948.

The exchange of energy between a system of nuclear spins immersed in a strong magnetic field, and the heat reservoir consisting of the other degrees of freedom (the lattice) of the substance containing the magnetic nuclei, serves to bring the spin system into equilibrium at a finite temperature. In this condition the system can absorb energy from an applied radiofrequency field. With the absorption of energy, however, the spin temperature tends to rise and the rate of absorption to decrease. Through this saturation effect, the spin-lattice relaxation time can be measured. [The *SCI*® indicates that this paper was cited 1,145 times in the period 1961-1975.]

---

Professor N. Bloembergen
Division of Engineering & Applied Physics
Harvard University
Cambridge, Massachusetts 02138

January 20, 1977

"When I arrived early in 1946 from the war-torn Netherlands as a graduate student at Harvard University with the objective of doing some kind of research leading to a Ph.D. thesis in physics, it was less than two months after E.M. Purcell, R.V. Pound and H.C. Torrey had demonstrated nuclear magnetic resonance in condensed matter experimentally. This work was stimulated by discussions of a group of physicists active in World War II radar development at the M.I.T. Radiation Laboratory. F. Bloch and W.W. Hansen, who were involved with radar work in other groups, independently and nearly simultaneously carried out a similar experiment, which they, together with M.C. Packard, called nuclear induction....

"It was my good fortune to arrive at the right time at the right place. Purcell, who had just been appointed as an associate professor of physics at Harvard, needed somebody to assist him in exploiting the discovery. Since Purcell and Pound were preoccupied with writing volumes for the monumental M.I.T. Radiation Laboratory series during the spring of 1946, I had time and opportunity, as Purcell's first graduate student, to become familiar with novel experimental techniques and catch up with my senior mentors in the field of magnetic resonance. After years of isolation in the Netherlands occupied by German forces who had closed the Dutch universities, it was a most stimulating experience to learn and discuss modern physics. The field of nuclear magnetic resonance proved to be enormously fertile, with numerous ramifications in atomic, molecular and solid state physics. Within 18 months I had plenty of material for a Ph.D. thesis, which I submitted at Leiden University in 1948, because I had already previously passed all required formal examinations in Holland. The same material proved the basis for the paper which appeared in *Physical Review* and is so often quoted as BPP.

"In retrospect, it remains a very basic and seminal paper. It deals with the relaxation times, $T_1$ and $T_2$, introduced by Bloch and Frenkel, in solids, liquids and gases. The sharp resonances which are based on the concept of motional narrowing are basic to NMR spectroscopy. The exploitation of this field by the chemists and biochemists, who are more numerous in numbers and more prolific in authoring papers than physicists, is undoubtedly responsible for the high incidence of citation. This does not fully explain, however, why the paper is still so much quoted in the period 1961-1975, 13 to 28 years after its original publication. Many comprehensive books on the subject of magnetic resonance and relaxation now exist, which certainly constitute an improvement on the naive early experimental and theoretical discussions of BPP. Perhaps new workers, confronted with the complexities of modern NMR and its applications, like the account of our early wrestling with some basic problems ....

"All three co-authors are still at Harvard University, although their research interests have diverged in different directions. E.M. Purcell is University Professor, and his research is focused on radioastronomy and astrophysics. R.V. Pound is Mallinckrodt Professor of Physics and has done fundamental work on Mossbauer spectroscopy, although he keeps an active program on nuclear magnetic relaxation in solid hydrogen. N. Bloembergen is Rumford Professor of Physics in the Division of Engineering and Applied Physics, and his present research is concerned with nonlinear optics."

| Number 4 | **Citation Classics** | January 23, 1978 |

Paul M A & Long F A. $H_O$ and related indicator acidity functions.
*Chem. Rev.* 57:1-45, 1957.

The authors review the literature pertaining to indicator acidity functions and their applications to elucidating the mechanisms of acid catalysis. [The *SCI®* indicates that this paper was cited 699 times in the period 1961-1976.]

Professor M.A. Paul (retired)
and Professor F.A. Long
Department of Chemistry
Cornell University
Ithaca, New York 14853

January 31, 1977

"This article, together with a companion article on the application of the $H_O$ acidity function to kinetics and mechanisms of acid catalysis,[1] reviewed twenty-five years of research stemming from the establishment by L.P. Hammett and A.J. Deyrup in 1932[2] of quantitative means of extending acidity measurements to highly acid solutions through a sequence of indicators with progressively decreasing base strengths, selected according to criteria minimizing extraneous medium effects not immediately related to acidity. Conspicuous among such medium effects are those associated with the charge type of the indicator; Hammett and Deyrup's $H_O$ acidity function (a logarithmic measure of acidity reducing to pH in dilute aqueous solutions) is derived from indicators of charge type: $B^O + H^+ \rightleftarrows BH^+$

"In concentrated aqueous solutions of strong acids, $H_O$ is observed to decrease (indicating increasing acidity) with increasing acid concentration C significantly more rapidly than does $-\log C$, and furthermore the $H_O$ values for the different strong acids are spread apart. Since the rates of certain acid-catalyzed reactions were found to correlate rather well with the behavior of $H_O$ whereas the rates of others showed poor correlation, interest grew in the possibility of using the presence or absence of such correlation as evidence for the particular mechanism of acid catalysis involved. Our review was prepared at a time when interest in acid catalysis had been renewed by the ideas of C.K. Ingold and his associates on the mechanisms of homogeneous organic reactions in general. By assembling and critically evaluating the published data on acidity functions and their applications to the study of acid catalysis (252 references), we evidently helped many investigators in further research on the subject. Indicator acidity functions of various types have since been determined and redetermined with improved precision by modern spectrophotometric methods, and found to be less general than was originally hoped. Also, many interesting complexities have been discovered in the mechanisms of acid-catalyzed reactions. An updated review of the field is now available, published in 1970 by C.H. Rochester.[3]

## REFERENCES

1. **Long F A & Paul M A.** Application of the $H_O$ acidity function to kinetics and mechanisms of acid catalysis. *Chem. Rev.* 57:935-1010, 1957.
2. **Hammett L P & Deyrup A J.** A series of simple basic indicators.
    1. The acidity functions of mixtures of sulfuric and perchloric acids with water; 2. Some applications to solutions in formic acid.
    *J. Am. Chem. S.* 54:2721-39, 4239-47, 1932.
3. **Rochester C H.** *Acidity functions.* New York: Academic Press, 1970. 310 pp.

# This Week's Citation Classic™

CC/NUMBER 6
FEBRUARY 6, 1984

Rottenberg H, Grunwald T & Avron M. Determination of ΔpH in chloroplasts.
1. Distribution of [$^{14}$C]methylamine. *Eur. J. Biochem.* 25:54-63, 1972.
[Department of Biochemistry, Weizmann Institute of Science, Rehovot, Israel]

A new method for the determination of ΔpH across chloroplast thylakoid vesicles which is based on the distribution of [$^{14}$C]methylamine was developed. It was demonstrated that under optimal conditions for light driven ATP synthesis a large ΔpH (acidic inside) is established and appears to be a competent intermediate of photophosphorylation. [The *SCI®* indicates that this paper has been cited in over 235 publications since 1972.]

---

Hagai Rottenberg
Department of Pathology
and Laboratory Medicine
School of Medicine
Hahnemann University
Philadelphia, PA 19102

January 11, 1984

"In the fall of 1969, after completing my PhD thesis at Harvard University, I arrived at the Weizmann Institute of Science for postdoctoral work with Mordhay Avron. My thesis work at Harvard, under the guidance of A.K. Solomon, was concerned with the mechanism of active potassium transport in mitochondria. During the course of this work, I became a 'convert' to Mitchell's chemiosmotic mechanism, and considered it important, therefore, to develop methods for the quantitative determination of the proton electrochemical gradient in mitochondria in order to test the predictions of the chemiosmotic hypothesis.[1]

"When I arrived at the Weizmann Institute, Avron, together with his then student Steve Karlish, was investigating light-induced proton transport in chloroplasts. It was apparent from their work that there was no quantitative relationship between the extent of proton uptake and photophosphorylation, but there was no way to measure the proton electrochemical gradient in chloroplasts. Since chloroplasts pump protons *into* the interthylakoid space, in contrast to mitochondria which pump protons out, the ion distribution methods which proved useful in mitochondria could not be employed. However, it seemed to us that the principle, namely, that the equilibrium distribution of permeable ions can be used to indicate the magnitude of the membrane potential and that of permeable acids (or bases) to indicate ΔpH, must also be valid in this case.

"Considering the polarity of the chloroplast, we reasoned, we should use anions (rather than cations) for the measurement of membrane potential, and amines (rather than carboxylic acids) for the measurement of ΔpH. Since it was already known that chloroplasts can pump ammonium in the light, we felt confident that our approach must work. It was our great fortune, at that point, to be joined by a very young and eager Dutch student, Tilly Grunwald (now Bakker-Grunwald) who came to spend a year with us. Grunwald used all her considerable experimental skill to perform the long series of experiments which proved our basic assumptions. We know today that we were also aided by lucky accidents and guesses. First was the lucky choice of $^{14}$C-methylamine, which was the only commercially radiolabeled aliphatic amine available then. As it turns out, it is also the best. Second was the use of $^{14}$C-sorbitol as a marker for the extra-thylakoid space, which turns out to be the best marker in this system. Third, it also happened that a few weeks before we started these experiments, the new Beckman microfuge arrived. With transparent cover and transparent tubes we had no trouble in illuminating the chloroplasts during centrifugation, which we know today to be essential for detection of a large ΔpH. After publishing a short report[2] on our initial experiments, we proceeded together with Simon Schuldiner, who had just arrived as a fresh graduate student, to develop faster methods for the measurement of ΔpH based on the same principles. Schuldiner is now at Hebrew University in Jerusalem.

"These studies resulted in a series of papers of which this work was the first.[3,4] In the decade that followed these studies, proton pumps have been found in many cells and organelles and have proved to play a central role in ATP synthesis and active transport. Moreover, in many of these systems, as well as in vesicle preparations and reconstituted pumps, protons are pumped into the internal space, simulating the case of the chloroplast. Hence, the methods which we developed for the measurement of ΔpH in chloroplasts have found wide use in all of these systems."[5,6]

---

1. **Rottenberg H.** ATP synthesis and electrical membrane potential in mitochondria. *Eur. J. Biochem.* 15:22-8, 1970.
2. **Rottenberg H, Grunwald T & Avron M.** Direct determination of ΔpH in chloroplasts and its relation to the mechanism of photoinduced reactions. *FEBS Lett.* 13:41-4, 1971.
3. **Schuldiner S, Rottenberg H & Avron M.** Determination of ΔpH in chloroplasts. 2. Fluorescent amines as a probe for the determination of ΔpH in chloroplasts. *Eur. J. Biochem.* 25:64-70, 1972.
4. **Rottenberg H & Grunwald T.** Determination of ΔpH in chloroplasts. 3. Ammonium uptake as a measure of ΔpH in chloroplasts and sub-chloroplast particles. *Eur. J. Biochem.* 25:71-84, 1972.
5. **Pick U & McCarty R.** Measurement of membrane ΔpH. *Meth. Enzymology* 69:538-47, 1980.
6. **Rottenberg H.** The measurement of membrane potential and ΔpH in cells, organelles, and vesicles. *Meth. Enzymology* 55:547-69, 1979.

# Citation Classics

Number 35 — August 29, 1977

Brunauer S, Emmett P H & Teller E. Adsorption of gases in multimolecular layers. *Journal of the American Chemical Society* 60:309-19, 1938.

This paper recognized that gases adsorb on solids in multilayers when the pressure is gradually increased at the boiling point of the gas being studied. It contains a derivation of an adsorption equation (the BET Equation) based on an extension to multilayers of the Langmuir treatment of monomolecular adsorption. [The *SCI*® indicates that this paper was cited 1,008 times in the period 1961-1975.]

---

Professor Paul H. Emmett
Chemistry Department
Portland State University
P.O. Box 751
Portland, Oregon 97207

June 17, 1977

"Speaking for my colleagues, Dr. Stephen Brunauer and Dr. Edward Teller and myself, I can say that we are all flattered to have our paper and names included in the 'most cited paper' list. Perhaps a little background of the history of the work leading to the BET method for plotting isotherms for multilayer adsorption and measuring surface areas of porous and finely divided substances should be given.

"After some preliminary work it had become evident that some method for measuring the internal surface area of the porous iron catalysts was needed. Means of differentiating between qualitative changes in the surface and quantitative extension of the surface area had to be found. Some time earlier I had furnished a sample of pure iron catalyst for adsorption measurements to Dr. A.F. Benton of the University of Virginia who first interested me in adsorption and catalysis and supervised my thesis work. The adsorption data that he obtained included adsorption isotherms for $N_2$ at $-191.5°$. The isotherm for $191.5°$ contained two distinct kinks, one at about 13 and the other at 48 cm. pressure. Dr. Benton suggested that the first of these might correspond to the formation of a statistical monolayer of adsorbed nitrogen and the other might mark the completion of a second layer. This seemed to me to be a good place to start so Brunauer immediately undertook an extended series of measurements of various gases near their boiling points.

"The first discovery we made was that the two kinks for the nitrogen isotherms disappeared when correction was made for the fact that nitrogen at $-195°C$ is about 5% imperfect at atmospheric pressure, the correction being a linear function of the pressure. Instead of a curve with kinks a smooth S-shaped isotherm was obtained with nitrogen at $-195°C$. The portion between about .05 and 0.35 relative pressure was linear. It joined a curved portion concave at the pressure axis below 0.05 to 0.1 relative pressure and a curved portion convex to the pressure axis above about 0.35 relative pressure.

"About this time Brunauer and I arrived at the conclusion that the linear part of the S-shaped isotherms corresponded to a completion of the first layer and the beginning of a second layer. We designated this part of the isotherm as 'point B' and proceeded to calculate the surface areas of the various adsorbents and catalysts by multiplying the number of molecules in the adsorbed monolayer by the estimated cross-sectional areas of the adsorbate molecules.

"While this work had been going on Dr. Edward Teller, a young Hungarian physicist, had joined the George Washington University in Washington, D.C. My Hungarian collaborator, Dr. Brunauer, suggested that he contact Teller and try to work out a theory for the S-shaped adsorption curves that might give more credence to the use of the isotherms for measuring the volume of gaseous adsorbate, $V_m$, needed to form a monolayer. Out of this collaboration came the equation cited above. Testing it against our data confirmed its reasonableness and led to $V_m$ values in good agreement with those selected by the 'point B' method.

"The question naturally arises as to why the BET paper made the 'most cited' list. I suspect that the principal favorable factors were the rationalization of the S-shaped curves by the theory which led to the BET equation, the obtaining of a value for $V_m$ much more reproducible than one obtained by estimating the point of juncture of a curve and a straight line by the point B method and, finally, a critical appraisal of the application of this method to readily reproducible isotherms. At any rate, the paper is generally recognized after nearly forty years of use as the best approach to the measurement of the surface area of catalysts and other finely divided and/or porous materials."

# Citation Classics

Number 38 — September 19, 1977

Dexter, David L. A theory of sensitized luminescence in solids.
*Journal of Chemical Physics* 21:836-50, 1953.

The probability of electronic energy transfer between atoms is computed for multiple and exchange interactions along the lines of the Forster theory[1] for dipole-dipole coupling. [The *SCI®* indicates that this paper was cited 531 times in the period 1961-1975.]

Professor David L. Dexter
Department of Physics & Astronomy
University of Rochester
Rochester, New York 14627

February 21, 1977

"In the summer of 1952, between jobs in Urbana and Rochester, I spent a few months with the Luminescence Section of the U.S. Naval Research Laboratory (NRL). At that time luminescence in inorganic crystals was studied at a dozen or two laboratories around the world. A small subdivision, investigated primarily by a group in The Netherlands and by J.H. Schulman's group at NRL, was 'sensitized luminescence,' which dealt with the luminescence by an impurity subsequent to excitation of a different impurity (or the host matrix). Schulman introduced me to the subject.

"In the late '40's Forster had developed a theory[1] which had been extremely useful in describing energy transfer between organic molecules in solution. It was based on Fermi's Golden Rule, where the perturbation is the induced dipole-dipole interaction ('near-zone' interaction) between the molecules, and the density of states is provided by nuclear motion and is expressed in terms of an energy overlap integral between the emission band of one molecule and the absorption band of the other. I think that Forster's theory was well known to chemists, but not so much so to solid state physicists in those days. It was clearly pertinent to the sensitized luminescence problem. In some cases of greatest interest, however, such as the Mn impurity, there was no measurable absorption spectrum for the energy acceptor. Of course 'forbidden' transitions are familiar, reductions in strength by $10^{-6}$ being not uncommon, so the weakness of absorption was no mystery. A little reflection on the nature of the near zone field shows that the degree of forbidden-ness in transfer probability is much less than in radiative transition probability, so that transfer via forbidden transitions is by no means precluded. Also, the electron exchange interaction may well be important when the dipole-dipole interaction vanishes. I adapted Forster's theory to the case of solids, incorporating ideas of electron-phonon interactions so as to include discussions of Stokes' shifts, temperature dependences, etc., and particularly including higher multipole and exchange interactions. It's rather a long paper, and covers a lot of territory.

"After a few years a lot of people in several disciplines began working on energy transfer problems. Many new effects, such as cooperative absorption, emission, and transfer, were discovered and interpreted in solids, glasses, and solutions, and the field is still very active. A large number of these investigations dealt with rare earth ions, triplet states in organic molecules, and other systems in which the electronic transitions were indeed forbidden by optical selection rules. I suppose people cite my paper as background for the later developments, and as a convenient label for the kind of energy transfer under discussion, e.g., more-or-less resonant transfer without the intervention of free charge carriers or (real) photons.

"In summary, then, Schulman told me what the problem was, Forster had shown how to solve it, and the intrinsic interest in and importance of energy transfer phenomena led to a lot of research activity in this area. Jim Schulman went on to become Director of NRL and is currently with the London Office of Naval Research, and Theodor Forster's untimely death in 1974 terminated a distinguished and still highly productive research career."

REFERENCE

1. Forster T. Zwischenmolekulare energiewanderung und fluoreszenz (Intermolecular energy transfer and fluorescence). *Annalen der Physik* 2:55-77, 1948.

# Citation Classics

**Number 29** — July 18, 1977

Wilson A J C. Determination of absolute from relative x-ray intensity data. *Nature* 150:151-2, 1942.

The average value of the intensity of an X-ray reflexion is given, and a comparison made between observed relative intensities of reflexion and the sum of the squares of the atomic scattering factors. This gives the approximate values of the conversion factor required to place the observed intensities on an absolute scale, and the overall temperature factor. [The *SCI®* indicates that this paper was cited 824 times in the period 1961-1975.]

---

Professor A.J.C. Wilson
Department of Physics
University of Birmingham
Birmingham B15 2TT
England

April 6, 1977

"Perhaps appropriately in view of its statistical nature, the existence as well as the popularity of this paper is largely a matter of luck. It originated as my 'referee's report' on the paper by Yü[1] immediately preceding it in *Nature*. If the editor of *Nature* had not sent Yü's paper to the Cavendish Laboratory for refereeing, if I had not been the most mathematically inclined crystallographer at the Cavendish at the time, and if my mind had not been full of statistical ideas concerned with mistake broadening in X-ray powder photographs of metals, it is unlikely that the problem (and even less likely that the solution) would ever have occurred to me. The paper appeared in the middle of World War II and attracted no interest at the time; the earliest citation known to me (and that by myself) is dated 1949.

"About 1948 D. Harker[2] and E.W. Hughes[3] rediscovered the essential idea, independently of me and of each other. Luckily for my citation rating, I noticed their publications, and immediately sent reprints of my paper to them and other crystallographers in key positions. Otherwise, it might be Harker or Hughes that would be heading the crystallographic popularity charts, rather than Wilson.

"Luck aside, why is it that this short paper has been cited about as frequently as all my other papers put together, and ranks sixth in total citations on the *SCI®* list of highly-cited physics, chemistry and mathematics papers published in the 1940s?[4] I think there may be four reasons.

"(1) It is short and, as such papers go, readily understandable.

"(2) It provides a method for obtaining quickly approximate values of two quantities of considerable use in the early stages of any structure analysis, and thus tends to be quoted in many structural papers.

"(3) It was the first paper, as far as I know, to apply statistical ideas in structural crystallography, and thus tends to be quoted by many of those who are using statistical methods in developing, for example, direct methods of structure determination (There are earlier statistical papers, including one of my own, in 'textural' crystallography.)

"(4) One phrase, almost a throw-away, 'certain coincidences can be predicted from the space group only,' contains the germ of several later papers by myself and by others.

"All this said, however, it seems to me that some of my other papers are more worthy of a high citation rating, and that perhaps some of those that in fact out-rank it [on the 1940s list] ought to come lower down!"

---

1. Yü S H. Determination of absolute from relative x-ray intensity data. *Nature* 150:151-2, 1942.
2. Harker D. Absolute intensity scale for crystal diffraction data. *American Mineralogist* 33:764-5, 1948.
3. Hughes E W. Limitations on the determination of phases by means of inequalities. *Acta Crystallographica* 2:34-7, 1949.
4. Garfield E. Highly cited articles. 36. Physics, chemistry and mathematics papers published in the 1940s. *Current Contents®* No. 10, 7 March 1977, p. 5-11.

Number 15 **Citation Classics** April 10, 1978

**Dalgarno A.** Atomic polarizabilities and shielding factors.
*Advan. Phys.* 11:281-315, 1962.

The author discusses methods of calculating the response of an atomic system to perturbation. Atomic polarizabilities and shielding factors are considered and the relationships between them are demonstrated. A formulation is presented of a self-consistent complex Hartree-Fock approximation in terms of which several other methods can be interpreted. [The *SCI*® indicates that this paper was cited 302 times in the period 1962-1976.]

A. Dalgarno
Center for Astrophysics
60 Garden Street
Cambridge, Massachusetts 02138

December 1, 1977

"This article was written in response to an invitation from Dr. B.H. Flowers (now Sir Brian Flowers), who had assumed responsibility for the editorship of *Advances in Physics* and who was, I believe, attempting to extend the range of subject matter published in the journal beyond its earlier devotion to solid state physics. The subject was not specified, but I understood that he was seeking a review of some aspect of atomic and molecular physics. I had worked for several years on atomic perturbation theory and I had been impressed by the confusion which attended the development of an accurate description of the response of an atomic or molecular system to the application of a static electric field. The subject was a simple one, and the confusion lay in the propagation of the inevitable inaccuracies in the description of the unperturbed system, into the effects of the perturbation. There were two obvious ways of proceeding: in one, the unperturbed and perturbed systems could be treated simultaneously at the same level of approximation; and in the other, the error in the description of the unperturbed system could be ascribed to an additional perturbation and double perturbation theory used to identify the sources of uncertainty in the calculation of the response. It was not difficult to organize the two viewpoints into a unified presentation and I thought that a review with this end would clarify my understanding and perhaps be more generally useful. Perturbation theory can also be expressed usefully in variational terms, and a review of some aspect of perturbation theory would allow me to emphasize the close relationship between the two apparently disparate approaches. The history of atomic polarizabilities had been a long one in quantum mechanics and the theory had retained its earlier lack of mathematical sophistication. The introduction of more recent angular momentum techniques scarcely merited an original paper but could be conveniently incorporated into a review.

"That my article is often cited is due largely, I believe, to the fact that in further extensions of the topic it was no longer necessary to seek out the appropriate reference in an early extensive but contradictory literature. The article was useful also in that its essential theme, a self-consistent theory of atomic and molecular perturbation, was later to be readily generalized to the description of frequency-dependent response functions, to the calculations of long range intermolecular interactions and to multiphoton processes. It is not without interest to note that the article was a review which had no original content."

Number 30 **Citation Classics** July 24, 1978

Ness N F, Scearce C S & Seek J B. Initial results of the IMP-1 magnetic field experiment. *J. Geophys. Res.* **69**:3531-69, 1964.

The authors present the first comprehensive survey and discovery by the IMP-1 satellite of the magnetosphere. The physical characteristics and locations of these boundaries, formed by the interaction of the magnetized solar wind plasma with the geomagnetic field, are studied. [The *SCI®* indicates that this paper was cited 270 times in the period 1964-1977.]

---

Norman F. Ness
Laboratory for Extraterrestrial Physics
National Aeronautics and
Space Administration
Goddard Space Flight Center
Greenbelt, MD 20771

January 6, 1978

"Identification of this paper as a 'citation classic' is a very pleasant way to be reminded of successful research conducted some years ago. My colleagues and I are suitably flattered by this long term result, although that does not compare with the excitement of discovery which permeated many of the results of this early experiment in space magnetometry. Our research was done just as the space age began, but prior to any accurate or comprehensive studies of the interplanetary magnetic field or the interaction of the solar wind with the geomagnetic field. The contemporary spirit of exploration of space was heightened by the competition with the USSR. The technological challenges of all spacecraft investigations were great because of the new environment and criteria by which such experiments were conducted.

"The IMP-1 spacecraft was the first of 10 such customized science laboratories to be launched into space during a particularly vigorous phase of exploration in space (1963-1973). One of them, Anchored IMP (AKA Explorer 35) was placed in lunar orbit and provided definitive comparison data for several ALSEP experiments. This series of spacecraft represents one of the better investments the NASA and the nation have made with respect to scientific return.

"A particularly critical but not unique aspect of space experiments had been the slow processing, analysis and interpretation of data obtained. I was well versed in the use of high speed digital computers at a time when such facilities and related experiences were in short supply. Challenged by the scientific and technical tasks of exploring our unknown terrestrial environment in space and the properties of magnetized plasmas associated with the earth and sun, I also accepted an assignment to provide initial processing of data for all other 8 experiments on the Interplanetary Monitoring Platform No. 1 (AKA Explorer 18). Thus, with other colleagues, we developed the IMP Information Processing System, and so the results of not only our experiment but all the others on the spacecraft benefited from its development and prompt distribution of data.

"IMP-1 was launched during a period of minimum solar activity, so that conditions in interplanetary space were relatively stationary. As a result, we were able to detect readily the regular ordering of the direction of the magnetic field in interplanetary space ('sectors') and to identify the field as being of solar origin, having been carried outward by the expanding solar corona ('solar wind'). Our first measurements of the detached bow shock wave, standing off from the distorted geomagnetic field, the magnetosphere, made a collisionless shock wave in astrophysical plasmas available for direct study. However, the most striking result of our studies was the discovery of an extended geomagnetic tail, directed away from the sun, much in the fashion of a cometary ion tail and due to the same cause, i.e., the solar wind flow. The tail is now known to be extremely important in the dynamics of the magnetosphere. But we still have much to learn before the dynamics of the magnetosphere are understood."

Number 48 **Citation Classics** November 28, 1977

Stewart R F, Davidson E R & Simpson W T. Coherent x-ray scattering for the hydrogen atom in the hydrogen molecule. *Journal of Chemical Physics* **42**: 3175-87, 1965.

The authors calculate the x-ray form factors for a bonded hydrogen in the hydrogen molecule for a spherical approximation to the bonded atom. [The *SCI*® indicates that this paper was cited 1879 times in the period 1965-1975.]

---

Professor Ernest Davidson
Department of Chemistry
University of Washington
Seattle, Washington 98195

February 2, 1977

"This particular paper was written in response to a recognized need. It contains a table of X-ray scattering factors appropriate for a bonded hydrogen atom. This table has been incorporated in most standard X-ray crystallography computer packages and generates an 'automatic' citation each time a new crystal structure containing a hydrogen atom is solved. The popularity of this table was greatly enhanced by the fact that R.F. Stewart is a crystallographer who at that time had a joint post-doctoral appointment with W.T. Simpson and L. Jensen. Consequently the table was generated in a format which could be readily incorporated in the standard crystal programs, and it was put to immediate use at the University of Washington.

"This paper was, in a sense, a finishing touch to a sequence of more fundamental papers. A few years previously Kolos and Roothaan[1] had generated an improved James and Coolidge-type wavefunction for the hydrogen molecule. Also Löwdin[2] had introduced the concept of natural orbitals and Shull and Löwdin[3] had showed their utility for understanding simpler wavefunctions for the hydrogen molecule. Finally Davidson[4] had developed a method for converting the James and Coolidge-type wavefunctions to natural orbital form and Davidson and Jones[5] had published the natural orbital expansion of the Kolos and Roothaan wavefunction. From the natural orbital expansion an accurate charge density for the molecule could be constructed. The fundamental conceptual problem remaining in calculating the X-ray scattering factors was how to partition the molecular charge density into the sum of two 'atomic' charge densities. In addition there was the difficult numerical problem of performing the Fourier transform of the atomic density in confocal elliptical coordinates, which was solved by R.F. Stewart following suggestions of Dr. George Hufford. After many false starts, the partitioning problem was finally solved in an elegant manner by showing the connection between a best least-squares fit to the density in coordinate and momentum space."

---

1. **Kolos W & Roothaan C C J.** Accurate electronic wave functions for the $H_2$ molecule. *Reviews of Modern Physics* **32**:219-32, 1960.
2. **Löwdin P O.** Quantum theory of many-particle systems. I. Physical interpretations by means of density matrices, natural spin-orbitals, and convergence problems in the method of configurational interaction. *Physical Review* **97**: 1474-89, 1955.
3. **Löwdin P O & Shull H.** Natural orbitals in the quantum theory of two-electron systems. *Physical Review* **101**:1730-9, 1956.
4. **Davidson E R.** Natural expansions of exact wave functions. I. Method. *Journal of Chemical Physics* **37**:577-81, 1962.
5. **Davidson E R & Jones L L.** Natural expansion of exact wave-functions. II. The hydrogen-molecule ground state. *Journal of Chemical Physics* **37**:2966-71, 1962.

# Citation Classics

Number 30 — July 25, 1977

Hanson H P, Herman F, Lea J D & Skillman S. HFS atomic scattering factors.
*Acta Crystallographica* 17:1040-4, 1963.

Atomic scattering factors calculated from Hartree-Fock-Slater (HFS) wave functions are presented for the neutral elements for $Z = 2$ to $Z = 100$ over a range of sin $\theta/\lambda$ values up to 6·0. It is asserted that this compilation should represent the most complete and accurate $f$ values available. [The *SCI*® indicates that this paper was cited 1,079 times in the period 1961-1975.]

---

Dr. Harold P. Hanson
Executive Vice President
University of Florida
Gainesville, Florida 32611

January 13, 1977

"Naturally it is gratifying that a work that one has produced has become a Citation Classic. This means that the work has been useful and it has had a significant impact on the scientific scene.

"However, a paper that qualifies as a Citation Classic is not necessarily one which represents a major scientific breakthrough. It is likely that most of the oft-cited papers are either (1) papers which outline a general method/technique that gets employed in numerous investigations or (2) papers which provide data, coefficients, parameters, etc., which are fundamental to a great number of experiments or studies. The paper I did with James Lea, Frank Herman, and Sherwood Skillman is in the latter category.

"The impact of the paper is in large part a result of timing. Herman and Skillman had just produced a generalized method of obtaining Hartree-Fock-Slater wave functions utilizing the newly-appreciated power of what was then the large computers. It was merely a matter of performing the indicated integration to produce a dense set of internally consistent atomic scattering factors for all elements of the periodic table. The wave functions themselves are of interest and significance to quantum theorists, but the atomic scattering factors have utility for a much wider group of investigators, viz., x-ray crystallographers, electron diffractionists, and even high-energy theorists.

"The question may be raised as to why these data continue to be used as extensively as they are at this late date. Atomic scattering factors which are presumably better have been produced in the meantime. However, the corrections to our numbers provided by more sophisticated wave functions are fairly small compared to the effects of bonding, geometry, etc. It is still probably true that the Hanson-Herman-Lea-Skillman atomic scattering factors will give about as good experimental residuals as any other values that are generally available, and there is no impetus for investigators to change to other values."

# Citation Classics

Number 33 — August 15, 1977

Jaffé, Hans H. A reexamination of the Hammett equation.
*Chemical Reviews* 53:191-261, 1953.

Since the publication of Hammett's book,[1] which proposed an empirical relation on the parallelism of the effects of substituents on the rate or equilibrium constants in many different side-chain reactions of benzene derivatives, many additional reactions have been investigated which permit the application of Hammett's equation. One of the aims of this review is to gather together and thus render accessible all material pertinent to the Hammett equation. The equation is reexamined for its range of application, usefulness, and precision. [The *SCI*® indicates that this paper was cited 1,667 times in the period 1961-1975.]

Dr. Hans H. Jaffé
Department of Chemistry
University of Cincinnati
Cincinnati, Ohio 45221

January 25, 1977

"In a major lecture at the 1951 Organic Mechanism Conference at Swarthmore, Pennsylvania, C. Gardner Swain said that if he just had a collection of some 3,000 pieces of data permitting application of the Hammett equation, he would like to undertake a statistical reevaluation of the σ constants. A brash young graduate student stuck up his hand and bragged that he had such a collection. After the lecture, H.C. Brown took self-same young man aside to tell him that such a collection should be made available to the scientific community. Next morning this same student just accidentally found a seat next to Professor Hammett, related the conversation with Professor Brown, and intimated that an invitation from a journal such as *Chemical Reviews* (on whose editorial board Professor Hammett was just serving), would make it infinitely easier for a government employee to prepare such a publication.

"The result of the events, coupled with much subsequent literature work and calculator punching led to the paper 'A Reexamination of the Hammett Equation' in *Chemical Reviews*. An interesting sidelight of this publication is the widespread use of the correlation coefficient in physical organic chemistry, which, it appears, was first introduced into this field in that paper.

"A further sidelight—Swain never did the statistical study he had suggested. We later attempted such a study, and after many nights on an IBM 650, concluded that the results are not sufficiently stable to be meaningful. The results change too much by addition or omission of a single data set, sometimes even of a single data point. We thus concluded that the values of McDaniel and Brown[2] are the most acceptable set of values."

1. Hammett L P. *Physical Organic Chemistry*. New York: McGraw-Hill Book Company, Inc., 1940.
2. McDaniel D H & Brown H C. An extended table of Hammett substituent constants based on the ionization of substituted benzoic acids. *Journal of Organic Chemistry* 23:420, 1958.

| Number 27 | **Citation Classics** | July 4, 1977 |

Fano, Ugo. Effects of configuration interaction on intensities and phase shifts.
*Physical Review* 124(6):1866-78, 15 December 1961.

The interference of a discrete auto-ionized state with a continuum gives rise to characteristically asymmetric peaks in excitation spectra. The earlier qualitative interpretation of this phenomenon is extended and revised. A theoretical formula is fitted to the shape of the $2s2p\,^1P$ resonance of He observed in the inelastic scattering of electrons. The theory can also give the position and intensity shifts produced in a Rydberg series of discrete levels by interaction with a level of another configuration. [The *SCI*® indicates that this paper was cited 647 times in the period 1961-1975.]

---

Professor Ugo Fano
Department of Physics
University of Chicago
Chicago, Illinois 60637

January 25, 1977

"The paper appears to owe its success to accidental circumstances, such as the timing of its publication and some successful features of its formulation. The timing coincided with a rapid expansion of atomic and condensed matter spectroscopy, both optical and collisional. The formulation drew attention to the generality of the ingredients of the phenomena under consideration. In fact, however, the paper was a rehash of work done 25 years earlier and its context still needs extension and clarification.

"It is well known that an atomic system can absorb only discrete amounts of energy as long as these amounts do not suffice to break it up; one observes then a line spectrum. A continuous spectrum is observed, instead, when the energy absorption can achieve ionization or dissociation of the system. Conspicuous discrete structures do nevertheless occur in continuous spectra when the absorbed energy runs initially into blind alleys thus allowing only a delayed break-up; the deeper the blind alley, the sharper is its influence. These spectral structures thus furnish evidence on energy migration within the absorbing system.

"In January, 1935, Emilio Segrè gave me some spectroscopy papers by H.A. Beutler as a fruitful subject of study. The Beutler spectra showed unusual intensity profiles which struck me as reflecting interferences between alternative mechanisms of excitation. Fermi then taught me sequentially within a few days how to formulate my interpretation theoretically; a paper was sent to the *Nuovo Cimento* quickly and I dropped the matter. I remained, however, sensitized to evidences of analogous phenomena, noticing, e.g., how they emerged through the influence of surface waves on the spectra of diffraction gratings.

"Late in 1960, R.L. Platzman called to my attention a strikingly asymmetric line profile in an unpublished spectrum of energy transfers in electron collisions by Lassettre and co-workers. This spectrum appeared analogous to those I had interpreted in 1935. My reply to Platzman provided the opportunity for a modernized formulation of the analytical treatment. He urged me to publish the new derivation; this I did, complementing it with illustrations and with fragmentary extensions and interpretations, with unexpected success. The amount of effort spent on this paper was, however, far larger than for its distant predecessor.

"By 1965 discrete structures had proved ubiquitous in the vacuum ultraviolet spectra of most materials. The theoretical framework I had utilized appeared, however, too restrictive. Only for isolated, sharp spectral features can we extract from spectral data well-defined quantitative information on the mechanisms that produce them. That is, we do well only when excitation energy can flow and ebb from a single blind channel before being dissipated away. Extensive efforts have been devoted to extending this analysis, over the years, by many physicists besides myself, but their results have remained fragmentary."

| Number 21 | **Citation Classics** | May 23, 1977 |

Van Hove L. Correlations in space and time and born approximation scattering in systems of interacting particles. *Physical Review* **95**:249-62, 1954.

The paper's central aim is to describe the distribution of momentum and energy transfers in a scattering process in terms of the correlation between scattering centers in space and time. The general method is described and illustrated for the case of slow neutron scattering by liquids, gases, and crystals. [The *SCI*® indicates that this paper was cited 582 times in the period 1961-1975.]

Dr. L. Van Hove
European Organization for
Nuclear Research
CERN
CH 1211 Geneve 23
Switzerland

March 15, 1977

"Although I knew that my 1954 paper was much read, I was very surprised to hear that it was among the most cited science articles of the last 15 years. Wondering why the paper was so often referred to, I got the impression that the general considerations which led my work at that time may have something to do with its later impact.

"In 1952-1953 I was working on the theoretical interpretation of slow neutron scattering experiments in solids and liquids, the aim being to extract information on structure properties of the solid or liquid from the scattering data. I was disturbed by the difficulty, with the methods then in use, to obtain a general physical insight into a problem which was basically the same, whether the scattering system was a solid, a liquid or a molecule. In the case of X-ray scattering this difficulty had been overcome as early as 1927 by F. Zernike and J. Prins by their description of the distribution of scattered X-rays in terms of the pair correlation function of the scattering system. My aim was to achieve something similar for the more general case where, contrary to X-rays, the energy transfer could not be neglected with respect to the incident energy of the particle to be scattered.

"After the question had been formulated in this way, the answer turned out to be both very simple and very general. Basically it consisted in realizing that the pair correlation concept had to be generalized to deal with both space and time co-ordinates of the scattering centres. The 1954 paper does no more than to work out this idea. It may well be that this combination of simplicity and generality is one of the reasons why the paper was so often quoted. For example, its basic idea can be found again in later work by many people on the time-dependent Green functions used in the field-theoretic treatment of solid-state physics problems.

"In addition, there was of course the circumstance that my article was written at the moment when slow neutron scattering began to be extensively exploited as a tool for structure studies of condensed matter. It was only natural for the many experimentalists working in this new domain to make reference to the paper when publishing their results."

| Number 34 | **Citation Classics** | August 21, 1978 |

Russell G A, Janzen E G & Strom E T. Electron-transfer processes. 1. The scope of the reaction between carbanions or nitranions and unsaturated electron acceptors. *J. Amer. Chem. Soc.* 86:1807-14, 1964.

This paper documents by electron spin resonance (ESR) techniques the widespread occurrence of electron transfer between mono- and di-anions, wherein the negative charge is centered on carbon or nitrogen, and a wide variety of unsaturated organic compounds, particularly those containing nitro, nitroso, azo or carbonyl functional groups, or easily reducible saturated compounds such as peroxides, disulfides, and alkyl or aryl halides. Electron transfer was also observed using a variety of organometallic reagents as the donors and various polynuclear- or hetero-aromatic compounds as the electron acceptors. [The *SCI*® indicates that this paper was cited 330 times in the period 1964-1977.]

Glen A. Russell
Department of Chemistry
Iowa State University
Ames, Iowa 50011

November 23, 1977

"Ed Janzen, Tom Strom and I are indeed flattered to have one of our publications included in the *SCI*® most-cited paper list. Frankly, at the time this diffuse paper was written in 1963, the editor and I had some reservations about its suitability. We had in hand some rather nice examples of the occurrence of electron transfer (ET) in organic systems. However, I felt that these hard-core examples of ET between diamagnetic substrates did not place in proper perspective the scope or implications of this process. Therefore, I tried to gather together various scraps from the older literature, various predictions about when ET was likely to occur, and a variety of illustrations from our own work to form a sort of summary paper as our initial full publication. Our basic premise was that one electron oxidation or reduction of an unsaturated organic system did not require classical agents such as the alkali or other metals, radical anions, transition metal ions, or electrode surfaces. Instead, diamagnetic carbanions ($\equiv C{:}^-$) or nitranions ($>N{:}^-$) could spontaneously ET to an unsaturated system ($\pi$) to yield a free radical ($\equiv C\cdot$ or $>N\cdot$) and the radical anion derived from the unsaturate ($\pi\cdot^-$), while carbocations ($\equiv C^+$) or dications could accept an electron from easily oxidized molecules. One variation in the overall scheme was to increase the electron affinity of the acceptor by photo-excitation. In many ways we simply extended basic concepts of inorganic and electroanalytical chemistry to the organic sphere. However, I feel we did perform a service for subsequent workers in this area by providing a key reference or benchmark from which they could develop their own discussions. I think the paper added another dimension to the organic chemist's outlook. At the time this paper appeared many chemists thought of reactions of organic anions or cations only as being electrophilic or nucleophilic in nature. Electrons were presumed to move in pairs, and perhaps it was a small but important step forward to demonstrate that electrons do not always move as pairs. In fact, I think the editor in accepting this paper missed this point completely but accepted the paper because he felt it had something to do with dihydropyridine chemistry, a subject of considerable interest but one which involves mainly electron pair chemistry.

"The circumstance which led me to the somewhat presumptuous attempt to summarize a diffuse area of chemistry in a research article can be traced to our application of new techniques and instrumentation. In the early 1960's my group was one of the first to apply electron spin resonance (ESR) spectroscopy to the study of organic reactions. We soon recognized that some of the standard reactions of organic chemistry required a new mechanistic formulation involving ET. One such process described in a later paper in this series was the base-catalyzed coupling of nitrosobenzene and phenylhydroxylamine to yield azoxybenzene. Our appreciation of ET led us to look for new examples of this process possessing utility. Thus, we subsequently discovered the ET chain mechanism for the substitution reaction between carbanions and $\alpha$-substituted nitroalkanes. This process, dubbed the $S_{RN}1$ reaction, was later shown by others to be an important general replacement reaction in a variety of aromatic systems."

# Citation Classics

Karle J & Karle I L. The symbolic addition procedure for phase determination for centrosymmetric and noncentrosymmetric crystals.
*Acta Crystallographica* 21:849-59, 1966.

The theoretical background and practical procedures are presented for the determination of the phases of x-rays scattered by a crystal. The phases are determined directly from the measured intensities. Knowledge of the phases permits the immediate calculation of the crystal structure. [The *SCI®* indicates that this paper was cited 945 times in the period 1961-1975.]

---

Drs. Jerome & Isabella Karle
Laboratory for the Structure of Matter
U.S. Naval Research Laboratory
Washington, D.C. 20375

January 11, 1977

"The development of the symbolic addition procedure for the direct determination of the phases of scattered x-rays provided the opportunity to solve readily the structures of a large class of crystals which had heretofore been inaccessible. This class consists of structures composed of atoms with essentially equal atomic numbers (neglecting hydrogen atoms), particularly those structures which crystallize in space groups which lack a center of symmetry. Earlier methods had already made it possible to address essentially equal atom centrosymmetric crystals, but not many applications were made.

"The alternative approach to solving crystal structures has been to take advantage of a heavy atom already present in the structure or one which is deliberately introduced. This can greatly simplify the analysis and lead to a solution in an entirely routine fashion. It is not, however, always possible or desirable to introduce a heavy atom.

"The analysis of crystal structures, facilitated by the symbolic addition procedure and the development of computers and automatic diffractometers has had a major impact on the progress of many scientific areas. For example, the availability of such capabilities has played a significant role in the practice of organic chemistry by aiding in the identification of reaction intermediates, final products and the clarification of reaction mechanisms....

"The problem of solving crystal structures afforded fascinating challenges from the beginning. As a theoretical problem it has gone through several stages. The problem was regarded as unsolvable because the phases of the scattered x-ray amplitudes are lost in the recording of the intensities. It was possible to show, however, that the values of the phases were still contained subtly in the measured intensities, so that a solution existed in principle. The next stage involved the development of mathematical formulas which had the potential for practical development. Finally, considerable effort was required to bridge the gap between theoretical formulas and practical procedures.

"The key role played in the development of the theory by the physical constraint imposed by the nonnegativity of the electron density is particularly noteworthy. In fact, the role of physical and mathematical constraints in the various fields of structure research and the development of practical procedures are subjects of particular interest and a detailed manuscript concerning these philosophical and mathematical aspects of the subject is being contemplated.

"There is one final word concerning the development of a facility for readily carrying out crystal structure analyses with good single crystals. Such analyses have too often been considered the sole concern of crystallographic research and so, with this development, the field has been considered to be largely without challenges. This is very far from the truth and several examples come immediately to mind. Crystallography is also concerned with structural problems for which good single crystals are often not available. The fields of earth sciences and materials science can attest to this. Actually, this is a familiar occurrence in all areas of application. There are also many types of structural problems for which diffraction data are extremely limited. In addition, there is the broad field of diffraction physics with a great variety of problems other than crystal structure analysis...."

# Citation Classics

Number 40 — October 2, 1978

Hughes E W. The crystal structure of melamine.
*J. Amer. Chem. Soc.* **63**: 1737-52, 1941.

The crystal structure of melamine was investigated by X-ray diffraction. A new method for refining atomic coordinates, based upon least-squares adjustment, was developed and used to find the positions of the atoms. The carbon-nitrogen distances in the cyanuric ring were found to be equal well within experimental error. [The *SCI®* indicates that this paper was cited 637 times in the period 1961-1977.]

Edward W. Hughes
Division of Chemistry &
Chemical Engineering
California Institute of Technology
Pasadena, CA 91125

January 5, 1978

"The paper covered two topics with only one mentioned in the title. The second described a new method for refining crystal structures, using the least-squares procedure. This 'hidden' part accounts for most of the citations since this method has become *the* standard way of refining crystal structures.

"Melamine, $C_3N_3(NH_2)_3$, is cyanuric triamide. It was then a rather rare chemical. I was surprised on the day that I mailed the paper to *The Journal of the American Chemical Society* to see in *Chemical and Engineering News* that 'Melamine is now available in carload lots.' The principal result was that the C—N distances in the cyanuric ring are all equal, within experimental error, not alternately short and long as suggested in one earlier examination of cyanuric triazide.

"Crystal structures solved during the first decade of such work involved only one or two atoms not fixed in the crystal lattice by symmetry and the variable coordinates could easily be calculated from X-ray diffraction amplitudes. Sir William Bragg in 1915 had pointed out the equality of the diffraction amplitudes and the Fourier coefficients for the density of scattering matter in crystals, but by 1935 no more than ten multiple-parameter structures had been determined this way, which involved evaluating two-dimensional Fourier series at hundreds of points with about a hundred terms per point. For melamine all possible two-dimensional Fourier pictures were incompletely resolved. A three-dimensional Fourier, in those days, was impossible, involving the evaluation of about a thousand trigonometric terms at about 5000 points.

"I decided to try least-squares adjustment of the atom coordinates in a two-dimensional projection, although I had never then heard of least-squares problems involving so many data. A projection for melamine involves eighteen atomic coordinates. The equations connecting these with the diffraction amplitudes are trigonometric, so for linear least-squares it is necessary to linearize the equations with respect to a trial structure. It took about two eight-hour days to set up the 105 linear observational equations in 18 unknowns. With some I.B.M. bookkeeping equipment (loaned rent-free by I.B.M., to whom thanks) these were reduced to 18 'normal equations' in 18 unknowns in 4 hours. These were solved by a standard iteration method in another 4 hours.

"With the advent of electronic computers the method could be applied to three-dimensional data and has become the usual way of getting optimum agreement between calculated and observed amplitudes. A modern computer would require less than a minute for one round of the 2-D melamine refinement, which in 1940 required 24 man-hours of very tedious calculation. Today at least 500 structure papers per year use this method but citations are usually to published computer programs."

# This Week's Citation Classic

Babb A L, Popovich R P, Christopher T G & Scribner B H.
The genesis of the square meter-hour hypothesis.
Trans. Amer. Soc. Artif. Internal Organs 17:81-91, 1971.
[Depts. Nuclear Engineering, Chemical Engineering, and Medicine, Univ. Washington, Seattle, WA]

In 1965, we believed that there may be uremic toxins that behaved as though they had molecular weights larger than urea (60 daltons). Based on a retrospective study of our first patients on dialysis, together with a new kinetic model of toxin transport and *in vitro* and *in vivo* experiments, we deduced that their apparent molecular weight was in the 2,000-5,000 dalton range. These insights led to the 'Square Meter-Hour Hypothesis.' [The *SCI*® indicates that this paper has been cited in over 245 publications since 1971, making it one of the three most-cited papers ever published in this journal.]

A.L. Babb
Departments of Nuclear Engineering and
Chemical Engineering
University of Washington
Seattle, WA 98195

January 12, 1983

"In 1961, B.H. Scribner, a nephrologist, invented the arteriovenous shunt for providing continuous access to the cardiovascular system of patients with end stage renal disease. At that time, a crude and costly refrigerated 400 liter stainless steel tank was used to contain the aqueous solution of chemicals recirculated through a hemodialyzer, which consisted of three parallel double layers of a cellophane membrane between which blood flowed by the pumping action of the heart. The aqueous solution, or dialysate, was pumped on the outside of the membranes and the uremic toxins passed from the blood to the dialysate through the membrane by the process of dialysis. The cleansed blood was then returned to the patient.

"The patients were treated for 12 hours twice weekly. There were so many candidates for treatment and so little money to provide therapy to all, that the so-called 'Who Shall Live?' committee was formed by the local medical society to select patients for treatment. Having been turned down for federal funds, in desperation Scribner called the chairman of the department of chemical engineering, R.W. Moulton, for some engineering assistance. Moulton recommended me.

"I met with Scribner and his colleagues shortly thereafter and was soon caught up in developing the lower cost continuous systems summarized in a recent review.[1] In a subsequent meeting, Scribner made the observation that 'in contrast to our early experience with hemodialysis, neuropathy was not seen in our first chronic peritoneal dialysis patients even though their urea nitrogen and serum creatinine levels were considerably higher than those of most hemodialysis patients who developed this complication.' He postulated that since the peritoneum leaked protein during peritoneal dialysis, it might also be passing toxins of higher molecular weight than urea relatively more efficiently than the cellophane membranes in hemodialyzers. My group then developed mathematical models, and designed protocols for *in vitro* and *in vivo* confirmation. One of the group, R.P. Popovich, stayed on for postdoctoral work and ultimately became a co-inventor of the Continuous Ambulatory Peritoneal Dialysis (CAPD) scheme[2] while at the University of Texas. Another member, T.G. Christopher, is now in private practice as a nephrologist.

"The initial results of our efforts resulted in the 'Square Meter-Hour Hypothesis' in which it was assumed that in uremia there were large molecular weight toxic solutes in the 2,000-5,000 dalton molecular weight range. For these solutes, it was also assumed that doubling the hemodialyzer surface area would cut the time in half for the equivalent so-called 'middle molecule' removal.

"This publication attracted worldwide attention because it was the first attempt to quantify the kinetics of the removal of potential uremic toxins of higher molecular weight and led to the concept that large surface area dialyzers could dramatically reduce treatment time. There is now a plethora of large surface area hemodialyzers on the market. Subsequent research led us to believe that the middle molecules were in the 500-2,000 dalton range and to enunciate the 'Middle Molecule Hypothesis.'[3] These two hypotheses triggered a worldwide search for evidence of middle molecules in uremic plasma.[4]

"Although we were unaware of early experimental work that suggested the existence of middle molecules in uremia, a group in Sweden successfully fractionated uremic plasma in 1976 and found that certain middle molecule fractions exhibited toxicity *in vitro*.[4] No definitive studies of *in vivo* toxicity of middle molecule compounds have yet been carried out.

"I believe that the publication cited here together with concomitant contributions to dialysis theory and technology were the major reasons for my election to the US National Academy of Engineering in 1972 and the Institute of Medicine of the National Academy of Sciences in 1982."

1. Babb A L & Scribner B H. Dynamics of hemodialysis systems. (Cooney D O, ed.) *Advances in biomedical engineering. Part I.* New York: Marcel Dekker, 1980. p. 93-141.
2. Popovich R P, Moncrief J W & Decherd J B. Continuous ambulatory peritoneal dialysis. *Ann. Intern. Med.* 88:449-56, 1978.
3. Babb A L, Farrell P D, Uvelli D A & Scribner B H. Hemodialyzer evaluation by examination of molecular spectra: implications for the square meter-hour hypothesis. *Trans. Amer. Soc. Artif. Internal Organs* 18:98-105, 1972.
4. Bergström J, Fürst P, Asaba M & Oules R. Dialysis and middle molecules. *Dialysis Transplant.* 7:344-7, 1978.

Chapter

# 7

# Chemical Analysis and Preparative Methods

The need to measure accurately and with great sensitivity the catecholamines epinephrine (adrenalin), norepinephrine (noradrenalin), and related compounds such as serotonin and histamine account for nearly one-half of the Citation Classics described in this chapter. Although the citation count for these Classics is distributed over unequal time periods and varies according to the publication dates of the original article and the "This Week's Citation Classic" (TWCC), the average number of citations exceeds 900. Two of the papers were cited more than 1,000 times, and two others were cited 995 and 975 times. That nine publications with the single goal to quantitate catecholamines should achieve citation status of this magnitude forces speculation as to its cause. Although some increase in sensitivity exists in the later methods, the authors do not claim this as the reason for the popularity of their articles. They state instead that their method is reproducible, is applicable to tissue analysis and histochemistry, or is novel in its use of a new fluorophor, or some such other explanation. Most of these procedures were intended to measure total catecholamines extractable from tissue. The different methods used for this purpose were slightly modified or significantly altered over time and illustrate a more general pattern of progress in methodology. Vogt was obliged, as were many early investigators in other fields, to rely on an inherently expensive and inaccurate bioassay. Later, several variations of fluorometric analysis with o-phthalaldehyde or trihydroxyindole (ninhydrin) were published (pp. 229–234). The last Classic on catecholamine assays (p. 235) is based on the sensitive radiometric–enzymatic assay, and even this has been replaced by electrochemical detectors coupled to chromatography. These procedures were used to evaluate the catecholamine content of different regions of the brain, and these stereotaxic–topographic data provided evidence of a second neurotransmitter system distinct from the cholinergic system (see the chapter on Neurobiology in volume 1).

It is noteworthy that many of these TWCCs contain references and the authors are quite faithful in citing previous articles on the subject of their Classic. For example, Häggendal cites the article by Bertler et al., and Shore cites Maickel et al., as do Curzon and Snyder. Missing from authorship among these Classics on the catecholamines are the names Brodie and Udenfriend, both prominent investigators on this subject at the National Institutes of Health. Their names are mentioned by several of the authors of these Classics.

The next Classics are randomly dispersed among assay methods for other prominent biomolecules and some preparative procedures. Four of these thirteen Classics exceeded 1,000 citations, and one has been used as a reference 970 times. The vital role of phosphorus in biological systems accounts for the vast popularity of Allen's modification (1,410 citations) of the Fiske and Subbarow method based on the reduction of phosphomolybdic acid. Later, ascorbic acid was used by Lowry (the same Lowry of the protein method) and Lopez as the reducing agent, with the attendant ability to quantitate esters that were labile in the stronger acid media used earlier. Ascorbic acid was also used by Chen in his brief two-page article that achieved 1,940 citations. Indeed, these three articles on phosphorus have been cited nearly 4,500 times.

Nash's defense of the popularity of the Hantzsch formaldehyde method (1,340 citations) based on its spelling, its century-old pedigree, and even its potability is a nice combination of English humor and modesty. Beverley Pearson Murphy's tale on protein binding of thyroxine is tinged with the humor that only expectant mothers could relate (see Murphy's other TWCC in Chapter 2, Lipids and Related Compounds).

The groundwork for the widely applicable affinity chromatography technique is described by Porath in his brief essay on the use of cyanogen bromide to link proteins to agarose or, of course, to other polymers. Cleland's TWCC on his search for a sulfhydryl protecting reagent is as nicely written as his TWCC on his earlier Classic (see Chapter 5, Enzymes). The first total synthesis of a small protein from its component amino acids added one at a time is the subject of the Classic by Bodanszky and du Vigneaud. This was an important step in protein synthesis, as was the preparation of protected amino acids by Schnabel that enabled development of the solid-phase method of protein synthesis. Merrifield's Nobel Prize in 1984 for this latter achievement was to a degree based on a method Schnabel was surprised to find acceptable for publication.

## This Week's Citation Classic™

CC/NUMBER 16
APRIL 16, 1984

Vogt M. The concentration of sympathin in different parts of the central nervous system under normal conditions and after the administration of drugs.
*J. Physiology* 123:451-81, 1954.
[Department of Pharmacology, University of Edinburgh, Scotland]

A map of the distribution of noradrenaline (NA) and adrenaline was obtained by bioassay of extracts of about 50 freshly dissected regions of the dog's brain and spinal cord. The NA concentration ranged from 2.0 to 0.01 µg/g fresh tissue. [The *SCI*® indicates that this paper has been cited in over 995 publications since 1955.]

---

Marthe Vogt
Institute of Animal Physiology
Agricultural Research Council
Babraham, Cambridge CB2 4AT
England

March 15, 1984

"The work was done in J.H. Gaddum's laboratory at the University of Edinburgh. It took about four years to carry out (1950-1953), but the manuscript was accidentally buried on the editor's desk until I finally had the courage to enquire about its fate. It was then published with great speed and without editorial corrections.

"The fact that cholinergic neurones (identified by the presence of choline acetyltransferase) had been shown[1] to be very unevenly distributed in the brain and practically absent from large regions suggested that acetylcholine could not be the only nervous transmitter in the central nervous system. The only other transmitter known at that time (1952) in peripheral nerves was noradrenaline (NA). A small amount of this transmitter was known to occur in the brain but was assumed to be localized exclusively in nerves to the blood vessels.

"A preliminary comparison of the concentration of NA in hypothalamus and cerebellar cortex showed that the NA was so much greater in hypothalamus than in cerebellar cortex (both highly vascularized regions) that the explanation of the difference by variable vascularity could not be correct. This observation led to my undertaking a survey of concentrations of NA and adrenaline in many brain regions with a method which used paper chromatography followed by biological assay, at the time the most sensitive method available. The results could hardly be explained by assuming that vasomotor nerves were the main source of cerebral NA, and suggested a characteristic uneven distribution of noradrenergic neurones in the brain tissue itself. Fluorescence microscopy carried out soon afterward by Swedish workers showed that catecholamine fluorescence was indeed present within neurones, many of which were found in the hypothalamus, but only a few in the cerebellum.

"Since dopamine (DA) is an obligatory precursor of NA and adrenaline, it is found in all catecholamine containing neurons. However, as shown by Carlsson,[2] there are large brain regions in which DA is the final product, and where it is therefore supposed to play the role of a transmitter or 'modulator' of impulses. Three examples are the nigro-striatal pathways, the arcuate nucleus, and certain layers in some cortical areas. There is particular interest in this last site, since the clinically beneficial effect of antagonists of DA (such as chlorpromazine) in schizophrenia can hardly be expected to be due to their action on the nigro-striatal system or on the arcuate nucleus, but a site of action in the cerebral cortex is a clear possibility."

---

1. **Feldberg W & Vogt M.** Acetylcholine synthesis in different regions of the central nervous system.
    *J. Physiology* 107:372-81, 1948. (Cited 265 times since 1955.)
2. **Carlsson A.** The occurrence, distribution and physiological role of catecholamines in the nervous system.
    *Pharmacol. Rev.* 11:490-3, 1959. (Cited 380 times.)

# This Week's Citation Classic

CC/NUMBER 49
DECEMBER 3, 1979

Bertler Å, Carlsson A & Rosengren E. A method for the fluorimetric
determination of adrenaline and noradrenaline in tissues.
*Acta Physiol. Scand.* 44:273-92, 1958.
[Dept. Pharmacol., Univ. Lund, Lund, Sweden]

This paper describes the first chemical method for the analysis of adrenaline and noradrenaline that proved sufficiently sensitive and specific for permitting accurate quantitative analyses of extracts of animal tissues in general. [The *SCI*® indicates that this paper has been cited over 975 times since 1961.]

Arvid Carlsson
Department of Pharmacology
University of Göteborg
S-400 33 Göteborg
Sweden

July 18, 1979

"Using a colorimetric assay, Hillarp and I had discovered, in 1956, that reserpine treatment resulted in the virtually complete disappearance of catecholamines from the rabbit adrenal medulla. We were eager to find out what happened to the catecholamines in other tissues. At this time, fluorimetry had started to focus general attention owing to its high sensitivity and to the recent development of the spectrophotofluorimeter by Bowman and his colleagues.[1] During a visit in 1955-1956 to the Laboratory of Chemical Pharmacology, National Heart Institute, Bethesda (head B.B. Brodie), I had become acquainted with and enthused by this instrument. Therefore, rather than trying one of the currently used bioassay procedures, we decided to develop a fluorimetric method for our purpose. Ehrlén had shown that adrenaline and noradrenaline can be conveniently converted into strongly fluorescent 'lutines,' but his method could be used only for pharmaceutical purposes.[2] An isolation technique had to be developed in order to apply the method for tissue extracts. A weak cation exchange resin had previously been used for this purpose but could not be used for tissues other than the adrenal medulla. Essentially, the new thing in our procedure was the introduction of a strong cation exchange resin (Dowex 50) for the column-chromatographic isolation of the catecholamines. When applying Ehrlén's method (somewhat modified) on our eluates, satisfactory results were obtained in terms of tissue blanks, internal standards, recoveries, and activation and fluorescence spectra. Moreover, the identification of adrenaline and noradrenaline in tissue extracts was secured by quantitative paper chromatography.

"To account for the widespread use of this method two additional points should be emphasized. Firstly, the paper describes a simple apparatus which greatly facilitates the chromatographic procedure. This apparatus consists of an all-glass syringe and a glass reservoir connected to each other and to the cation exchange column via a three-way stopcock. Secondly, the method could be further developed to permit the separation and quantitative analysis of a large number of amines and several of their precursors and metabolites in one single chromatographic procedure. This development has recently been reviewed by Atack.[3]

"The work reported in our paper was performed at the University of Lund, Sweden, where I held a position as associate professor of pharmacology. My co-authors Åke Bertler and Evald Rosengren were medical students. Their contribution was considerable. They did the main part of the experimental work, together with the technical assistant Carin Larsson, and they took a very active part in planning the work and in the construction of the apparatus. The paper forms part of Rosengren's dissertation in 1960, and both Rosengren's and Bertler's dissertations (in the same year) report results obtained with the new method.[4,5] At about the same time our paths separated. I became professor of pharmacology at the University of Göteborg. Several years later Bertler moved to Linköping, where he is professor of clinical pharmacology at the Medical School. Rosengren is associate professor of pharmacology at the University of Lund.

"Today our method has several competitors, utilizing not only fluorimetry but also mass fragmentography, isotope techniques based on enzymatic methylations, high-pressure liquid chromatography, etc. It is hard to predict how long the method will survive. In our, and presumably several other laboratories, it is still the dominant method for the analysis of catecholamines."

1. **Udenfriend S.** *Fluorescence assay in biology and medicine.* New York: Academic Press, 1962. 505 p.
2. **Ehrlén I.** Fluorimetric determination of adrenaline II. *Farm. Revy* 47:242-50, 1948.
3. **Atack C.** Measurement of biogenic amine using cation exchange chromatography and fluorimetric assay. *Acta Physiol. Scand.* **101** (Suppl. 451):1-99, 1977.
4. **Rosengren E.** Studier över katekolaminer i vävnader. Thesis. Lund, 1960. 21 p.
5. **Bertler Å.** Inverken av reserpin och andra läkemedel på katekolaminomsättningen. Thesis. Lund, 1960. 23 p.

# This Week's Citation Classic

CC/NUMBER 40
OCTOBER 5, 1981

Shore P A, Burkhalter A & Cohn V H, Jr. A method for the fluorometric assay of histamine in tissues. *J. Pharmacol. Exp. Ther.* **127**:182-6, 1959.
[National Heart Institute, National Institutes of Health, Bethesda, MD]

Histamine reacts with o-phthalaldehyde (OPT) in basic solution to form a fluorophore which, on acidification, rearranges to a stable, highly fluorescent product. Extraction of histamine from tissues or body fluid and reaction with OPT allows the estimation of submicrogram amounts of the biogenic amine. [The *SCI*® indicates that this paper has been cited over 1,345 times since 1961.]

P.A. Shore
Department of Pharmacology
University of Texas
Health Science Center
Dallas, TX 75235

August 25, 1981

"In the late 1950s, there was a rapidly expanding interest in the biogenic amines, and methods were being developed for their estimation in tissues. Thanks to the interest and drive of Bernard Brodie, Sidney Udenfriend, and Robert Bowman at the National Heart Institute (NHI), the first working spectrofluorometer was developed and new or improved fluorescent methods were developed for catecholamines and serotonin. I was fortunate to be at the NHI at the time, and thought that a sensitive fluorometric method for histamine would be a boon for better understanding the functions of this still mysterious amine.

"Histamine is not a natural fluorophore, but couples readily with various aldehydes to form a Schiff base product. This was not much help since such a coupling with a fluorescent aldehyde occurs with many amines, and one had also to cope with the presence of the excess fluorescent aldehyde. Scrutiny of the chemical literature revealed that histamine and pyridoxal form an acid stable condensation product,[1] and while this product was not fluorescent, it prompted my colleagues and me to search for a non-fluorescent aldehyde which might condense with histamine to form a fluorophore. We felt we needed to work with aromatic rather than aliphatic aldehydes and tried several, to no avail. After almost giving up, we attempted to obtain o-phthalaldehyde (OPT) for a final effort, but were unsuccessful for some time until we did obtain a small sample of a tarry mess that was reputed to be OPT. After consulting Beilstein we managed to purify it. To our joy, the reaction of OPT with histamine produced an immensely fluorescent product which, although apparently proceeding initially via a Schiff base, rearranged in acid to a stable fluorophore. Working out the details of a method was then simple.

"Subsequent work by us and others showed that OPT reacts with a variety of amines, each under quite specific conditions.[2] At the present time, there are specific conditions for fluorophore formation with histamine, catecholamines, serotonin, and m-hydroxyphenylethyl amines.

"The paper has been cited frequently because it was an innovation in the estimation of this biogenic amine and also because quantitative measurement of histamine release is frequently used in research on immediate hypersensitivity reactions. Many adaptations have been made of the initial procedure and, needless to say, highly purified OPT is now available from numerous sources. W. Lovenberg and K. Engelman have recently published a review in this field."[3]

1. Heyl D, Luz E, Harris S A & Folkers K. Chemistry of vitamin $B_6$. VII. Pyridoxylidene- and pyridoxylamines. *J. Amer. Chem. Soc.* **70**:3669-71, 1948.
2. Maickel R P, Cox R H, Jr., Saillant J & Miller F P. A method for the determination of serotonin and norepinephrine in discrete areas of rat brain. *Int. J. Neuropharmacol.* **7**:275-82, 1968. [Citation Classic. *Current Contents/Life Sciences* **24**(27):18, 6 July 1981.]
3. Lovenberg W & Engelman K. Assay of serotonin, related metabolites, and enzymes. (Glick D, ed.) *Analysis of biogenic amines and their related enzymes.* New York: Wiley, 1971. p. 1-34.

# Citation Classics

Anton, Aaron H & Sayre, David F. A study of the factors affecting the aluminum oxide-trihydroxyindole procedure for the analysis of catecholamines. *Journal of Pharmacology and Experimental Therapeutics* 138:360-75, 1962.

This paper presents a reliable, quantitative, highly sensitive, adaptable method for the estimation of catecholamines in diverse biological material from various vertebrate species. This method involves the selective adsorption of the catecholamines onto a constant amount of aluminum oxide, elution with constant volume of perchloric acid (0.05 N), and their measurement by the formation of a fluorescent trihydroxyindole derivative in the presence of potassium ferricyanide and alkaline (10 N alkali) ascorbate. [The *SCI®* indicates that this paper was cited 1,087 times in the period 1961-1975.]

---

Aaron H. Anton, Ph.D.
Department of Anesthesiology
School of Medicine
Case Western Reserve University
Cleveland, Ohio 44106

January 20, 1977

"Peer recognition of one's efforts is certainly a gratifying and pleasant experience, especially when it is unexpected. I was not aware that our method for the analysis of catecholamines was being used to the extent indicated in this citation. On the one hand, its wide use reflects the protean nature of the catecholamines which have been implicated in the physiology of diverse biological systems ranging from plants to man. On the other hand, each variable in the procedure, and there are many, was thoroughly examined and sufficient details are given in the publication so that it can be readily followed. Another reason for its frequent citation could be that it contains a useful table listing the distribution of norepinephrine and epinephrine in man and various laboratory animals.

"I became interested in the catecholamines in 1960 in relation to their controversial role in essential hypertension. It soon became apparent that the controversy was partly due to methodological problems. Dave Sayre, a superb technician, joined me about that time and we carried out a detailed analysis of the factors involved in various published procedures. This was a long, tedious, and at times, frustrating experience since we wanted a versatile method that could be used with a variety of biological material including urine, plasma, tissue and cerebrospinal fluid. The present method evolved from that study and we still use it essentially as described in the original publication.

"Catecholamine methodology is recognized as being difficult because: (1) the parent compounds are unstable if not properly preserved, (2) many technical factors, e.g., reagents, timing, temperature, can influence the results, and (3) their relatively low concentration *in vivo* requires careful attention to details in order to obtain reproducible results. We are very pleased with this citation since it indicates that our work was worthwhile, resulting in a method for the analysis of these important metabolites that apparently has been used successfully by many investigators."

# This Week's Citation Classic

CC/NUMBER 27
JULY 6, 1981

Maickel R P, Cox R H, Jr., Saillant J & Miller F P. A method for the determination of serotonin and norepinephrine in discrete areas of rat brain.
*Int. J. Neuropharmacol.* 7:275-82, 1968.
[Lab. Psychopharmacology, Depts. Pharmacology and Psychology, Indiana Univ., Bloomington, IN]

A method is described for the quantitative measurement of levels of serotonin and norepinephrine in geographically specific portions of the rat brain by utilization of a selective solvent extraction procedure and specific spectrophotofluorometric techniques. [The *SCI*® indicates that this paper has been cited over 505 times since 1968.]

Roger P. Maickel
Department of Pharmacology
and Toxicology
School of Pharmacy and
Pharmacal Sciences
Purdue University
West Lafayette, IN 47907

May 27, 1981

"The fact that this paper has been labeled a *Citation Classic* is proof that the phrase 'build a better mousetrap and the world will beat a path to your door' holds true even in scientific research! But, after all, what is a new analytical procedure other than a 'better mousetrap'—an improved means for doing a necessary job? The aura attached to a major scientific breakthrough often hides the fact that much of the work involved is not glorious and spectacular. So it was with the development of this method, now labeled as a much-referenced citation.

"My undergraduate training as a chemist was followed by graduate and postdoctoral research in chemical pharmacology in the laboratory of B.B. Brodie at the National Institutes of Health during the period 1955-1965. This was an exciting era, especially since that laboratory was one of the world centers in brain biogenic amine research. Such research demanded new, more sensitive analytical methods; the technology we now know as spectrophotofluorometry was born there. Methods emanated in a virtually unending stream: 5HT, 5HIAA, norepinephrine, epinephrine, tryptamine, histamine. The procedures were similar: homogenization, extraction, back-extraction, and conversion to a fluorophore.

"In 1963, I was intrigued by the fact that o-phthalaldehyde (OPT) reacted with amino acids to form colored derivatives, but produced a powerful fluorophore with histamine. More than two years later (interrupted by a move to Indiana University) the research effort culminated; a number of 3,5-disubstituted indoles reacted with OPT under rigorous conditions (10 N HCl, 100°) to yield highly fluorescent products. By combining this procedure with an iodine oxidation of norepinephrine to a trihydroxyindole, homogenizing the brain tissue directly in acidified n butanol, extracting into dilute acid, and scaling down volumes, the 'mousetrap' was designed. Two additional years of repetitive testing and retesting, modifying and discarding modifications, culminated in this publication. The procedure was convenient, sensitive, specific, and cost-effective, and was thoroughly use tested before publication, thus accounting for the article's frequent citation.

"I would be remiss if I did not acknowledge Raymond H. Cox, Jr., and Francis P. Miller (then graduate students in my laboratory); without their tireless efforts and long hours, the trial and error process would have been significantly more lengthy and considerably less effective. In addition, Jean Saillant served as our technical help, performing many repetitive tasks with skill and precision. One final comment must be made. Although the 'mousetrap' was completed in 1967, at least two dozen modifications and improvements have been reported in the intervening 13 years, most recently in the *Journal of Chromatography* and in *Pharmacology, Biochemistry, and Behavior*.[1,2] Obviously, science, as time, marches on."

1. Ogasahara S, Mandai T, Yamatodani A, Watanabe T, Wada H & Seki T. Simple method for the simultaneous determination of noradrenaline, dopamine and serotonin by stepwise elution from a short column of weak cation-exchange resin.
   *J. Chromatography* **180**:119-26, 1979.
2. Jacobowitz D M & Richardson J S. Method for the rapid determination of norepinephrine, dopamine, and serotonin in the same brain region. *Pharmacol. Biochem. Behav.* **8**:515-19, 1978.

# This Week's Citation Classic

CC/NUMBER 37
SEPTEMBER 13, 1982

Häggendal J. An improved method for fluorimetric determination of small amounts of adrenaline and noradrenaline in plasma and tissues.
*Acta Physiol. Scand.* **59**:242-54, 1963.
[Department of Pharmacology, University of Göteborg, Göteborg, Sweden]

Several modifications of the trihydroxyindole method for the fluorimetric determination of adrenaline and noradrenaline were introduced. These included changes in the ion-exchange procedure for purification, stabilization of blank values, and reduction of the volumes of reagents. The result was an appreciable improvement in sensitivity and reproducibility. [The *SCI*® indicates that this paper has been cited in over 585 publications since 1963.]

---

C. Jan Häggendal
Department of Pharmacology
University of Göteborg
S-400 33 Göteborg
Sweden

April 27, 1982

"After I had completed my medical studies at the end of the 1950s, I wanted to carry on with the thesis for my PhD degree within the area of mental illness, psychotropic drugs, and monoamines. Therefore I became interested in quantitative methods for estimation of small amounts of noradrenaline and adrenaline. A few years earlier, a sensitive method for estimation of catecholamines had been described by Bertler, Carlsson, and Rosengren.[1] Columns of strong cation exchange resin were used for the purification and concentration of catecholamines in the tissue extracts. The catecholamines were subsequently fluorimetrically determined after they had been converted to strongly fluorescent trihydroxyindoles by oxidation followed by stabilization.

"On the basis of these principles I started to adapt the method for increased sensitivity. A detailed investigation of different factors affecting the separation was performed. After that I tried to increase the sensitivity of the estimation procedure. This was achieved, e.g., by reducing the negative effect of too large volumes of reagents. Furthermore, the different chemical steps in the procedure were stabilized and particular attention was paid to the question of various blanks. The method that finally came out permitted reproducible noradrenaline and adrenaline assays in rather small pieces of tissue and small blood plasma volumes. However, in order to be sensitive and reproducible the method demanded a large amount of carefulness and precision.

"The most important reason why this paper has been relatively frequently quoted is in all probability that the method was found to be useful in a field of research that rapidly expanded during the 1960s and 1970s. In this connection it should be noted that several different methods were available for catecholamine assay during that period of time. The other main methods were either based on fluorescence techniques or radiochemical methods.

"At present, however, the most sensitive and reproducible methods for quantitative catecholamine determinations are either radiochemical methods, or methods based on high-pressure liquid chromatography followed by electrochemical detection.[2] The role of procedures based on fluorescence appears now to be declining.

"I think that methodologic work can be of considerable importance for the student's training in science. Thus, the detailed knowledge about this method gave me self-confidence to use it as a tool for studies on basic mechanisms in the sympathetic noradrenergic neurons, where the results sometimes came out to be unexpected or even incompatible with what was earlier accepted.

"This may be illustrated by the studies on the axonal transport and life span of amine storage granules, together with A. Dahlström,[3] and also by the studies on the amount of noradrenaline released per nerve impulse from the varicosity of the adrenergic nerve ending, together with B. Folkow and B. Lisander.[4] In this connection the early suggestion of a local feedback mechanism for noradrenaline release[5] may be mentioned."

---

1. Bertler Å, Carlsson A & Rosengren E. A method for the fluorimetric determination of adrenaline and noradrenaline in tissues. *Acta Physiol. Scand.* **44**:273-92, 1958.
   [Citation Classic. *Current Contents/Life Sciences* **22**(49):12, 3 December 1979.]
2. Hallman H, Farnebo L O, Hamberger B & Jonsson G. A sensitive method for determination of plasma catecholamines using liquid chromatography with electrochemical detection. *Life Sci.* **23**:1049-52, 1978.
3. Dahlström A & Häggendal J. Studies on the transport and life-span of amine storage granules in a peripheral adrenergic neuron system. *Acta Physiol. Scand.* **67**:278-88, 1966.
4. Folkow B, Häggendal J & Lisander B. Extent of release and elimination of noradrenaline at peripheral adrenergic nerve terminals. *Acta Physiol. Scand.* **307**(Suppl.):1-38, 1967.
5. Häggendal J. Some further aspects on the release of the adrenergic transmitter. (Kroneberg G & Schümann H J, eds.) *New aspects of storage and release mechanisms of catecholamines.* Berlin: Springer-Verlag, 1970. p. 100-9.

# This Week's Citation Classic

CC/NUMBER 4
JANUARY 25, 1982

Curzon G & Green A R. Rapid method for the determination of 5-hydroxytryptamine and 5-hydroxyindoleacetic acid in small regions of rat brain.
*Brit. J. Pharmacol.* 39:653-5, 1970.
[Dept. Chemical Pathology, Inst. Neurology, London, England]

A rapid and sensitive method for measuring 5-hydroxytryptamine (5HT) and 5-hydroxyindoleacetic acid (5HIAA), using o-phthalaldehyde (OPT) and L-cysteine, is presented, enabling both compounds to be measured in small areas of rat brain. [The *SCI®* indicates that this paper has been cited over 590 times since 1970.]

Gerald Curzon
Department of Neurochemistry
Institute of Neurology
London WC1N 2NS
England

October 9, 1981

"From 1966 onward, Richard Green was a PhD student in my laboratory and was determining 5HT and its metabolite 5HIAA in rat brain. The best fluorometric methods then available needed more or less a whole rat brain per determination. Regional work necessitated pooling material from a number of brains. This was a big restriction on 5HT studies. The large numbers of animals needed was a particular problem for us as our animal housing was limited to a boxed-in space under a bench! However, in 1968 Maickel et al.[1] described a 5HT method sensitive enough for brain regions of a single rat. This was based on solvent extraction, heating at 100° with OPT in 10N hydrochloric acid, and fluorometry. It was clear from the literature that the extraction could be modified to separate 5HT from 5HIAA and as both fluoresced with OPT we thought that the method could be adapted for both substances. An afternoon's work early in 1970 confirmed this but the sensitivity was rather disappointing.

"Fortunately, at the bench next to Green a technician (Devi Kantamaneni) was trying out a new OPT method for 5HIAA in cerebrospinal fluid.[2] Its authors had noted that instead of forming the fluorescent complex, most of the 5HIAA was destroyed by the hot acid. They prevented this by cysteine. Green therefore repeated his determinations with cysteine present. Sensitivities for 5HT and 5HIAA were increased fourfold so that both substances were determinable in regions of a single rat brain.

"I had considerable misgivings about writing up the method as it had little originality, being essentially a combination of two previous papers.[1,2] Also, while we were testing it out, Maickel's group described a development of their method which permitted determination of both 5HT and 5HIAA in regions.[3] In the end, we decided that as we had found our procedure useful so also might others, and we sent it to the *British Journal of Pharmacology*.

"Why is the method cited so often? (a) It is more sensitive than other fluorimetric procedures and gives good recoveries. (b) It is robust, rapid, and simple. (Someone once told me that they interpreted failure to get it to work as a sign of incompetence.) (c) The residue of the extract containing 5HT can be used to determine tryptophan, its precursor. (d) Interest in 5HT remains high.

"Has the method had a crucial influence? I don't think so. If it had not existed, other methods would have been used more, though perhaps their lower sensitivity and/or greater laboriousness would have led to more effort being needed for the same return. Anyway, if we had not put these papers[1,2] together, others would probably have done so before long.

"Over the years, the method has been improved in various details. The present version was recently described and compared with high pressure liquid chromatography (HPLC).[4] To my relief, values correlate well. We still use the OPT method though we often use HPLC instead. Green is now at the MRC Clinical Pharmacology Unit, Oxford."

1. Maickel R P, Cox R H, Saillant J & Miller F P. A method for the determination of serotonin and norepinephrine in discrete areas of rat brain. *Int. J. Neuropharmacol.* 7:275-81, 1968.
   [Citation Classic. *Current Contents/Life Sciences* 24(27):18, 6 July 1981.]
2. Korf J & Valkenburgh-Sikkema T. Fluorimetric determination of 5-hydroxyindoleacetic acid in human urine and cerebrospinal fluid. *Clin. Chim. Acta* 26:301-6, 1969.
3. Miller F P, Cox R H, Snodgrass W R & Maickel R P. Comparative effects of p-chlorophenylalanine, p-chloroamphetamine and p-chloro-N-methylamphetamine on rat brain norepinephrine, serotonin and 5-hydroxyindole-3-acetic acid. *Biochem. Pharmacol.* 19:435-42, 1970.
4. Curzon G, Kantamaneni B D & Tricklebank M D. A comparison of an improved o-phthalaldehyde fluorometric method and high pressure liquid chromatography in the determination of brain 5-hydroxyindoles of rats treated with L-tryptophan and p-chlorophenylalanine. *Brit. J. Pharmacol.* 73:555-61, 1981.

# This Week's Citation Classic

CC/NUMBER 42
OCTOBER 17, 1983

Snyder S H, Axelrod J & Zweig M. A sensitive and specific fluorescence assay for tissue serotonin. *Biochem. Pharmacol.* 14:831-5, 1965.
[Lab. Clinical Science, Natl. Inst. Mental Health, Natl. Insts. Health, Bethesda, MD]

A sensitive and specific method for estimating serotonin in biological materials was developed based on the reaction between serotonin and ninhydrin when the mixture is heated. The resultant fluorescence is eight times more intense than the native fluorescence of serotonin in strong acid. [The *SCI*® indicates that this paper has been cited in over 520 publications since 1965.]

Solomon H. Snyder
Department of Neuroscience
Johns Hopkins University
School of Medicine
Baltimore, MD 21205

June 23, 1983

"My scientific career began as a research associate with Julius Axelrod a year after Richard Wurtman had joined the laboratory. Wurtman and Axelrod had just completed epochal studies establishing melatonin as a hormone of the pineal gland. Axelrod suggested I think about doing some research involving the pineal gland. I was impressed with the report of Wilbur Quay[1] of an extraordinary diurnal rhythm in the serotonin content of the rat pineal gland with levels ten times higher at noon than midnight, clearly the most dramatic biochemical diurnal variation I had ever seen. Quay measured serotonin using the native acid fluorescence assay reported by Udenfriend and colleagues[2] in their pioneering development of the spectrophotofluorometer as a powerful tool for measuring biogenic amines and numerous other body chemicals. Unfortunately, the rat pineal gland weighs only 1 mg. Though its serotonin content is rather high, Quay still required dozens of pineal glands for a single determination making detailed studies of the rhythm almost hopeless.

"About this time I noticed a publication by Vanable[3] showing that when serotonin and ninhydrin are heated together in $H_2O$ solution, an intense fluorescent product results. However, the paper did not indicate whether the reaction was specific for serotonin and no attempts had been made to assess the feasibility of using the fluorescence to assay tissue serotonin.

"Axelrod and I, with the assistance of Mark Zweig, a summer medical student, modified the technique to produce maximal fluorescence and then found that we could extract serotonin from tissues into organic solvents, return the serotonin to an aqueous phase, heat with ninhydrin, and measure the resultant fluorescence. We found that the reaction was quite specific for serotonin and that tissue levels monitored by the ninhydrin technique were essentially the same as those measured by the method of Bogdanski *et al.*[2] Most exciting was that the ninhydrin technique gave fluorescence eight to ten times more intense than that of serotonin in strong acid solution. We could detect serotonin in tissues such as the adrenal gland and heart where it was not apparent by previous methods.

"More importantly, with the ninhydrin technique we could measure serotonin in one or two rat pineal glands. This permitted a series of studies showing that the pineal serotonin rhythm behaves like a biological clock, remaining intact even in complete darkness or in blinded rats. However, exposing the rat to light would abruptly block the nocturnal decline in serotonin levels.[4] We then found that light regulated pineal gland serotonin in neonatal rats directly through the skull even in blinded neonates, suggesting that in the rat, a mammal, the pineal still can function, as it does in reptiles, as a third eye.[5]

"A large number of citations to the paper is no mystery. Any methods paper which describes an improved technique is likely to be well cited. However, no technique remains indispensable for long. Three years after our paper appeared, Roger Maickel and colleagues[6] reported another method for measuring serotonin based on its condensation with o-phthalaldehyde, providing yet a further improvement in the assay of tissue serotonin."

---

1. Quay W B. Circadian rhythm in rat pineal serotonin and its modification by estrous cycle and photoperiod. *Gen. Comp. Endocrinol.* 3:473-9, 1963.
2. Bogdanski D F, Pletscher A, Brodie B B & Udenfriend S. Identification and assay of serotonin in brain. *J. Pharmacol. Exp. Ther.* 117:82-8, 1956. (Cited 1,005 times.)
3. Vanable J W. A ninhydrin reaction giving a sensitive quantitative fluorescence assay for 5-hydroxytryptamine. *Anal. Biochem.* 6:393-403, 1963.
4. Snyder S H, Zweig M, Axelrod J & Fischer J E. Control of the circadian rhythm in the serotonin content of the rat pineal gland. *Proc. Nat. Acad. Sci. US* 53:301-5, 1965.
5. Zweig M, Snyder S H & Axelrod J. Evidence for a nonretinal pathway of light to the pineal gland of newborn rats. *Proc. Nat. Acad. Sci. US* 56:515-20, 1966.
6. Maickel R P, Cox R H, Jr., Saillant J & Miller F P. A method for the determination of serotonin and norepinephrine in discrete areas of rat brain. *Int. J. Neuropharmacol.* 7:275-82, 1968.

[Citation Classic. *Current Contents/Life Sciences* 24(27):18, 6 July 1981.]

# This Week's Citation Classic™

CC/NUMBER 51
DECEMBER 17, 1984

Coyle J T & Henry D. Catecholamines in fetal and newborn rat brain.
*J. Neurochemistry* 21:61-7, 1973.
[Laboratory of Clinical Science, National Institute of Mental Health, Bethesda, MD]

This paper described a sensitive and specific radiometric enzymatic assay for catecholamines in brain extracts. The method was applied to fetal rat brain and demonstrated the early developmental appearance of functioning catecholamine neurons. [The *SCI*® indicates that this paper has been cited in over 750 publications since 1973.]

Joseph T. Coyle
Division of Child Psychiatry
Departments of Psychiatry, Neuroscience,
Pharmacology, and Pediatrics
Johns Hopkins University
School of Medicine
Baltimore, MD 21205

November 8, 1984

"This study was a product of my collaboration with David P. Henry when we were both research associates in the Laboratory of Clinical Science at the National Institute of Mental Health. David was working in the laboratory of Irwin Kopin, and I was in the laboratory of Julius Axelrod. David was involved in a project designed to measure catecholamines in serum under a variety of physiologic and pathologic conditions. My studies were directed at characterizing the development of brain noradrenergic and dopaminergic neuronal systems by quantitative neurochemical methods. Although my studies had previously demonstrated that the synthetic enzymes for catecholamines were present in the rat brain as early as 15 days of gestation, it was not possible with the existing and rather insensitive fluorometric techniques to reliably quantify catecholamine levels in rat brain before birth.

"To measure the catecholamines in the fetal brain, the radiometric enzymatic assay being developed by David was modified so that it was applicable to brain extracts. The remarkable sensitivity of the assay resulted from the availability at the time of [$^3$H]-S-adenosyl methionine (SAMe) of high specific radioactivity (4.5 mCi/μmol). The assay's considerable specificity derived from the use of the partially purified enzyme, catechol-O-methyltransferase (COMT). COMT catalyzes the inactivation of catecholamines by transferring a methyl group from SAMe to one of the ring hydroxyl groups on the catecholamine. A particularly attractive feature of the assay was that it allowed the separate determination of norepinephrine and dopamine without reliance on cumbersome procedures, such as thin-layer chromatography.

"The study demonstrated that both norepinephrine and dopamine were detectable in the fetal rat brain as early as 15 days of gestation, when the brain weighed less than two percent of that of the adult. Furthermore, pharmacologic manipulations revealed that the catecholamines in the fetal rat brain behaved in a fashion similar to those in adult brain, thereby indicating that the neurotransmitters were localized in a dynamic, functionally relevant pool. This study provided the first quantitative evidence of the remarkably early formation and functional activity of catecholaminergic neurons in the developing brain, a conclusion that was supported by subsequent histochemical and immunocytochemical studies.

"The reason for the frequent citation of this report may be that it was the first description of a sensitive, specific, and relatively simple method for measuring catecholamines in brain tissue. The overarching strategies involved in the assay — use of a partially purified methyltransferase, [$^3$H]-S-adenosyl-L-methionine, and differential organic extraction techniques — were exploited by other members of the Laboratory of Clinical Science to develop assays for dopamine-beta-hydroxylase, serotonin, tryptamine, and tyramine. Radiometric enzymatic assays for catecholamines have recently been eclipsed by the development of techniques coupling high performance liquid chromatography with electrochemical detection.[1] This latter method is less expensive and more rapid than, but as sensitive as, the radiometric enzymatic assay.[2]

"It is noteworthy that this article was selected as a *Citation Classic* this year when Julius has formally retired from the National Institute of Mental Health. This study by two young postdoctoral fellows directly issued from the conceptual approaches and the ambience of collaborative interactions that characterized Julie's laboratory."

1. Keller R, Oke A, Mefford I & Adams R N. Liquid chromatographic analysis of catecholamines: routine assay for regional brain mapping. *Life Sci.* 19:995-1003, 1976. (Cited 405 times.)
2. Zaczek R & Coyle J T. Rapid and simple method for measuring biogenic amines and metabolites in brain homogenates by HPLC-electrochemical detection. *J. Neural Transm.* 53:1-5, 1982.

## This Week's Citation Classic

CC/NUMBER 39
SEPTEMBER 27, 1982

Allen R J L. The estimation of phosphorus. *Biochemical J.* 34:858-65, 1940.
[Low Temperature Station for Research in Biochemistry and Biophysics,
University of Cambridge, Cambridge, England]

In the Fiske and Subbarow[1] method for estimating phosphorus by reduction of phosphomolybdic acid the intensity of the blue colour so produced varies with time and temperature. A method is described in which these difficulties are eliminated and which is applicable also to turbid and coloured solutions. [The *SCI®* indicates that this paper has been cited in over 1,410 publications since 1961.]

---

Russell J. L. Allen
63 Abbotsbury Close
London W14 8EQ
England

April 27, 1982

"In my early days as a graduate student at the University of Melbourne I became familiar with the Duboscq colorimeter, much favoured by biochemists at that time for colorimetric analyses. In this device, the colour densities generated in test and reference samples by the chosen analytical procedure were compared by finding the depths of solution that produced equal intensities of colour when viewed by transmitted light in a split field eyepiece. The concentration of the substance under analysis present in the test solution could then be calculated.

"When I arrived in Cambridge in 1936 to work on the phosphorus metabolism of the higher plants, I found that the colorimeter was being replaced by the more refined photometer. Here the optical density of the test solution was measured directly by a photocell connected to a sensitive galvanometer and scale calibrated against a graded series of standard solutions. But when I started to use a photometer for estimating phosphorus by the popular Fiske and Subbarow[1] method I soon encountered problems. First, the colour density went on increasing for about 20 minutes after the reagents were added to the test solution. One had to wait for at least that time in order to avoid errors. This was inconvenient, but more serious was the real danger that organic phosphorus compounds in the materials I was working with would be hydrolysed in the strongly acid solution during the waiting period and thus appear as free phosphate. Secondly, colour density at any particular time depended on temperature: as the environmental temperature rose, the strength of the colour increased. Neither effect had mattered when a colorimeter was used because the test solution was compared with a reference solution prepared simultaneously at the same temperature. I eventually solved both problems by substituting 2:4-diaminophenol hydrochloride (amidol) for the 1-amino-2-naphthol-4-sulphonic acid used by Fiske and Subbarow as a reducing agent. Amidol had been considered for use by previous workers but never seriously investigated. The density of the blue colour produced with this reagent is for all practical purposes independent of time and temperature.

"I was able, by an extension of the amidol method, to overcome another vexing problem. The estimation of phosphorus in many of the plant extracts with which I was working was made difficult by the presence in these of turbidity and extraneous colour. I found that the blue colour could be extracted by *isobutyl* alcohol and measured in the photometer free from any interference.

"I have not worked in this field for more than 40 years.[2] The amidol method has been highly cited because it became available just when everyone was changing over from colorimeters to photometers and was no doubt welcomed by other investigators who were experiencing the same problems that I had encountered. But the survival of this old method into the age of analysis by black boxes and other sophisticated gadgetry is as surprising to me as it is gratifying."

1. Fiske C H & Subbarow Y. The colorimetric determination of phosphorus. *J. Biol. Chem.* 66:375-400, 1925.
2. No review of methods determining phosphorus has been published recently.

# This Week's Citation Classic

CC/NUMBER 31
AUGUST 3, 1981

Lowry O H & Lopez J A. The determination of inorganic phosphate in the presence of labile phosphate esters. *J. Biol. Chem.* **162**:421-8, 1946.
[Div. Nutrition and Physiol., Public Health Res. Inst., New York, NY]

Earlier colorimetric methods for inorganic phosphate, based on phosphomolybdate reduction, caused rapid cleavage of labile organic phosphates such as phosphocreatine or acetyl phosphate. This problem was solved by using a more powerful reducing agent, ascorbic acid, at a much less acid pH. [The *SCI*® indicates that this paper has been cited over 970 times since 1961.]

---

Oliver H. Lowry
Department of Pharmacology
School of Medicine
Washington University
St. Louis, MO 63110

June 9, 1981

"Herman Kalckar was stranded in the US during World War II and was hired by Otto Bessey to work with us in the newly formed Public Health Research Institute of New York City. There Kalckar discovered that hepatic nucleosidase, thought to be a simple hydrolase, was actually a phosphorylase,[1] and that one product, ribose-1-P, is very unstable under conditions for $P_i$ assay by the standard Fiske and Subbarow[2] method. This made it impractical to measure the enzyme by $P_i$ appearance or disappearance. I had been fiddling with the analytical conditions of the $P_i$ assay (to try to increase sensitivity) and bet Herman that I (meaning of course Jeanne Lopez) could easily find conditions that would overcome this problem. Jeanne and I won (I forget what the bet was) but it took several months and a notebook full of experiments.

"Most early $P_i$ methods depended on the fact that the phosphomolybdate complex is much more readily reduced (to give a blue color) than is molybdate itself. Bell and Doisy[3] were probably the first to use this principle to design a practical biological method, but the Fiske and Subbarow procedure became better known (and certainly would have been a *Citation Classic* in those days—and perhaps still is?). I knew both Fiske and Subbarow quite well, and Fiske had told me (between long pauses to puff on his pipe) some of the ways this exceedingly complex and versatile system can be manipulated.

"Four variables affect the reduction of molybdate and phosphomolybdate: pH, temperature, molybdate concentration, and the reducing agent. The first three affect hydrolysis of most organic phosphates. We started by tampering with pH, using the reducing agent of Fiske and Subbarow (sulfite with aminonaphtholsulfonic acid catalyst). When the pH was raised from that recommended (0.65) to the range between 0.8 and 1.8, molybdate itself was reduced. Above pH 3.2, phosphomolybdate was only slowly reduced. This left a pH window between 1.8 and 3.2 that could have been used with some gain in organic phosphate stability. However, by substituting ascorbic acid as the reducing agent, the useful window was shifted even further, to pH 3.5-4.5, and permitted the use of a lower molybdate concentration. (Ascorbic acid had been employed earlier by Ammon and Hinsberg,[4] but only in strong acid.) These changes increased stability 15-fold for P-creatine, and over 100-fold for ribose-1-P and acetyl-P and made it easy to measure $P_i$ in their presence.

"The generous number of citations of this paper merely reflects the tremendous importance of phosphate transactions in all manner of metabolic systems. However, this paper may be cited much less in the future. This is because of the discovery that phosphomolybdate gives 20 or 30 times more color when complexed with malachite green than when it is reduced.[5] Moreover, the reaction is nearly instantaneous. This probably means that even though strongly acid conditions are required, with a little more tinkering the absorption can be read in the presence of most phosphate esters before disturbing hydrolysis has occurred."

---

1. **Kalckar H M.** Enzymatic synthesis of a nucleoside. *J. Biol. Chem.* **158**:723-4, 1945.
2. **Fiske C H & Subbarow Y.** The colorimetric determination of phosphorus.
   *J. Biol. Chem.* **66**:375-400, 1925. [The *SCI*® indicates that this paper has been cited over 10,300 times since 1961.]
3. **Bell R D & Doisy E A.** Rapid colorimetric methods for the determination of phosphorus in urine and blood. *J. Biol. Chem.* **44**:55-67, 1920.
4. **Ammon R & Hinsberg K.** Colorimetrische Phosphor- und Arsensäurebestimmung mit Ascorbinsäure. *Z. Physiol. Chem.* **239**:207-16, 1936.
5. **Itaya K & Ui M.** A new micromethod for the colorimetric determination of organic phosphate. *Clin. Chim. Acta* **14**:361-6, 1966.

# Citation Classics

Number 9 — February 28, 1977

Chen P S, Toribara T Y & Warner H. Microdetermination of phosphorus.
*Analyt. Chemistry* 28:1756-58, 1956.

This paper presents a method for phosphorus determination sufficiently sensitive to dispense with microtechniques and special glassware and apparatus. The procedure also possesses the advantages of stability, constancy, and linearity. An ascorbic acid method for the reduction of phosphomolybdate is applied to the determination of phosphorus in whole blood, plasma, serum, and urine with a sensitivity about 8 times greater than that of the aminonaphtholsulfonic acid method. [The SCI® indicates that this paper was cited 1,984 times in the period 1961-1975.]

Dr. Philip S. Chen, Jr.
Assistant Director for Intramural Affairs
National Institutes of Health
Bethesda, Maryland 20014

December 22, 1976

"This method was worked out in early 1954 immediately after the writing had been completed on my Ph.D. thesis in pharmacology at the University of Rochester. Entitled 'Studies on the Renal Excretion of Calcium,' the thesis describes calcium excretion by the dog as a function of its physico-chemical state in the serum (i.e., ultrafilterability) and as influenced by the serum level of various ions (calcium itself, strontium, phosphate) and of various pharmacologic agents (complexing agents, metabolic inhibitors, etc.). The dependent relationship of serum calcium and inorganic phosphate was an important variable in these studies. During my thesis research, I used the well-known Fiske and Subbarow colorimetric method to determine inorganic phosphate in blood serum and urine.

"As a graduate student, I was greatly influenced and aided by Dr. Taft Toribara, an analytical chemist who had trained under the renowned H.H. Willard at the University of Michigan. Taft's wise counsel on such problems as ultrafiltration and calcium analysis contributed greatly to my own research style.

"By the end of 1953, my thesis was written and submitted to my thesis advisor, Dr. William Neuman; and other members of my graduate committee for review prior to the final oral examination. With some leisure time on my hands, and a happy feeling of having completed an arduous period of training, I happily perused a wide range of publications in the medical school library. Early in 1954, I read the publication by Lowry et al.[1] describing their microanalytical studies of a variety of constituents of brain tissue. Their article revived my interest in the analytical problems relating to phosphate determination in which I had a continued interest.... On the basis of a number of experiments run in the spring of 1954, it appeared that ascorbic acid reduction of the phosphomolybate complex would not only result in a blue color with much greater molar absorbancy than the product produced by aminonaphtholsulfonic acid in the Fiske-Subbarow method, but offered a number of other advantages as well....

"During the summer of 1954, Huber Warner, who had just finished his freshman year at Williams College, came to work as a summer employee in Dr. Toribara's laboratory. He was assigned the task of comparing the new ascorbic acid method with the Fiske-Subbarow method. Thus by the end of the summer of 1954, the basic characteristics of the procedure had been worked out.... The paper was finally written and submitted to *Analytical Chemistry* about the time that I left the University of Rochester to assume a position at the National Heart Institute, National Institutes of Health.

"Over the years, I have received a number of personal testimonials from friends and colleagues who had found the method useful in determining phosphorus in a wide variety of biological samples. I believe the basic reaction has proven itself over the course of time to be a most reliable and useful technique and the authors are most gratified at the kindly reception which it has received from a broad spectrum of users."

1. Lowry O H, Roberts N R, Leiner K Y, Wu M L & Farr A L. The quantitative histochemistry of brain. I. Chemical methods. *J. Biol. Chem.* 207:1-17, 1954.

# This Week's Citation Classic

CC/NUMBER 14
APRIL 6, 1981

Nash T. The colorimetric estimation of formaldehyde by means of the Hantzsch reaction. *Biochemical J.* 55:416-21, 1953.
[Air Hygiene Lab., Public Health Laboratory Service, London, England]

A colour reaction for low concentrations of formaldehyde is described which depends on the formation of diacetyl-dihydrolutidine from acetylacetone and formaldehyde in the presence of excess ammonium salt. Conditions can be mild enough for the survival of biological material. [The *SCI*® indicates that this paper has been cited over 1,340 times since 1961.]

T. Nash
Chemical Defence Establishment
Porton Down
Salisbury, Wiltshire SP4 0JG
England

March 17, 1981

"I am indebted to Dr. (now Professor Sir Robert) Williams for bringing me into the team [The (British) PHLS Committee on Formaldehyde Disinfection] which investigated the use of formaldehyde for disinfecting fabrics. He gave me the job of finding a non-toxic neutralizing agent with which to treat infected fabric after exposure to formaldehyde vapour, in order that the bacterial viability results should not be invalidated by any bacteriostatic action of residual formaldehyde.

"In the course of this work, which had a fairly successful conclusion,[1] I tested various single compounds and then went on to test two-component mixtures as suggested by a study of Robinson's classical researches into alkaloid synthesis in plants, where formaldehyde is a key intermediate.[2] In such reactions, aptly termed 'syntheses under physiological conditions' quite complicated condensations can occur in cold dilute aqueous solution at around neutral pH. Typical of these is the Hantzsch dihydropyridine synthesis by reaction between ammonia or a primary amine, an aldehyde, and two molecules of a beta-diketone or beta-ketoester. I found that the three simplest possible precursors, acetylacetone, ammonia, and formaldehyde, gave a coloured product, diacetyl-dihydro-lutidine (DDL). The reaction was of no value as regards the initial aim of the work, but was successfully followed up as a possible basis for a colorimetric method.

"The reagent was particularly useful in the estimation of formaldehyde vapour concentration over diluted commercial formalin, which gives off a non-bactericidal gas (probably methylene dimethyl ether) which reacted as formaldehyde towards, e.g., the chromotropic acid reagent, but was unreactive towards mine. As regards other methods of estimating DDL, its rather weak fluorescence in solution was not important at the time, but with modern instrumentation, fluorimetry is now a feasible alternative.

"DDL is readily obtained pure, and an interesting later development was the use of a saturated aqueous solution (about 100 micromolar) as a selective colorimetric reagent for ozone.[3] Working out various possible structures on paper, the ozonolysis product of DDL looks like a promising chelating agent for some metals, but I have not followed this up. The popularity of the Hantzsch reagent is probably due to (1) the spelling, (2) the impeccable century-old pedigree, (3) the fact that it can be spilt about and even imbibed with little harm, in contrast to the hot strong acids or alkalies of other reagents."

1. **Nash T & Hirch A.** The revival of formaldehyde-treated bacteria. *J. Appl. Chem.* 4:458-63, 1954.
2. **Robinson R.** The structural relations of some plant products. *J. Roy. Soc. Arts* 96:795-808, 1948.
3. **Nash T.** Colorimetric determination of ozone by diacetyl-dihydrolutidine. *Atmos. Environ.* 1:679-87, 1967.

# This Week's Citation Classic

CC/NUMBER 21
MAY 21, 1984

Murphy B E P & Pattee C J. Determination of thyroxine utilizing the property of protein-binding. *J. Clin. Endocrinol. Metab.* 24:187-96, 1964.
[Clin. Invest. Unit, Queen Mary Veterans' Hosp., and Dept. Investigative Med., McGill Univ., Montreal, Canada]

This article described the first practical assay for serum thyroxine (T4), with data demonstrating its diagnostic validity in hyper- and hypothyroid patients and its freedom from the effects of iodine. [The *SCI®* indicates that this paper has been cited in over 545 publications since 1964.]

---

Beverley E. Pearson Murphy
Montreal General Hospital
Montreal, Quebec H3G 1A4
Canada

April 20, 1984

"The work in this paper formed part of my PhD thesis[1] done under the supervision of Chauncey Pattee. Since the gestation of this paper paralleled that of my second child, conception of both occurring in the fall of 1962, my memories of them remain associated.

"I had recently (June 1962) presented a paper describing a new method for the determination of serum corticoids based on their binding to human corticosteroid-binding globulin[2,3] and I had a hunch that a similar technique would be applicable to the specific determination of thyroxine since a comparable sort of protein—thyroxine-binding globulin (TBG)—existed for thyroxine.

"In the fall of 1962, I began doing preliminary experiments—I obtained thyroxine and its analogues and then attempted to determine thyroxine in precisely the same way used for serum corticoids. The necessary steps were to precipitate the proteins to free the hormone, then to measure it according to its competition with $^{131}$I-thyroxine for sites on a fixed amount of TBG. The second step involved the separation of the protein-bound and unbound fractions; for this I used Sephadex gel filtration, a method we had recently substituted for dialysis in the determination of corticoids.

Development of the thyroxine method was a matter of working backward, i.e., first establishing the separation technique, then getting a satisfactory standard curve, and finally measuring the hormone in serum.

"The initial experiments went well, with the setting-up of 12 small Sephadex columns which successfully separated the bound and free hormone. However, the standard curve was unsatisfactory, not behaving at all as I had expected, and quite useless for assay purposes. One day in January 1963, feeling ill with a cold, nauseated due to my pregnancy, and discouraged with my results, I decided to go home to bed. On my way out, I picked up Antoniades's *Hormones in Human Plasma* to read. Between trips to the bathroom, I made my way through the chapter on thyroid hormones by Ingbar and Freinkel[4] and suddenly realized what was wrong. I had neglected to consider the effects of prealbumin, another, but weaker, T4 binder in blood. My reading that afternoon also provided the remedy—simply alter the buffer from phosphate to barbital. The following day, I obtained an excellent standard curve and from then on all went smoothly (with both method and pregnancy). Ably assisted by technician Sorel Cohen, we rapidly accumulated data and in March we were able to submit an abstract; in May, a manuscript. In June, I presented our findings in London, Ontario. Three weeks later, two professors from Ottawa visited the laboratory to learn the technique and invited me to lunch (dining out is one of Montreal's delights). Declining reluctantly, I instead made my way over to the maternity hospital and reported back to the lab a little later that it was a girl. During that summer, we revised the paper, which was accepted in October.

"This paper owes its popularity to describing the first validated method for the specific determination of thyroxine, demonstrating a rapid means of diagnosing with considerable accuracy the two common diseases, hypo- and hyperthyroidism.

"Currently, serum thyroxine is often determined by radioimmunoassay, a technique which is similar in principle but replaces TBG with an antibody to thyroxine."[5]

---

1. **Murphy B E P**. *Some aspects of the protein-binding of corticosteroids and thyroxine in human blood*. PhD thesis. McGill University, 1964. 205 p.
2. ─────. Some studies of the protein-binding of steroids and their application to the routine micro and ultramicro measurement of various steroids in body fluids by competitive protein-binding radioassay. *J. Clin. Endocrinol. Metab.* 27:973-90, 1967.
3. ─────. Citation Classic. Commentary on *J. Clin. Endocrinol. Metab.* 27:973-90, 1967. *Current Contents/Life Sciences* 24(3):17, 19 January 1981.
4. **Ingbar S H & Freinkel N**. Thyroid hormones. (Antoniades H N, ed.) *Hormones in human plasma*. Boston: Little, Brown, 1960. p. 515-79.
5. **Refetoff S**. Thyroid function tests. (De Groot L H, Cahill G F, Odell W D, Martini L, Potts J T, Jr., Nelson D H, Steinberger E & Winegrad A I, eds.) *Endocrinology*. New York: Grune and Stratton, 1979. Vol. 1. p. 387-428.

| Number 19 | **Citation Classics** | May 8, 1978 |

**Simon E J & Shemin D.** The preparation of S-succinyl coenzyme A.
*J. Amer. Chem. Soc.* 75:2520, 1953.

---

The authors present a method for the synthesis of succinyl coenzyme A (succinyl CoA) by addition of succinic anhydride to CoA in the presence of a weak base. The anhydride method is applicable to the synthesis of other acyl derivatives of CoA. [The *SCI®* indicates that this paper was cited 621 times in the period 1961-1976.]

---

Professor Eric J. Simon
New York University Medical Center
550 First Avenue
New York, New York 10016

December 2, 1977

"Late in 1951, with a fresh Ph.D. in Organic Chemistry, I arrived in the laboratory of Dr. David Shemin at Columbia College of Physicians and Surgeons. The object of my post-doctoral fellowship was to train in biochemistry, a field which appeared to me much more exciting than organic chemistry.

"Coenzyme A (CoA) had been discovered rather recently in Dr. Fitz Lipmann's laboratory and 'active acetate' had been shown to be acetyl CoA by Dr. Feodor Lynen. It had become quite clear that acyl CoA derivatives were the substrates of a large number of enzymes important to intermediary metabolism. Dr. Shemin had just proved that the first step in porphyrin synthesis was the condensation of succinyl CoA with glycine. It occurred to us that a simple chemical synthesis of succinyl CoA, preferably one applicable to other acyl CoA derivatives would therefore be extremely useful to many investigators.

"Since CoA was very expensive, all initial experiments were done with glutathione (GSH). It soon became evident that the sulfhydryl groups disappeared rapidly when succinic anhydride was added to GSH solution in the presence of a weak base. Only a slight excess of anhydride was needed for an essentially quantitative reaction. The product gave a hydroxamic acid test under conditions that suggested the presence of a thioester. Repetition of the experiment with CoA gave similar results.

"A crucial question remained: Did the succinic anhydride esterify other functional groups present in CoA and thereby destroy its activity? This question was best answered by determining whether our product was biologically active.

"I contacted Dr. Charles Gilvarg, who at that time was a post-doctoral fellow in Dr. Severo Ochoa's laboratory at NYU. Professor Ochoa and his collaborators, Gilbarg, Seymour Kaufman, Minor Coon and the late Joseph Stern, were anxious to test our synthetic material. They were working with several enzymes for which succinyl CoA was a substrate. Our material was quickly found to be biologically active. We went on to show that the method was applicable to the acylation of CoA by just about any acid for which an anhydride was available.

"In discussing the work with a number of distinguished biochemists, it became evident that many of them had thought of this simple approach but dismissed it since 'anhydrides would surely react with all kinds of functional groups present in the CoA molecule.' The biological innocence of an organic chemist and the carefree approach of youth overcame the hesitations, and the experiment that 'could not work' was carried out.

"The timing of this work was perfect since the demand for acyl CoA derivatives was world-wide. Publication of this extremely simple but useful synthesis provided a fine start for a young organic chemist who wanted to embark on a career in biochemistry. My everlasting thanks go to my mentor, David Shemin, whose encouragement, stimulating discussions and wide knowledge of biochemistry made this work possible and started me on a satisfying and exciting career."

## This Week's Citation Classic

Porath J, Axén R & Ernback S. Chemical coupling of proteins to agarose.
*Nature* 215:1491-2, 1967.
[Institute of Biochemistry, University of Uppsala, Sweden]

In alkaline media, cyanogen bromide reacts with hydroxyl groups of agarose and other polysaccharides. The polymers are converted to highly reactive derivatives. Such 'activated gel derivatives' are then used to prepare a variety of adsorbents for bioaffinity chromatography, diagnostics, and immobilized enzymes. [The *SCI®* indicates that this paper has been cited in over 860 publications since 1967.]

---

Jerker Porath
Biomedical Center
Institute of Biochemistry
University of Uppsala
S-751 23 Uppsala
Sweden

May 7, 1984

"As a continuation of our developmental work on molecular sieving gels (Sephadex, etc.), Per Flodin and I, together with Lennart Rhodén, attempted to prepare biospecific adsorbents for isoagglutinins. Isocyanate groups were introduced into Sephadex to make possible thiourea bridging with blood group substances. Due to unforeseen circumstances, the work was interrupted. Several years later, Rolf Axén and I published a similar immobilization procedure[1] which paved the way for the introduction of the first gel-based radioimmunoassay technique.[2] We soon realized, however, that a different, and perhaps better, immobilization technique was highly desirable to improve and extend our newly introduced method.

"Axén joined Bernhard Witkop at Bethesda to learn new techniques for selective cleavage of protein. Upon his return to the Institute of Biochemistry at Uppsala, he suggested a new approach to our project, *viz.*, the coupling of substances containing primary amino groups to cyanamide-Sephadex. The idea was suggested as a research project to Sverker Ernback, one of our first-year research students. The cyanamide-Sephadex was prepared by treating amino-Sephadex with cyanogen halide. To my surprise, the yield of coupled amino acid exceeded 100 percent! I urged Ernback to make blind experiments using unsubstituted Sephadex. Indeed, Sephadex after cyanogen halide treatment was found to immobilize protein in excellent yields![3]

"Cyanogen bromide coupling replaced the isothiocyanate procedure for the synthesis of immobilized antigens and antibodies to be used in radioimmunoassays. It is still the most commonly used method for preparing gel- and paper-based immunodiagnostics (RIA, RAST, PRIST, etc.).

"Our discovery of the cyanogen halide coupling method was followed by hectic work in several directions. Enzymes and enzyme inhibitors were immobilized onto a variety of hydroxylic supports. Not much later, our work on activation of agarose for enzyme immobilization was described, and this work initiated an almost explosive development in (bio-)affinity chromatography.

"We interpreted the activation to involve the formation of cyanate followed by its rapid conversion to imino carbonate. The final coupling products were thought to be gels containing mixtures of imino carbonic acid esters, carbonic acid esters, and carbamate substituents, and, somewhat later, isourea linkages were also considered. Evidence for this interpretation was obtained from IR-spectra including also some model compounds.

"The complicated scheme of reactions is now fairly well understood, thanks to Meir Wilchek[4] and others. The nucleophilic displacement of the ligands with ammonia is particularly interesting: original amino groups are converted into guanidino groups. By using the cyanogen bromide activated support as an organic reagent, Wilchek converted insulin into 'superinsulin.'

"Our original suggestions have been essentially confirmed, but the recent work has shed light on some important limitations. In improved form, the cyanogen bromide coupling is still the preferred method for the preparation of most biospecific adsorbents used in (bio-)affinity chromatography, and agarose is by far the most commonly employed support."

---

1. Axén R & Porath J. Chemical coupling of amino acids, peptides and proteins to Sephadex.
   *Acta Chem. Scand.* 18:2193-5, 1964. (Cited 35 times.)
2. Wide L & Porath J. Radioimmunoassays of proteins with the use of Sephadex-coupled antibodies.
   *Biochim. Biophys. Acta* 130:257-60, 1966. (Cited 570 times.)
3. Axén R, Porath J & Ernback S. Chemical coupling of peptides and proteins to polysaccharides by means of cyanogen halides. *Nature* 214:1302-4, 1967. (Cited 1,655 times.)
4. Kohn J & Wilchek M. The determination of active species on CNBR and trichloro-s-triazine activated polysaccharides. (Gribnau T C J, Visser J & Nivard R J F, eds.) *Affinity chromatography and related techniques: theoretical aspects industrial and biomedical applications.* Amsterdam: Elsevier Scientific Publishing, 1982. p. 235-44.

# This Week's Citation Classic

CC/NUMBER 15
APRIL 12, 1982

Cleland W W. Dithiothreitol, a new protective reagent for SH groups.
*Biochemistry* 3:480-2, 1964.
[Department of Biochemistry, University of Wisconsin, Madison, WI]

Because of its low redox potential, dithiothreitol (and its isomer, dithioerythritol) is capable of maintaining monothiols completely in the reduced state and of reducing disulfides quantitatively. It proved much superior to the thiols then used as protective reagents for sulfhydryl groups. [The *SCI®* indicates that this paper has been cited over 810 times since 1964.]

---

W. Wallace Cleland
Department of Biochemistry
University of Wisconsin
Madison, WI 53706

November 25, 1981

"When I was hired as an assistant professor at Wisconsin in 1959, they thought they were hiring a lipid biochemist. My interest in enzyme kinetics (which is now the major thrust of research in my lab) developed after I came to Wisconsin, and the theoretical work I published in *Biochimica et Biophysica Acta* in 1963 was an earlier Citation Classic.[1] But for eight years I had a National Institutes of Health grant to study the substrate specificity for acyl-CoA thioesters in the acylation of glycerophosphate to phosphatidic acids. We prepared acyl-CoA's by chemical acylation of CoA, which in those days was only 75 percent reduced. Since it cost $600 per gram, we wanted some way to reduce the rest and keep it reduced.

"Thiols then, as now, were most readily kept reduced by having an excess of another thiol present, since thiols readily exchange with disulfides. However, the equilibrium constant for thiol-disulfide interchange is near unity. It occurred to me that if one used a dithiol of suitable chain length, that the mixed disulfide of the dithiol and CoA would cyclize internally to give a cyclic disulfide and reduced CoA. Because the reaction produced two products from one reactant, it should go to completion. A cyclic disulfide with a six-membered ring seemed the best bet, so I bought a bottle of 1,4-butanedithiol. Opening the bottle convinced me instantly that this was not the reagent of choice (it smells like skunk oil, and the deep freeze in which this bottle was stored still reeks of the stuff 20 years later). I figured a couple of hydroxyl groups would cut down on the stench (as well as make the compound water soluble), and a trip to *Chemical Abstracts* showed that indeed 1,4-dithiothreitol and the erythro isomer had been made as part of a search for a better antiarsenical than 1,2-dithioglycerol.[2] I made the compounds, and to my delight they worked exactly as expected, and had a bad smell only at very close range.

"With the publication of the cited article, I began to receive requests for the compounds, and it was rapidly clear that I either had to start a pilot plant run, or convince a commercial vendor to make the stuff. Calbiochem was the first to market dithiothreitol (calling it 'Cleland's Reagent'), and they have continued to advertise the virtues of the compounds, and have distributed bibliographies on their use.[3]

"This paper has been highly cited for the following reason. In the 18 years since this paper was published, these dithiols have become the reagents of choice for keeping thiols, and the thiol groups of enzymes, in the reduced state. CoA became available commercially in the fully reduced state, and the yields of many enzyme isolations improved drastically. It is very satisfying to know that a pair of compounds designed on paper to accomplish a specific purpose did exactly what they were supposed to do; not all of one's ideas are so successful!"

---

1. **Cleland W W.** The kinetics of enzyme-catalyzed reactions with two or more substrates or products. I. Nomenclature and rate equations. *Biochim. Biophys. Acta* 67:104-37, 1963.
   [Citation Classic. *Current Contents* (28):8, 11 July 1977.]
2. **Evans R M, Fraser J B & Owen L N.** Dithiols. Part III. Derivatives of polyhedric alcohols. *J. Chem. Soc.* 1949: 248-55, 1949.
3. Cleland's reagent: a current bibliography prefaced with Professor Cleland's original paper describing dithiothreitol, a new protective reagent for SH groups. La Jolla, CA: Calbiochem-Behring Corp., 1979. 27 p.

## This Week's Citation Classic

CC/NUMBER 28
JULY 14, 1980

Bodanszky M & du Vigneaud V. A method of synthesis of long peptide chains using a synthesis of oxytocin as an example. *J. Amer. Chem. Soc.* 81:5688-91, 1959.
[Dept. of Biochemistry, Cornell Univ., Ithaca, NY]

The previously generally accepted method for the construction of peptides, combination of shorter segments of their chain, is replaced by a new approach to chain building, the addition of single amino acids. Unequivocal incorporation of the (protected) amino acids was achieved by the application of nitrophenyl esters. [The *SCI®* indicates that this paper has been cited over 615 times since 1961.]

Miklos Bodanszky
Department of Chemistry
Case Western Reserve University
Cleveland, OH 44106

June 27, 1980

"Early in 1957 my wife, our five-year old daughter, and I arrived as refugees in New York City. I came to join du Vigneaud in his studies (recognized a year earlier by the Nobel Prize) on the chemistry of the peptide hormones oxytocin and vasopressin. Our newly acquired freedom, together with the absence of material possessions such as home, car, or television set, allowed me to concentrate on the task at hand: a new synthesis of oxytocin. At the same time, I was still somewhat obsessed by thoughts about my new procedure for the coupling of amino acids to each other, developed before I left Hungary: the nitrophenyl ester method.[1] Thus, the idea to incorporate only protected-activated amino acids in the synthesis of oxytocin, rather than to follow the classical approach of combining segments of the peptide chain, presented itself quite naturally. The stepwise addition of single residues allowed systematic lengthening of the chain, without endangering the optical purity of the amino acid constituents. Du Vigneaud enthusiastically approved the project and supported it with his tremendous knowledge of the problems surrounding oxytocin, his 'baby protein.' The progress of the synthesis appeared breathtakingly fast. Within a short time we had a fairly large sample of oxytocin in our hand in a yield far exceeding the yields of previous syntheses. The purity of the intermediates, all isolated in crystalline form, was also encouraging. The simplicity of the handling of the prefabricated reactive intermediates, p-nitrophenyl esters of benzyloxycarbonyl amino acids, suggested that this could be a general method for the synthesis of any long peptide chain. This view was expressed also in the title of our paper. The repetitiveness of the operation seemed to lend itself to mechanization and automation,[2] and the stepwise strategy indeed acted as a stimulus in the invention of techniques for the facilitation of peptide synthesis.

"It remains to be seen whether or not all peptides and proteins can be synthesized by a single process. The specificity of the biological activities of individual peptides is directly related to differences in their chemistry. Therefore, an entirely universal approach could be near to impossible. Yet, stepwise synthesis is often practical and we could successfully apply it for the first synthesis of the gastrointestinal hormone secretin and, most recently, for the avian vasoactive intestinal peptide. Several laboratories adopted the stepwise strategy. Some of the publications referred to this much cited paper, but gave credit for the details of the reactive intermediates (which we described years earlier[1]) rather than for the general idea proposed in it. Thus, we read these references with mixed feelings. Nevertheless, it is very gratifying to write these comments. They bring back the memory of our first years in the United States, the excitement of New York City, the congenial atmosphere of the du Vigneaud laboratory, the colleagues (many distinguished peptide chemists by now: J. Meienhofer, J. Stouffer, R. Studer, A. Light, W. Cash, P. Katsoyannis, C. Ressler, V.V.S. Murti, P. Fitt, to mention a few), a time of great expectations, and a time when we all were very much alive. That a 'Citation Classic' resulted from this period, shows that some of the expectations were fulfilled: it is not easy to make even a small dent on the surface of chemistry. Our time was not spent in vain."

1. **Bodanszky M.** Synthesis of peptides by aminolysis of nitrophenyl esters. *Nature* (London) 175:685, 1955.
2. **Bodanszky M.** Stepwise synthesis of peptides by the nitrophenyl ester method.
   *Ann. NY Acad. Sci.* 88:655-64, 1960.

# This Week's Citation Classic

CC/NUMBER 48
NOVEMBER 28, 1983

Schnabel E. Verbesserte Synthese von tert.-Butyloxycarbonyl-aminosäuren durch pH-Stat-Reaktion. (Improved synthesis of tert.-butyloxycarbonyl-amino acids by pH-stat reaction.) *Liebigs Ann. Chem.* **702**:188-96, 1967.
[Pharmazeutisch-Wissenschaftlichen Laboratorium der Farbenfabriken Bayer AG, Elberfeld, Federal Republic of Germany]

The tert.-butyloxycarbonyl derivatives of free and partially protected amino acids are obtained in good yields by acylation with t-butylazidoformate at controlled pH. Sterically hindered amino acids also react smoothly. By variation of the pH values, the yield and the speed of the acylations can easily be determined for each amino acid derivative. [The *SCI®* indicates that this paper has been cited in over 540 publications since 1967.]

Eugen Schnabel
Institut für Biochemie
Bayer AG
D-5600 Wuppertal 1
Federal Republic of Germany

August 9, 1983

"In 1957, the acid labile tert.-butyloxycarbonyl group (BOC) was introduced for N-protection of the amino groups of amino acids in peptide synthesis by McKay and Albertson[1] and independently by Anderson and McGregor.[2] After Schwyzer and his co-workers[3] had developed a generally feasible synthesis for BOC-amino acids using t-butylazidoformate[4] and magnesium oxide in water-dioxane as the solvent, these compounds found increasing application in peptide synthesis.

"When I started working for Bayer at Wuppertal in 1964, I was faced with the problem of how to synthesize analogues of the peptide hormone physalaemin. Our approach was to use extensively BOC-protected amino acids in conventional fragment condensation utilizing as many common intermediates as possible. Employing Schwyzer's method, most of the BOC-derivatives could readily be prepared but some were obtained in moderate yields only—especially BOC-Tyr(Bzl), BOC-Cys(Bzl), BOC-Ile, and BOC-Asp. Poor solubility and steric hindrance seemed to be the cause. Better yields of BOC-Tyr(Bzl) were indeed obtained with sodium hydroxide as base, but, during the reaction, Tyr(Bzl) precipitated. I next used an autotitrator to run the acylation at a constant pH of 10.4 in order to prevent precipitation. The reaction could easily be followed by the uptake of base and came out beautifully. BOC-Tyr(Bzl) was obtained in excellent yield and purity. On trying other amino acids whose acylation with t-butylazidoformate had been problematic, equally good results were obtained. Even the N-alkylamino acids reacted smoothly. It was amazing to watch the reaction of t-butylazidoformate with Pro. At pH 9.1, it proceeded with warming and was virtually complete within a few minutes. Even at pH 7.9, the acylation went to completion, though it took 12 hours. Next, my technician, J. Stoltefuss, suggested the use of buffers. He tried a few. The results looked promising but did not match those of autotitration.

"For several years, the autotitrator method was widely used for the preparation of the BOC-amino acids. Recently, di-tert.-butyldicarbonate[5,6] has become the reagent of choice for their synthesis, since it is commercially available, easy to handle, and an efficient acylating agent.

"I was really surprised at the acceptance of this paper: the use of t-butylazidoformate for the synthesis of BOC-amino acids was known,[3] autotitration was an already long known principle, too, and most of the derivatives had been synthesized previously. I, therefore, had hesitated for some time to publish this paper. Progress in peptide synthesis and especially Merrifield's solid phase method[7] produced a big demand for BOC-amino acid derivatives, of which only a few were commercially available at that time, and any procedure giving better yields was more than welcome. In addition, the paper contained a table with the physical constants of most BOC-amino acids and many colleagues have cited that data. In my opinion these are the reasons why this paper has become a *Citation Classic.*"

1. McKay F C & Albertson N F. New amine-masking groups for peptide synthesis.
   *J. Amer. Chem. Soc.* **79**:4686-90, 1957.
2. Anderson G W & McGregor A C. *t*-Butyloxycarbonylamino acids and their use in peptide synthesis.
   *J. Amer. Chem. Soc.* **79**:6180-3, 1957. (Cited 310 times.)
3. Schwyzer R, Sieber P & Kappeler H. Zur Synthese von N-t-Butyloxycarbonyl-aminosäuren.
   *Helv. Chim. Acta* **42**:2622-4, 1959. (Cited 255 times.)
4. Carpino L A. Oxidative reactions of hydrazines. IV. Elimination of nitrogen from 1,1-disubstituted-2-arenesulfonhydrazides. *J. Amer. Chem. Soc.* **79**:4427-31, 1957.
5. Pozdnev V F. Application of di-tert-butyl pyrocarbonate for preparation of N-tert-butyloxycarbonyl derivatives of amino acids. *Khim. Prir. Soedin.* SSSR **1974**:764-7.
6. Moroder L, Hallett A, Wünsch E, Keller O & Wersin G. Di-tert.-butyldicarbonat—ein vorteilhaftes Reagenz zur Einführung der t.-Butyloxycarbonyl-Schutzgruppe. *Hoppe-Seylers Z. Physiol. Chem.* **357**:1651-3, 1976.
7. Merrifield R B. Solid phase peptide synthesis. I. The synthesis of a tetrapeptide.
   *J. Amer. Chem. Soc.* **85**:2149-54, 1963. (Cited 1,340 times.)

# This Week's Citation Classic

CC/NUMBER 13
MARCH 29, 1982

Chen R F. Removal of fatty acids from serum albumin by charcoal treatment.
J. Biol. Chem. 242:173-81, 1967.
[Laboratory of Technical Development, National Heart Institute, Bethesda, MD]

The paper described a new and simple method for removing fatty acids and other impurities from serum albumin; namely, treatment with activated charcoal at low pH. Physical tests showed that the method did not denature the protein. [The $SCI^®$ indicates that this paper has been cited over 1,070 times since 1967.]

---

Raymond F. Chen
National Heart, Lung, and
Blood Institute
National Institutes of Health
Bethesda, MD 20205

December 4, 1981

"This work was a by-product of studies on protein fluorescence. In the early-1960s, as a result of work by Gregorio Weber,[1] there was great interest in fluorescence spectroscopy as a new tool for studying proteins. Unfortunately, most laboratories did not have the expertise to construct the requisite fluorescence instrumentation. To remedy this situation, the chief of my laboratory at the National Heart Institute (NHI), Robert L. Bowman, encouraged me to see how commercial instruments could be adapted to measure quantum yields, corrected spectra, polarization, and lifetimes. Eventually, I was able to show how these data could be obtained with available instruments, thus contributing to making fluorescence a generally accessible technique. Some measurements, such as quantum yields, required standards, and I was led to examine the suitability of many compounds. For protein quantum yields, I naively thought that the best standard would be another protein, and what better choice could there be than common serum albumin?

"However, I soon discovered that because of impurities, no two samples of serum albumin had the same emission properties. It turns out that albumin binds nearly everything and is more stable when complexed with fatty acids, certain alcohols, or N-acetyltryptophan. These or similar additives are often deliberately incorporated into commercial samples. Impurity removal required prolonged solvent extraction or other tedious procedures.

"The idea of using charcoal to purify serum albumin probably came to me for two reasons. First, our storeroom at the NHI was chronically obstructed with an enormous amount of activated charcoal. Theodor Kolobow, in another section of Bowman's multifaceted laboratory, had been studying charcoal facilitation of metabolite transport across silicone rubber membranes used in artificial lungs and kidneys. In response to his request for a free sample, Atlas Chemical Industries had shipped him (to his horror) several hundred pounds of charcoal. Second, while working on my dissertation a few years earlier under Gerhard W.E. Plaut at the University of Utah, I was studying the stereochemistry of the reaction catalyzed by the DPN-linked isocitrate dehydrogenase of cardiac mitochondria. In order to see if tritium was transferred to the $\alpha$- or $\beta$-side of $DPN^+$, Plaut suggested isolating the tritiated DPNH by charcoal adsorption, a technique which I found worked despite the presence of protein. (Incidentally, the enzyme uses the $\alpha$-side of DPNH.[2])

"As perhaps the most studied protein, serum albumin serves as a model for other proteins and has great importance and interest in its own right. There are many clinical, commercial, and laboratory applications which require purified serum albumin. These factors contribute to the frequency of citation of my paper.

"Incidentally, serum albumin turns out not to be a particularly good fluorescence standard, because even after charcoal treatment, there are minor differences between samples[3] for reasons which continue to be obscure."

---

1. Weber G. Rotational Brownian motion and polarization of the fluorescence of solutions.
   Advan. Prot. Chem. 8:415-59, 1953.
2. Chen R F & Plaut G W E. Studies on the hydrogen transfer mediated by the DPN-linked isocitric dehydrogenase of heart. Biochemistry 2:752-6, 1963.
3. Chen R F & Koester V J. Fluorescence properties of human serum albumin: effect of dialysis and charcoal treatment. Anal. Biochem. 105:348-53, 1980.

Number 14 **Citation Classics** April 3, 1978

**Roberts J D & Mazur R H.** Small-ring compounds. 4. Interconversion reactions of cyclobutyl, cyclopropylcarbinyl and allylcarbinyl derivatives.
*J. Amer. Chem. Soc.* 73: 2509-20, 1951.

This paper was the first to provide a basis for understanding the almost bewildering variety of product mixtures obtained in reactions starting with the simple $C_4H_7X$ derivatives which have cyclobutyl, cyclopropylcarbinyl, and the allylcarbinyl carbon skeletons. In addition, it was shown that cyclopropylcarbinyl and cyclobutyl derivatives were very unusually reactive, far more so than the theories of organic reactions could accommodate at the time the research was done. [The *SCI®* indicates that this paper was cited 214 times in the period 1961-1976.]

John D. Roberts
California Institute of Technology
Div. of Chemistry & Chemical Engineering
Pasadena, CA 91125

November 23, 1977

"Of the some 400 papers I have published before and after 1951, none has given me more satisfaction than this one. The reasons are several. First, R.H. Mazur was one of the first graduate students to elect to work with me at MIT, and he was absolutely outstanding—a man of great intellectual power, with a deep interest in mathematics, who for reasons not clear to me, then or now, wished to do experimental work in organic chemistry. It was a continuing pleasure to watch Mazur's scientific development. Whenever I suggested he try an additional experiment, he said, 'That was done last week, and here are the results.' After completing his PhD at MIT, Mazur went on to become a well-known research chemist at G. D. Searle where, among other things, he and his coworkers developed the low-calorie peptide sweetener, aspartame.

"The second reason was that, although the research involved rather simple, four-carbon, organic compounds and simple reactions, the problem was a challenging one, because we had no good clues at all to what the final outcome would be. Of course, the dilemmas posed by the previously reported work in the area could have had trivial solutions we were just not smart enough to anticipate. However, it did turn out that we were breaking new ground, with implications for organic chemistry in a rather broad way. But even more, the work brought forth a whole new set of chemical puzzles associated with the patterns of reactivity of the substances we investigated. Attempts to solve these puzzles account for the twenty-two follow-up papers we published, and for the large number of times the article has been cited by others. Interestingly, the most important of the new puzzles has still not been resolved 26 years later, and is the subject of active experimental and theoretical investigation in several laboratories.

"Finally, the paper aroused controversy, particularly because one of the important conclusions was seriously questioned by H. C Brown (Purdue), who claimed it was based on faulty experimental data. The reason was that our results did not agree with those he and his coworkers had obtained on the key compound described in our paper. It was indeed true that Mazur had carried on his research with relatively unsophisticated techniques compared to what we might use on the same problem today. However, Mazur was a very careful worker and seemed incapable of the error attributed to him. It was therefore very satisfying when a re-investigation of the samples and procedures used by Mazur, ten years after completion of his thesis, verified his findings in every detail and further determined that Brown had analyzed his own data incorrectly.

"That the key problem posed by our 1951 paper regarding the structures and ease of interconversion of the evanescent cationic intermediates involved in the reactions of cyclopropylcarbinyl, cyclobutyl and allylcarbinyl derivatives has still not been resolved, even after so many years of intensive study, strongly suggests that some of our basic approaches to determine and formulate organic structures, even for very simple molecules, need further revision and strengthening."

Number 11 **Citation Classics** March 13, 1978

Cram D J & Elhafez F A A. Studies in stereochemistry. 10. The rule of 'steric control of asymmetric induction' in the synthesis of acyclic systems.
*J. Amer. Chem. Soc.* 74:5825-35, 1952.

A rule was formulated that correlates and predicts, on the basis of steric effects, the direction of stereochemical bias in addition reactions of aldehydes and ketones in which one chiral carbon is generated in the presence of an attached chiral carbon. [The *SCI*® indicates that this paper was cited 265 times in the period 1961-1976.]

---

Professor Donald J. Cram
Department of Chemistry
University of California
Los Angeles, California 90024

December 2, 1977

"In 1952 we published 'The Rule of "Steric Control of Asymmetric Induction" in the Syntheses of Acylic Systems.' This hypothesis resulted from our syntheses of organic chemical systems needed to investigate our newly discovered 'phenonium ion.' Our rule evolved from earlier observations of the Frenchman, M. Tiffeneau, and the Scot, A. McKenzie. These early 20th century investigators observed a pattern in addition reactions where one chiral (handed) carbon center was constructed while attached directly to a second chiral carbon. Two diastereomerically related compounds were produced. These authors found that the predominant isomer depended on the order in which substituents were introduced into the products.

"A simple analogy illustrates the evolution of our rule. Imagine that Tiffeneau and McKenzie were assembling diads of N or И letters out of parts, ⊳ , turned in any direction. Assume that two three-quarter finished assemblies, N ⊳ and N ⊳, were in hand which needed one part to complete the diads. These early authors noticed that addition of one part to one assembly gave predominantly one diad, and addition of another part to the other assembly gave predominantly a different diad. The diads had the structure, N N and N И, but Tiffeneau and McKenzie did not know which was which. They were unable to correlate the patterns obtained in one set of reactions with the patterns observed in other sets of reactions.

"Our contribution consolidated the many patterns into a single rule with predictive power. In effect, we found that adding ⊲ to N ⊳ gave mainly N N, and adding ⊲ to N ⊳ gave N И. We correlated the structures of the stereoisomeric products in terms of steric effects in the transition states leading to these products. We suggested a reaction mechanism as an explanation for the rule. Our explanation stimulated controversy! My co-workers and colleague-competitors the world over joyously suggested alternative explanations for why the rule worked.

"Our first manuscript on the rule, a Communication to the Editor, was rejected, but our full paper was accepted. At about the same time, V. Prelog in Switzerland investigated a similar problem, except that the chiral centers were 1,4 rather than 1,2 to one another. His and our results were similar in form, as were our rationalizations. Special exceptions to our rule were designed by J.W. Cornforth in Britain and by H. Felkin in France.

"Our publication is much quoted for several reasons. It deals with a problem encountered in many syntheses. The rule is useful in designing synthetic sequences. The explanation offered is reasonable, arbitrary, unprovable, and provocative. It has therefore stimulated experiments, alternative explanations, calculations, and even semiquantitative expression in algebraic language. In my judgment, the reason why the rule works is still not understood. Probably it works in different systems for different reasons which, by chance, can be summarized in terms of our single generalization."

# Chapter 8

# Statistics

The emergence of biostatistics has created an essential companion science to all the life sciences. Discoveries made in the infancy of any compartment of science are generally descriptive in nature and unneeding of mathematical manipulation. Even early quantitative studies are often quite fairly evaluated on a simple plus–minus scale. As more and more subtle features are examined, or when the magnitude of samples studied clouds interpretations of scientific experiments, statistical analysis becomes correspondingly more important. No other defense is needed for the inclusion of commentaries on these sixteen Citation Classics in a volume on the Life Sciences.

It is noteworthy that three of these Classics are textbooks. This possibly expresses the feeling of inadequacy that many biologists experience in the statistical interpretation of their data, particularly when more and more sophisticated tactics are required. Possibly this only reflects the reading habits of biologists who seldom consult journals primarily devoted to the mathematical sciences because of the reading burden in their own specialty. But whatever the cause for their popularity, these three volumes have also been proved as highly worthy. If life scientists do not have one of these three volumes in their personal libraries, they will usually have sufficient judgment to request the librarian to maintain the preferred one on the permanent reserve list. Of these three textbooks that by Snedecor, before his coauthorship with Cochran, was first published in 1937 and that by Steel and Torrie not until 1960. During the period 1961–1975, these two volumes differed little in their citation record. Since Finney's text was evaluated over a seven-year longer span, it should not be compared with the others in terms of numbers. Probit analysis, though a special statistical method, has the advantage of a broad range of application, and the popularity of Finney's book undoubtedly relates to the use of probit analysis in educational testing, in the arrangement of trials in clinical medicine, in the evaluation of the voluminous amounts of data now generated by clinical laboratories, and in similar situations.

Of the thirteen research papers grouped here, only two were published after 1960. The implication that Classics in statistics vanished after that time is erroneous. The emergence of other volumes of *Current Contents* covering the social and physical sciences made them equally as appropriate as the life sciences for recording mathematical advances in statistics.

Nothing is more pleasing to the biologist who must do his own statistics than a well-constructed nomograph that will lead him to a safe conclusion, and this surely must account, in part, for the popularity of the Litchfield-Wilcoxon article. The end dose-50% ($ED_{50}$) calculation so common in pharmacology, microbiology, virology, physiology, etc., is clearly a reason for the popularity of Weil's Classic.

The multiple range test is the topic for the next three Classics, followed by two written by Nathan Mantel, one on relative risk and the other on ranking. Relative risk has become a useful expression in subjects where the actual incidence of a character is too small to allow the usual statistical methods. For example, the correlation of a disease state with the presence of a particular transplantation antigen, most of which are distributed at a low incidence in any population, is evaluated by calculating the relative risk. Ranking, because of its simplicity, has become popular largely for that reason. Five additional Classics complete this chapter, and that by Kaplan and Meier has exceeded the coveted 1,000 citations threshold.

Biostatistics, relative to the life sciences as a whole, is a small field, but the recurrence of Wilcoxon, Duncan, Mantel, and Haenszel as authors of these Classics is unlikely to be due to chance alone. Perhaps this should be evaluated statistically.

Although not strictly statistical in content, the Classic written by Pelz and Andrews provides a suitable conclusion to this volume. Their analysis of the working environment of scientists in industrial, governmental, and academic settings identified, and objectively illustrated by means of charts and tables, the elements essential to research success. These may be summarized in the phrase "creative tension"—a symbiosis of security and pressure to succeed, a combination that is equally applicable to research teams and individuals.

# Citation Classics

Number 19 — May 9, 1977

Snedecor G W & Cochran W G. *Statistical methods applied to experiments in agriculture and biology.* 5th ed. Ames, Iowa: Iowa State University Press, 1956.

This book is presently in its 6th edition, under the title *Statistical Methods* (Iowa State University Press, 1967). Dr. George W. Snedecor, who wrote the original work in 1937, died in February, 1974, at the age of 92. [The *SCI*® indicates that this book was cited 1,688 times in the period 1961-1975.]

---

William G. Cochran
Professor of Statistics Emeritus
Harvard University
Cambridge, Massachusetts 02138

March 8, 1977

"Variation is a fundamental characteristic of living beings. Consequently, basic problems for the investigator in agriculture and biology are how to conduct investigations and how to analyze the collected data in such a way that these variations do not obscure what he or she is trying to learn and thus lead to wrong conclusions. It happens that the best methods we have been able to devise for analyzing data subject to variation depend on results in the rather abstract theory of probability. Many of the specific techniques currently found most useful were developed by one man--Sir Ronald Fisher--at the Rothamsted Experimental Station [England] in agriculture in the 1920's and early 1930's.

"Fisher's ideas were described by him in his book *Statistical Methods for Research Workers,* first edition 1925 [14th ed. Hafner, 1973]. But Fisher had no natural gift for teaching, and successive editions of his book have always been found heavy going.

"Professor Snedecor of Iowa State College (now University) was one of the first to realize the importance of Fisher's ideas and to begin presenting them in his lectures. In 1924, at the suggestion of Henry A. Wallace, later Secretary of Agriculture, Snedecor started a series of Saturday afternoon seminars for research workers in agriculture, in which Wallace participated.

"Snedecor's book *Statistical Methods Applied to Experiments in Agriculture and Biology,* first edition 1937, was itself pioneering. It was both an introductory text for students and a reference source for research workers. It was written clearly in an informal style with a minimum of mathematical symbolism or sophistication, making Fisher's ideas widely usable. Drawing on his long experience as a statistical consultant, Snedecor filled the book with illustrations from agricultural and biological research. For some years it was without a competitor, the third edition appearing in 1940 and the fourth in 1946.

"In time, other good introductory books with similar aims appeared on the market. But simultaneously the importance of sound statistical methods began to be appreciated in engineering, in the social sciences, in business, and in medicine and public health. Since statistical methods are to a considerable extent transferable from one field to another, Snedecor's *Statistical Methods* was extensively quoted as a reference in these fields also. This may help to explain its good performance in the citation race.

"I first became involved in 1955, when George Snedecor asked me to write a final chapter on the planning and analysis of sample surveys for the fifth edition, which appeared in 1956. Dr. Snedecor was then 75. Soon afterwards he asked me to prepare a sixth edition when this became due. It appeared in 1967. In this edition the phrase 'applied to experiments in agriculture and biology' was dropped from the title. This was for two reasons. Examples from other fields had been brought in, and a number of the newer techniques are not directed primarily at data gathered in experiments.

"Incidentally, I wonder if the statistics profession realizes the debt that it owes to the agronomists. Beginning before 1800, the astronomers were, I suppose, the first to apply probability ideas to the handling of their variation. But the agronomists, early in this century, were the first to make intensive studies of the nature of the variation in plant and tree crop yields with a view to finding the best methods for conducting and analyzing experiments. Perhaps more important to us, they were the first to realize that statisticians are useful and must eat. Many of the outstanding professional statisticians in the world today found either their first job, or an early job before they were well known, in agriculture."

# Citation Classics

Number 39     September 26, 1977

Steel, Robert G D & Torrie, James H. *Principles and procedures of statistics.* New York: McGraw, 1960.

---

This book is a valuable reference text for all scientific fields, especially the experimental sciences. James H. Torrie, coauthor of the book, died in 1976. [The *SCI*® indicates that this book was cited 1,381 times in the period 1961-1975.]

---

Professor Robert G.D. Steel
Department of Statistics
School of Physical & Mathematical Sciences
North Carolina State University
Raleigh, North Carolina 27607

February 1, 1977

"The taste of success is sweet! I believe I speak for James H. Torrie, as well as myself, when I say that the wide acceptance of *Principles and Procedures* has been very satisfying to its authors. The satisfaction is all the greater because it was unexpected.

"Perhaps Torrie and I should have been more optimistic. Very many factors were in our favor. Foremost, I think, is that ours was a happy association at all times and the rare differences of opinion were quickly and easily resolved. A second important factor was that we had a testing period of four or five years, during which time we, he on the Wisconsin campus and I at Cornell, taught from earlier drafts of the book in mimeo form with each student having his own copy. We listened to the criticisms of upwards of a thousand students and faculty and the resulting input caused many drastic changes in content and organization as we progressed through successive revisions. Closer to publication, suggestions by conscientious reviewers were carefully studied and appraised and they, too, made a vital impact. One 'criticism' provided a small laugh for the two of us when a reviewer stated, rather positively we thought, that he could recognize which chapters were the work of which author, especially as we ourselves could no longer make that distinction, so deeply buried in revisions was the original. I must acknowledge the great contribution of Cornell in providing time, typing and mimeoing without limitation.

"How did it come to be written? My memory is a short one. Perhaps a book salesman making his rounds suggested it. In any event, Jim and I were co-workers from 1949 to 1952 at the University of Wisconsin. He was already firmly established as a plant breeder and applied statistician; I was fresh from Iowa State with a Ph.D. in statistics. Both of us had felt for some time that the field of statistics needed a new text, for teaching as a reference source for workers in subject matter areas. Our proposed text would be a real alternative to others already in existence and provide a new approach to the study of statistics, particularly at the graduate level. If you can forgive the conceit, we thought we knew what that approach should be and what material should be included. A year later, when I left for Cornell, I took with me the first plan of the book and a few chapters in manuscript form.

"Over the years, the book grew slowly because of the geographic separation of the authors and the priorities we had to give to our university commitments, but it did grow. In fact, it grew too much and the publisher said 'this many pages and no more.' We were compelled to control our enthusiasm. We became more selective about techniques and exposition was tightened up. Eventually we were reduced to condensing every paragraph. We felt we had won the battle when one reviewer remarked that 'every sentence has to be read with care.'

"The time arrived when we had to have closer collaboration and I spent a Sabbatic leave at the Mathematics Research Center at the University of Wisconsin. On winter nights of that year, seven years after its conception, *Principles and Procedures* was put into shape for the publisher.

"A few years after *Principles and Procedures* appeared, it became clear that a revision was needed. Mistakenly or not, we chose instead to concentrate our energies on a new text at the undergraduate level. James H. Torrie lived long enough to see this in print, for which I am extremely grateful, but I regret that he did not have the additional satisfaction of knowing that *Principles and Procedures* is a highly-cited 'classic'."

# This Week's Citation Classic

Finney D J. *Probit analysis: a statistical treatment of the sigmoid response curve.*
Cambridge: Cambridge University Press, 1947. 256 p.
[University of Oxford, Oxford, England]

This book gives a systematic account of maximum likelihood estimation for quantal response data, with comparison of methods. Special problems considered include relative potencies, bivariate tolerance, mathematical formulations for the tolerance of mixtures of poisons, and aspects of experimental design. [The *Science Citation Index®* (*SCI®*) and the *Social Sciences Citation Index®* (*SSCI®*) indicate that all editions of this book have been cited in over 2,930 publications since 1961.]

David J. Finney
Department of Statistics
University of Edinburgh
Edinburgh EH9 3JZ
Scotland

·April 2, 1982

"About 1940, as a young statistician at Rothamsted Experimental Station, I had to continue with the help, previously provided by W.G. Cochran, on research on insecticides. The little I knew about probits I had learned a year or two previously from W.L. Stevens at the Galton Laboratory. When I realized the range of theoretical and computational complications, I suggested to Stevens that he should write a book on the subject. 'Why don't you?' he replied. The outcome was one of the first statistical texts that presented a group of relatively advanced techniques rather than a standard introduction to basic methods. I wrote of techniques that were unfamiliar even to most professional statisticians, but I discussed experimental data in a detail intended to make analyses accessible to biologists. In particular, I showed computations step-by-step for the calculators of the day; I emphasized strongly the practical interpretation of the results.

"The book clearly met a need. It showed the methods as applicable not only to insecticides but to estimation of drug potencies, psychometric data, educational tests, and other problems. It also demonstrated that iterative maximum likelihood computations were practicable for a biologist, provided that he organized them with care. Perhaps few scientists under the age of 45 can imagine a time when even simple mechanical calculators were rarely available to biologists. This placed a high premium on a standardized arithmetical routine, easily followed by technicians, and made popular various ingenious types of approximate analysis. The computer revolution has completely changed the situation; with good software, full maximum likelihood estimation is as easy and inexpensive as the approximations. Today, the latter methods have no more than historic interest. My second[1] and third[2] editions introduced new applications and updated the computational schemes; though the classical maximum likelihood probit iteration is still presented, recommended procedures are now entirely in terms of comprehensive computer software. My own program, BLISS (named for the originator of probits), is used in illustration, but of course many others would serve.

"I believe that *Probit Analysis* was one stimulus to the subsequent production of many books on special topics in statistical science. More personally, it initiated my own broader interest in biological assay, which has remained one of my major interests in statistical practice for 35 years and on which I later published another and possibly better book.[3] Though I was certainly not the first statistician active in this field, I believe I have made a substantial contribution to its systematic presentation and terminology. Fifteen years ago, I thought that practical applications had gone as far as would ever be necessary; since then, the explosive growth of immunoassay techniques has called for methods very close to those of probits, and the need for generalizations and improved design and computation continues. At least, in 1982, confusion with 'probate' seems almost to have vanished! Yet, even within the last few weeks at a conference on immunoassay, a biochemist remarked to me, 'Is it not strange that you, a statistician, should have developed an interest in data?'"

1. Finney D J. *Probit analysis: a statistical treatment of the sigmoid response curve.* Cambridge: Cambridge University Press, 1952. 318 p.
2. --------------. *Probit analysis.* Cambridge: Cambridge University Press, 1971. 333 p.
3. --------------. *Statistical method in biological assay.* London: Griffin, 1978. 508 p.

## Citation Classics

Litchfield J T & Wilcoxon F A. A simplified method of evaluating dose-effect experiments. *J. Pharmacol. Exp. Ther.* 96:99-113, 1949.

Due to the widespread use of statistical methods for evaluating biological data, it was necessary to solve a dose-percent curve. This paper presents a rapid graphic method for approximating the median effective dose and the slope of dose-percent effect curves. This method is able to estimate the confidence limits for any conventional probability or for a dose other than the median effective dose. [The *SCI*® indicates that this paper was cited 2,238 times in the period 1961-1975.]

---

John T. Litchfield, Jr., M.D.
Hardscrabble
P.O. Box 85
Heathsville, Virginia 22473

December 14, 1976

"Frank Wilcoxon, a complete stranger to me, learned early in 1945 that I had accepted a position with the same laboratories where he was working. He had already concluded that the Litchfield-Fertig method for 'Graphic solution of the dosage-effect curve'[1] was logical and attractive but capable of improvement. His enthusiastic wish to collaborate with me on this project was such that he left explicit instructions with the reception desk that he was to be called there immediately upon my arrival. As a consequence, I met Frank and began discussing the method before I even saw my employer or went to the personnel department.

"I had spent many hours computing dose-per cent effect curves using a mechanical calculator in order to devise a 15-20 minute approximation to replace a 2-3 hour task. In view of Frank's intense interest in the method, my considerable ego was abruptly deflated when I learned that he considered the method too laborious, confusing, and time consuming. However, he rapidly convinced me that most biologists found logarithms and probits incomprehensible but were accustomed to dosages and per cent effects or responses. By using specially ruled graph paper (logarithmic-probability), the data could be plotted directly and a line fitted by eye without need to convert to logarithms and probits. In addition, he had devised a $Chi^2$ test to tell whether or not the trial line was a reasonable one. This included a nomograph which eliminated the calculations necessary for determining $Chi^2$ values.

"From that first meeting on we concentrated on the construction of the other nomographs needed to eliminate calculations and on the set of instructions for the revised and simplified test. As these were perfected we undertook 'clinical trials' with less and less sophisticated subjects in order to discover and clarify any points which were confusing. Our product was judged satisfactory when it was used successfully by our colleagues whose speciality was organic chemistry. It was then submitted for publication.

"Although formally trained as an organic chemist, Frank was also an expert in plant physiology, mathematics, and statistics. His intense interest in short, quick statistical methods led to his discovery and the opening up of the field on nonparametric statistics. Had it not been for his intense interest there might never have been the Litchfield-Wilcoxon method."

1. Litchfield J T & Fertig J W. Graphic solution of the dosage-effect curve. *Bull. J. Hopk. Hosp.* 69:276-86, 1941.

## This Week's Citation Classic

NUMBER 39
SEPTEMBER 24, 1979

Gehan E A. A generalized Wilcoxon test for comparing arbitrarily singly-censored samples. *Biometrika* 52:203-13, 1965.
[Birkbeck College, London, and National Institutes of Health, Bethesda, MD]

A statistical test was proposed extending the Wilcoxon test to arbitrarily right-censored samples. The test is particularly applicable to comparing two survival distributions in a clinical trial when there is a mixture of individuals who have died and others who are still alive in each group. [The *SCI*® indicates that this paper has been cited over 230 times since 1965.]

Edmund A. Gehan
University of Texas System
Cancer Center
M. D. Anderson Hospital
and Tumor Institute
Texas Medical Center
Houston, TX 77030

November 28, 1978

"This paper was written while on a special fellowship from the National Cancer Institute at Birkbeck College, University of London. During the previous four years (1958-1962), I worked as a consulting statistician at the National Cancer Institute and identified the problem of the paper. In collaboration with Dr. E. J. Freireich et al. on a clinical trial comparing 6-MP to placebo in the maintenance of complete remissions in children with acute leukemia, there were 21 pairs of patients with 21 times to relapse on placebo, and 9 times to relapse and 12 times to censoring on 6-MP.[1] Times to censoring were lengths of remission in individuals who had not relapsed at time of analysis. Clearly, 6-MP was significantly superior to placebo, but there was no statistical test utilizing all the data for demonstrating this.

"The fellowship provided a wonderful opportunity to work on the problem and I was most fortunate to work in collaboration with Professor David R. Cox who was then head of the small Department of Statistics of Birkbeck. In addition, I had married my wife, Brenda, shortly before leaving for London so it was indeed a 'honeymoon experience.' The test developed was a natural generalization of the Wilcoxon test for comparing two samples. Professor Cox was of great help on approaches to determining the theoretical characteristics of the test. He should have co-authored the paper, but was only willing to be acknowledged. Brenda helped by her encouragement and hand-calculating over 1,000 examples of the test which demonstrated that a normal approximation worked well in small samples.

"The paper has received many references in the statistical literature because it led to many additional theoretical developments for survival studies with censored data, most notably a paper by Cox on regression methods.[2] There are also many references in the medical literature as justification for the statistical test performed to compare survival times in two groups—some even using the term 'Gehan test,' though Wilcoxon developed the original test. It is still amazing that the problem was not considered much earlier, since it arises so commonly in clinical trials. John Gilbert of Harvard University recognized the problem at about the same time and developed an identical test in an unpublished Ph.D. thesis. I've always been thankful for using the example of acute leukemia (I seriously considered using hypothetical data with censored observations in both samples), since many subsequent authors have used the same data. What may not be known by many statistical authors is that the medical paper by Freireich et al. was a breakthrough—it was the first to demonstrate that maintaining patients on treatment in complete remission was beneficial."

1. Freireich E, Gehan E, Frei E, III, Schroeder L, Wolman I, Anbari R, Burgert E, Mills S, Pinkel D, Selawry O, Moon J, Gendel B, Spurr C, Storrs R, Haurani F, Hoogstraten B, and Lee S. The effect of 6 mercaptopurine on the duration of steroid-induced remissions in acute leukemia. *Blood* 21:699-716, 1963.
2. Cox D. Regression models and life tables. *J. Roy. Statist. Soc.* B 34:187-220, 1972.

# This Week's Citation Classic

NUMBER 38
SEPTEMBER 17, 1979

Weil C S. Tables for convenient calculation of median-effective dose (LD50 or ED50) and instructions in their use. *Biometrics* 8:249-63, 1952.
[Mellon Institute, Pittsburgh, PA]

The acute toxicity of chemicals is most accurately expressed as the median-effective dose, LD50, e.g., the dose to kill half the animals. The tables permit easy calculations of this median and its confidence interval, using a minimum number of animals. [The *SCI*® indicates that this paper has been cited over 410 times since 1961.]

Carrol S. Weil
Carnegie-Mellon Institute
of Research
Carnegie-Mellon University
Pittsburgh, PA 15213

November 25, 1978

"Prior to the appearance of these tables, the calculation of the median-effective dose, ED50, LD50, LC50, LT50, etc., required approximately 30 to 60 minutes of time, using the only method then available, mechanical calculators, graphing, tables, curve fitting, and goodness of fit calculations. The probit procedure involved attempting to straighten the sigmoid dose-response mortality data and, therefore, calculate its midpoint. However, to 'prove' the goodness of fit many animals and dosage levels had to be used.

"When the method of moving averages was published, I found it gave almost identical results as the much more complicated probit method, with much less computational time and with no assumption about curve fitting. However, I found that when the same mortality ratios resulted, e.g., 0 of 5, 1 of 5, 3 of 5, and 5 of 5 (0,1,3,5) the calculation could be made more simple by use of constants, that I then calculated and put in table form. With the use of my tables and a table of logarithms one could determine the LD50 and its 95% confidence limits in 3 to 5 minutes. Now, with electronic calculators or computers, the time is seconds to a minute.

"At the start, I suggested the use of four dosage levels in a geometric series (a constant number of animals per level). We found that, the majority of the time, with a factor of 2 between dosage levels, we had mortality ratios below and above the median with only 2 to 3 dosage levels; e.g., 1 of 5 and 3 of 5 or 0 of 6, 3 of 6, and 5 of 6; (1,3) or (0,3,5). Not only were these sufficient for calculation but they saved animals. The published tables are for 4 levels; if you write me I will send you others to use with 2 or 3 dosage levels.

"We, and many other toxicologists, have used these tables for many years. They have proved valuable in interlaboratory and intralaboratory tests of consistency. We have found, with LD50 assays using only a total of 10 to 15 rats each, that acute LD50s were consistent in our laboratory over a 12-year period, testing the same 26 chemicals each year. We have also run hundreds of joint-toxic estimates using this method, which would have been impractical, more time and energy consuming, and more costly because of the larger number of animals necessary using the probit methods. And, as beforementioned, this non-parametric method should be preferred in these toxicity assays to the parametric, curve-fitting methods."

# Citation Classics

**Duncan D B.** Multiple range and multiple *F* tests. *Biometrics* 11:1-42, 1955.

This new multiple range test, for determining the homogeneity of a set of *n* values in an analysis of variance in a population, combines the proposals by Newman in 1939 and Keuls in 1952 with the author's earlier multiple comparison tests. A series of tests paralleling the methods of multiple range tests have been termed multiple *F* tests, which use "protection levels based on degrees of freedom." An *F* test alone, it is demonstrated, "falls short of satisfying all of the practical requirements." One of several test procedures examined is termed the *least-significant-difference* (or *L.S.D.*) *test*. [The SCI® indicates that this paper was cited 3,610 times in the period 1961-1975.]

---

Professor David B. Duncan
Department of Biostatistics
Johns Hopkins University
Baltimore, Maryland 21205

November 1, 1976

"I am naturally glad and encouraged to have a 'most cited' paper and to learn of the frequent use of the DMR (my multiple range) rule.

"The 1955 DMR rule was a modified version of my earlier 1951 DMF (multiple F) rule. Both of these rules ranked in conservatism and power between the less conservative 1935 FLSD (Fisher LSD) rule and the more conservative 1939 NMR (Newman MR) rule. By using F tests, the DMF rule could be used to test comparisons (subsequently called contrasts by Scheffe) as well as pairwise differences. However, the multiple use of F tests was more cumbersome than that of range tests, and the DMF rule received much less attention.

"In between these rules came the TLSD (Tukey LSD) rule of 1952 and the SLSD (Scheffe LSD) of 1953. Both of these were based on the use of experimentwise levels making them much more and very much more conservative and less powerful than even the NMR rule.

"The frequent use of the DMR rule has been encouraging in the support it has given for my less conservative approach.

"In 1955 when I published the DMR rule I had not been able to finish a multiple decision theory approach to the problem which I had started in my thesis in 1947.

"In this kind of approach, which can be held to be the ideal, it seemed reasonable from the start to choose an additive loss function. That is, a loss function whereby, roughly speaking, the seriousness of the error made by any joint decision is scored in proportion to the number of individual differences about which it is wrong. In presenting the DMF and DMR rule I had been influenced by this approach but had also had to resort to arguments of a more *ad hoc* nature.

"On taking up the decision theory approach again in 1962, I was able to show that, by using a super normal population prior model $\mu_i \sim N(\mu_0, \sigma_\mu^2)$ for the true treatment means, the optimal rule was a 'k-ratio' LSD rule with a LSD which depends on the variance ration $r\sigma_\mu^2 / \sigma_e^2$, $\sigma_e^2$ being the error variance. Subsequently in 1965 it was exciting to find that, by putting conjugate $\chi^2$ priors on $\sigma_\mu^2$ and $\sigma_e^2$, the LSD for this rule could be made to depend directly on the *observed* F ratio.

"Joined by R.A. Waller in 1969, we were able, by switching the prior on $\sigma_\mu^2$ to being an independent conjugate $\chi^2$ on $\sigma_T^2 = r\sigma_\mu^2 + \sigma_e^2$, to derive and table the precise t values for the k-ratio LSD Bayes rule for all small sample values of the numerator and denominator degrees of freedom of the observed F ratio.

"By having its t value depend on the F ratio the k-ratio LSD rule is adaptive. It can vary, in an intuitively pleasing way, all of the distance from being conservative like an experimentwise rule when the F is small to being less conservative and more powerful than even a comparisonwise rule when F is large. This rule, which I showed later in 1975 can be used on all contrasts as well as pairwise differences,[1] is the one I now recommend in place of all of the earlier rules including the cited DMR rule."

---

1. **Duncan D B.** T tests and intervals for comparisons suggested by the data. *Biometrics* 31:339-59, 1975.

| Number 44 | **Citation Classics** | October 31, 1977 |

**Kramer C Y.** Extension of multiple range tests to group means with unequal numbers of replications. *Biometrics* 12:307-10, 1956.

The author shows how means with unequal numbers of replications can be grouped into homogenous subgroups to determine which means are different from other means. [The *SCI*® indicates that this paper was cited 539 times in the period 1961-1975.]

---

Professor Clyde Y. Kramer
Department of Statistics and
Statistical Laboratory
Virginia Polytechnic Institute
and State University
Blacksburg, Virginia 24061

February 2, 1977

"It is indeed an honor and both extremely satisfying and flattering to be a most cited author. To have a statistical paper in the list of most cited papers certainly shows the utility of statistical methods in all subject matter fields. This paper, as most of my many papers, was written because there were researchers in all subject matter areas facing this problem.

"As an Experiment Station Statistician, my primary duties were to consult with research workers in all areas. This included talking over the proposed research, providing the optimum statistical design or sampling plan, analyzing the results, and helping the research worker interpret his results. Most of the time the experiments were set up to have an equal number of observations on each treatment, but about as many times as not when the data came in for analysis there were missing observations. The missing observations were caused by, I guess you could say 'Mother Nature' because any research conducted is merely a game against Mother Nature.

"Many papers dealt with grouping equal means but these were no help to my clients and, of course, one could not just compare two treatments at a time by a student t-test, so I began to work on the problem. After many approaches I found a very simple relationship that turned out to give very satisfactory results.

"Before I published the paper, I took many experiments in all areas that had equal numbers and at random deleted observations in the treatments and applied my new procedure. I found that only in a very few instances did the groupings change, which was very gratifying. This was during the time when one only had desk digital calculators and it was very time consuming but I felt necessary.

"The procedure proved very easy to use and from the number of citations, I guess it was just what the subject matter people needed to draw their inferences. It is now incorporated as a procedure in Statistical Analysis System 1976 and probably will continue to be used more and more.

"I think of myself as an applied statistician and a consultant, in that I work on problems that have been brought to me by my clients. The researcher workers at Virginia Polytechnic Institute and State University should share in this honor because they are the ones that brought the problem to my attention and published their results using my method even before I published the paper."

## This Week's Citation Classic
CC/NUMBER 47
NOVEMBER 19, 1979

Harter H L. Critical values for Duncan's new multiple range test.
*Biometrics* 16:671-85, 1960.
[Aeronautical Research Laboratories, Wright-Patterson Air Force Base, OH]

The author tabulates critical values for Duncan's new multiple range test, at significance levels $\alpha$ [protection levels $P = (1-\alpha)^{p-1}$], of p successive values out of n ordered means, for $\alpha$ = 0.10, 0.05, 0.01, 0.005, 0.001; p = 2(1)n; n = 2(1) 20(2) 40(10) 100 and degrees of freedom $v$ = 1(20), 24, 30, 40, 60, 120,∞. [The *SCI*® indicates that this paper has been cited over 150 times since 1961.]

H. Leon Harter
32 South Wright Avenue
Dayton, OH 45403

February 14, 1979

"In 1954 I began a study of error rates and sample sizes for multiple comparison tests, especially those based on the studentized range. The results of this study were published in a 1957 journal article.[1] During the course of this work, I discovered sizeable errors in the published tables of critical values for Duncan's new multiple range test, and smaller ones in those of percentage points of the studentized range used as critical values for Tukey's studentized range test, the Newman-Keuls test, and Tukey's compromise X procedure.

"In order to correct and extend these tables it was necessary to compute new tables of the probability integral of the studentized range, more extensive and more accurate than those of Pearson and Hartley (1943).[2] This in turn required the computation of more extensive and more accurate tables of the probability integral of the range than those of Pearson and Hartley (1942).[3] The new tables of percentage points of the studentized range were published in another 1960 journal article.[4] The complete results, including the tables of probability integrals, were published in two 1959 technical reports. These results, together with corrected and extended tables of error rates, were reproduced in a 1970 book.[5]

"This work was performed at the Aeronautical (later Aerospace) Research Laboratories (ARL), with the active encouragement of Paul R. Rider, then chief statistician at ARL. The programming for the Univac Scientific (ERA 1103A) computer was done by Donald S. Clemm and Eugene H. Guthrie, based on numerical analysis by Gertrude Blanch. These and other colleagues deserve much credit for successful completion of the mammoth undertaking.

"I attribute the frequent citation of this paper to the fact that it provided the first accurate critical values for the widely used new multiple range test proposed by David B. Duncan in an even more frequently cited 1955 paper featured earlier in their series."[6]

1. **Harter H L.** Error rates and sample sizes for range tests in multiple comparisons.
*Biometrics* **13**:511-36, 1957.
2. **Pearson E S & Hartley H O.** Tables of the probability integral of the 'studentised' range.
*Biometrika* **33**:89-99, 1943.
3. **Pearson E S & Hartley H O.** The probability integral of the range in samples of $\eta$ observations from a normal population. *Biometrika* **32**:301-10, 1942.
4. **Harter H L.** Tables of range and studentized range. *Ann. Math. Statist.* **31**:1122-47, 1960.
5. **Harter H L.** *Order statistics and their use in testing and estimation. Volume 1: Tests based on range and studentized range.* Washington, DC: US Government Printing Office, 1970. Out of print, but paper-bound and microfiche copies available from NTIS. (AD-A058262).
6. **Duncan D B.** Multiple range and multiple *F* tests. *Biometrics* **11**:1-42, 1955.

# This Week's Citation Classic
CC/NUMBER 26
JUNE 29, 1981

Mantel N & Haenszel W. Statistical aspects of the analysis of data from retrospective studies of disease. *J. Nat. Cancer Inst.* **22**:719-48, 1959.
[Biometry Branch, Natl. Cancer Inst., NIH, Public Health Serv., US Dept. Health, Education, and Welfare, Bethesda, MD]

The relationship of the retrospective study to the forward-type study and other practical issues are discussed. Methods of statistical analysis controlling on confounding factors are provided. A chi-square test for significance of any observed association is given. One of several summary measures of relative risk is recommended. [The *SCI®* indicates that this paper has been cited over 815 times since 1961.]

Nathan Mantel
Statistics Department
Biostatistics Center
George Washington University
Bethesda, MD 20014

May 27, 1981

"This paper was the conception of its junior author, William Haenszel, who had the practical familiarity with the problems of retrospective studies. My experience had been largely in the application of statistics and statistical thinking to laboratory investigations and Haenszel suggested that I augment his own work by any statistical concepts I thought appropriate. Those concepts were, in a way, simple and I was not satisfied to give them only as mathematical formulas. In the end, there was a blending of Haenszel's practical ideas with my own—Haenszel, in his generosity, suggested that the order of authorship be reversed.

"In a way, our work was an extension of still earlier work by Jerome Cornfield who had suggested the effective utilization of retrospective studies.[1,2] He illustrated his concepts with the homogeneous case, a single-stratum population, but was aware of and had published on the heterogeneous case. Haenszel and I went into the heterogeneous case more thoroughly and more formally.

"The high frequency of citation of our paper comes about from a number of reasons. It may be cited because of its generally useful ideas, though most likely in relation to retrospective or other observational studies. It may be cited for the chi-square test it provides and/or for its summary measure of relative risk. It may be cited for the emphasis that it puts on stratified analysis, or it may be cited for no good reason that I can see.

"Since the publication of the paper, both observational studies and clinical trials have been on the increase, with a consequent rising frequency of citation to our paper. Also contributing has been the growing awareness of the statistical community of the relevance of the paper. The statistical methods of the paper have become subjects of statistical investigation in their own right. For my own part, I have written several papers extending the concepts of the initial Mantel-Haenszel paper in interesting ways.[3,4] To the extent that these derivative papers are increasingly cited, there may be reducing citation of the original paper. Citations by statistical writers frequently emphasize advantageous properties of some aspect of the Mantel-Haenszel package (stratification + significance test + summary relative risk or relative odds measure), but citations of a critical nature are by no means uncommon. I have recently published several articles in this field."[5-7]

1. Cornfield J. A method of estimating comparative rates from clinical data. Applications to cancer of the lung, breast, and cervix. *J. Nat. Cancer Inst.* **11**:1269-75, 1951.
2. ............., A statistical problem arising from retrospective studies. *Proc. Third Berkeley Symp. Math. Statist. Probab.* **4**:135-48, 1956.
3. Mantel N. Chi-square tests with one degree of freedom; extensions of the Mantel-Haenszel procedure. *J. Amer. Statist. Assn.* **58**:690-700, 1963.
   [The *SCI®* indicates that this paper has been cited over 285 times since 1963.]
4. ............., Evaluation of survival data and two new rank order statistics arising in its consideration. *Cancer Chemother. Rep.* **50**:163-70, 1966.
   [The *SCI®* indicates that this paper has been cited over 300 times since 1966.]
5. Mantel N, Bohidar N R & Chalmers J L. Mantel-Haenszel analyses of litter-matched time-to-response data, with modifications for recovery of interlitter information. *Cancer Res.* **37**:3863-8, 1977.
6. Mantel N. Tests and limits for the common odds ratio of several 2x2 contingency tables: methods in analogy with the Mantel-Haenszel procedure. *J. Statist. Plan. Infer.* **1**:179-89, 1977.
7. Mantel N & Fleiss J L. Minimum expected cell size requirements for the Mantel-Haenszel one-degree-of-freedom chi-square test and a related rapid procedure. *Amer. J. Epidemiol.* **112**:129-34, 1980.

# This Week's Citation Classic

Mantel N. Evaluation of survival data and two new rank order statistics arising in its consideration. *Cancer Chemother. Rep.* **50**:163-70, 1966.
[Biometry Branch, Natl. Cancer Inst., NIH, Public Health Service, US Dept. Health, Education, and Welfare, Bethesda, MD]

The Mantel-Haenszel procedure is adapted to the comparison of survival or time-to-response curves. Separate 2 x 2 contingency tables are formed by considering the numbers at risk and responding in each time interval. For infinitesimal time intervals, the method becomes a ranking procedure. The two rank order statistics described include the final chi square and the maximal interim chi square. Philosophical aspects of the problem are discussed, including the need for having a value function. [The *SCI®* indicates that this paper has been cited in over 500 publications since 1966.]

Nathan Mantel
Department of Mathematics, Statistics, and Computer Science
American University
Bethesda, MD 20814

December 14, 1982

"A footnote in this paper indicates that it was presented at a Conference of Regional Statisticians in Bethesda in May 1965. That actually was not the case—I had presented another paper but submitted the instant one for publication, a case of false pretenses.

"Sometime in 1964, I followed through on a suggestion in my 1963 extension[1] to the original 1959 Citation Classic by myself and Haenszel.[2] The methodology I suggested was extremely simple—just treat the data arising in each time interval as constituting simply a separate fourfold contingency table. Also, by allowing the durations of the time intervals to become infinitesimal, it was apparent that I was dealing with a ranking procedure.

"Philosophical considerations, often overlooked by others, also arose, which I brought out. If we could not decide which was the better of two survival curves which we knew exactly, but which crossed each other, then how could we decide which was the better if we had only estimates of those curves?

"Shortly after having prepared my own draft, I got involved in a competitive procedure. Ed Gehan had, in press or in preparation, two manuscripts on generalizing the Wilcoxon procedure for censored data, which he brought to my attention.[3,4] Subsequently, I saw how Gehan's apparently complex procedure could be greatly simplified, and my own resulting manuscript was a *tour de force*.

"It was this alternative manuscript which I presented at the conference. It was a manuscript with much more substance than the one I had previously prepared, and would make for a much better presentation. Yet, also, it was one which I had little doubt about getting published. But I had misgivings about getting my future Citation Classic published, and it was that one which I did submit in connection with the conference. (At the actual conference, I made clear the existence of the two manuscripts.)

"My *tour de force*, however, did not have clear sailing. Gehan wanted very much to see it in print, but his own editor judged otherwise. I had recourse to another journal and, after a year of waiting, inquired. Apparently, the reviewer, Wilcoxon, had died in the course of the year and the paper had been forgotten. The repentant editor then accepted it for publication without changing a word—he apparently recognized its special merits.[5]

"Substantial impetus for the popularity of my original work on survival data has come from the 1972 publications of Cox[6] and of Peto and Peto.[7] For one thing, the method I gave could be recognized as a special simple case of an ingenious procedure by Cox for avoiding time model assumptions. Also, Peto and Peto gave a name, log ranks, to the method provided, with the first of the Petos continuing to give the method publicity in his writings. But it is the general usefulness of the method, its applicability to a wide range of important problems, and its theoretical appeal which have mattered in the long run."

1. **Mantel N.** Chi-square tests with one degree of freedom: extensions of the Mantel-Haenszel procedure.
   *J. Amer. Statist. Assn.* **58**:690-700, 1963.
2. **Mantel N & Haenszel W.** Statistical aspects of the analysis of data from retrospective studies of disease.
   *J. Nat. Cancer Inst.* **22**:719-48, 1959.
   [Citation Classic. *Current Contents/Life Sciences* **24**(26):19, 29 June 1981.]
3. **Gehan E A.** A generalized Wilcoxon test for comparing arbitrarily singly-censored samples.
   *Biometrika* **52**:203-23, 1965. [Citation Classic. *Current Contents/Life Sciences* **22**(39):12, 24 September 1979.]
4. ─────. A generalized 2-sample Wilcoxon test for doubly censored data. *Biometrika* **52**:650-3, 1965.
5. **Mantel N.** Ranking procedures for arbitrarily restricted observation. *Biometrics* **23**:65-78, 1967.
6. **Cox D R.** Regression models and life tables. *J. Roy. Statist. Soc. Ser. B Metho.* **34**:187-220, 1972.
7. **Peto R & Peto J.** Asymptotically efficient rank invariant test procedures.
   *J. Roy. Statist. Soc. Ser. A Gener.* **135**:185-206, 1972.

# Citation Classics

Number 49     December 4, 1978

**Ebel R L.** Estimation of the reliability of ratings. *Psychometrika* **16**:407-24, 1951.

When several judges rate several products, questions about the reliability of their ratings, i.e., the consistency of the ratings across judges, may arise. This paper considers several procedures for estimating that reliability, and recommends one that is generally most satisfactory. [The *Science Citation Index*® *(SCI*®*)* and the *Social Sciences Citation Index*™ *(SSCI*™*)* **indicate that this paper was cited a total of 219 times in the period 1961-1977.**]

Robert L. Ebel
Professor of Education and Psychology
Michigan State University
East Lansing, Michigan 48824

January 30, 1978

"As part of the program of general education launched at the University of Iowa in the early 1940's, students were required to demonstrate or develop skill in writing and speaking, among other abilities. The themes they wrote and the speeches they gave were rated by professors in the Communication Skills Program. Directors of the program were concerned that the ratings should be consistent across raters, not only in fairness to the students, but in pursuit of agreement among the professors on the elements of quality in a theme or speech. The ratings were analyzed in the Examinations Service of the University.

"My predecessor as director of that Service, the late Professor Paul Blommers, had worked out a routine for calculating the extent of agreement in the ratings, based on R. A. Fisher's intraclass correlation coefficient. There were, I found, two other formulas which appeared to be applicable. But when the three formulas were applied to the same sets of ratings they gave a somewhat discrepant result. Something was wrong, and I set out to find what it was. The discovery might just possibly put me one step closer to a firm grasp of the ideas developed by R.A. Fisher, Charles C. Peters, or Paul Horst.

"Taking some simple numerical hypothetical examples of possible ratings, I applied the three formulas and studied the results. It became apparent that the different results sometimes yielded by the formulas were due to differences among the formulas in two characteristics: (1) whether the over-all within-raters variance was the arithmetic mean or the geometric mean of the separate within-raters variances, and (2) whether the between-raters variance was included or excluded in the error term.

"If examples are chosen in which the arithmetic mean is the same as the geometric mean, and if between-raters variance is always included as error, the three formulas will give identical results.

"Good mathematicians mistrust general inferences from specific numerical examples. Harold Gulliksen, reviewing an early draft of the paper, suggested that I support the conclusions I had reached from the numerical examples with a generalized algebraic derivation. Substituting persistence for brilliance, I managed to do this. Still, I do find simple numerical examples helpful in suggesting generalizations worthy of algebraic validation.

"Why has the paper been cited often? Not, I think, because many others are concerned with the discrepancy that attracted my attention. Rather, it may be because the paper included a fairly simple explanation, with simple examples, of the use of analysis of variance methods in solving some frequently encountered problems of reliability estimation."

# Number 13     Citation Classics     March 27, 1978

Cronbach L J. Coefficient Alpha and the internal structure of tests.
*Psychometrika* 16:297-334, 1951.

The author defends as an appropriate index of text equivalence a general formula estimating the correlation $a$ between two random samples of items from a universe of items like those in the text. Comparison is made to indices proposed by Guttman and Loevinger. [The *Science Citation Index*® (*SCI*®) and the *Social Sciences Citation Index*™ (*SSCI*™) indicate that this paper was cited a total of 238 times in the period 1961-1976.]

             Lee J. Cronbach
             School of Education
             Stanford University
             California 94305
                     November 1977

"I am sure the paper is cited mostly because I put a brand name on a commonplace coefficient. Thousands of investigations report such coefficients, and some label it 'Cronbach's $a$.'

"Testers judge instruments by examining the consistency of scores over items or half-tests. Only recent theory[1] provides a *direct* rationale for putting the ancient intraclass correlation to this use. As a research assistant in 1939 I was taught to use a 'Kuder-Richardson Formula 20' developed by experts who worked in the office next door. None of us realized it, but KR20 is the ancient intraclass correlation, specialized to fit items scored 1/0 (pass/fail). The KR paper made heroic assumptions. It was an easy and intriguing exercise to derive the formula from variant assumptions, and that exercise became, as someone said, the second-favorite indoor sport of psychometricians. In 1941 Cyril Hoyt had presented another rationale leading to the general coefficient I christened $a$. Hoyt's proof was exotic, hence his paper attracted no following.

"My paper connected the specialized and variant formulas with Hoyt's and examined debated interpretations. (Others later identified it as an intraclass correlation.) Not incidentally, my paper reacted to Jane Loevinger's powerful monograph attacking the whole psychometric tradition including KR20. I think history sided with me, in that the Loevinger-Guttman techniques faded out of psychology and educational research. The debate is more forgotten than it should be, given the current attraction of Rasch's variant of Guttman-Loevinger.

"More personally: in 1949 I presented to a regional meeting twenty minutes' worth of the dissent from Loevinger. Philip Dubois of Washington University invited me to come from Urbana to consult with him and Jane on a project using her ideas. The visit did not come to pistols at high noon; Goldine Gleser, the junior member of the project team, found a bridging formula that Jane and I could each see as a victory for our surely incompatible principles. Out of that encounter grew a 20-year collaboration and two Cronbach-Gleser books.

"As for the name ' $a$ ', my paper required a symbol for the coefficient (as reached by whatever computation). I had fantasies of companion analyses. $a$ describes consistency 'of persons, over items, holding occasion fixed'. The obvious permutations ('over occasions...,' etc.) lead to five more coefficients. By the time we had the multifacet theory[2] we distrusted the pseudosymmetry and were no longer centrally interested in coefficients. 'Alpha' was put into my title to set this paper off from 'Coefficient Beta' and four other papers never written."

1. Cronbach L J, Gleser C G, Rajaratnam N & Nanda H. *The dependability of behavioral measurements.* New York: Wiley, 1972. 410 pp.
2. Gleser G C, Cronbach L J & Rajaratnam N. Generalizability of scores influenced by multiple sources of variance. *Psychometrika* 30:395-418, 1965.

# This Week's Citation Classic

CC/NUMBER 24
JUNE 15, 1981

Armitage P. Restricted sequential procedures. *Biometrika* 44:9-26, 1957.
[Statistical Res. Unit, Medical Res. Council, London Sch. Hygiene and Tropical Medicine, England]

A family of closed sequential plans is described, providing control over error probabilities for a two-sided hypothesis test. The results are based on diffusion theory, but provide useful approximations for various distributional forms. Applications to clinical trials are proposed. [The *Science Citation Index®* (*SCI®*) and the *Social Sciences Citation Index®* (*SSCI®*) indicate that this paper has been cited over 115 times since 1961.]

---

Peter Armitage
Department of Biomathematics
University of Oxford
Oxford OX1 2JZ
England

April 28, 1981

"In 1947, I joined the Medical Research Council's Statistical Research Unit under A. (now Sir Austin) Bradford Hill, whose advocacy of the randomized controlled trial for the comparison of rival therapeutic measures was strong and influential. Within a few years I started to explore the possible use of sequential methods for the design and analysis of clinical trials. The motivation seemed clear: if the results of a trial were analysed as they became available, one could stop the trial early if treatment differences were clearly emerging. I had done some fragmentary work on sequential analysis during the war, in connection with sampling inspection. The theory was dominated by the important work of Abraham Wald, published later in his book,[1] and I was at first inclined to think that Wald's methods could be applied fairly directly to clinical trials. However, it later seemed more realistic to incorporate 'closure' (an upper limit on the number of observations) as an integral feature of the plans. Some *ad hoc* plans of this type were published by Bross[2] at about this time.

"The 1957 paper represented one of the first attempts to provide a general theory. The name 'restricted' was chosen as I needed something more specific than 'closed,' and 'truncated' had been used by Wald for a rather different approach. The plans that emerged seemed to have the right sorts of characteristics for clinical trials. There would typically be a cumulative sum plotted against the number of observations, with two divergent boundaries, the crossing of which indicated an advantage for one treatment over another. If no boundary had been crossed before a certain sample size N, the trial was closed. The theory was based on a diffusion approximation to the distributions of sums of Gaussian variables, but this could be regarded as a 'normal approximation' to other situations. For the common case of binary observations, the approximation could be checked by exact calculations.

"The paper formed the basis of the first edition of my book,[3] and the methods were used a good deal, particularly in the 1960s. In the second edition[4] I advocated use of curved boundaries corresponding to the repeated use of standard significance tests, but the plans were really very little altered. I suppose that the general interest in sequential trials since 1960 is the reason for the citation of my (now rather outdated) paper. Practical interest has now moved away from the idea of sequential analysis after every observation towards that of interim analyses after each of a small number of stages.[5] This shift of emphasis is probably related to the wider implementation of very large multicenter chronic disease trials, where interim analyses are conveniently done for the periodic meetings of investigators. The techniques involved[5] spring directly from the earlier work on sequential analysis, thus (I suppose) providing a continuing trickle of citations. Theoretical work by others on the original restricted procedures has taken the theory well beyond that in my paper."

1. **Wald A.** *Sequential analysis.* New York: Wiley, 1947. 212 p.
2. **Bross I.** Sequential medical plans. *Biometrics* 8:188-205, 1952.
3. **Armitage P.** *Sequential medical trials.* Oxford: Blackwell, 1961. 105 p.
4. ——————. *Sequential medical trials.* Oxford: Blackwell, 1975. 194 p.
5. **Pocock S J.** Group sequential methods in the design and analysis of clinical trials. *Biometrika* 64:191-9, 1977.

# This Week's Citation Classic

CC/NUMBER 24
JUNE 13, 1983

Kaplan E L & Meier P. Nonparametric estimation from incomplete observations.
*J. Amer. Statist. Assn.* 53:457-81, 1958.
[Univ. California Radiation Laboratory, CA and University of Chicago, IL]

The product-limit formula estimates the proportion of organisms or physical devices surviving beyond any age t, even when some of the items are not observed to die or fail, and the sample is rather small. The actuarial and reduced-sample methods are also studied. [The *Science Citation Index®* (*SCI®*) and the *Social Sciences Citation Index®* (*SSCI®*) indicate that this paper has been cited in over 1,495 publications since 1961.]

---

Edward L. Kaplan
Department of Mathematics
Oregon State University
Corvallis, OR 97331

April 15, 1983

"This paper began in 1952 when Paul Meier at Johns Hopkins University (now at the University of Chicago) encountered Greenwood's paper[1] on the duration of cancer. A year later at Bell Telephone Laboratories I became interested in the lifetimes of vacuum tubes in the repeaters in telephone cables buried in the ocean. When I showed my manuscript to John W. Tukey, he informed me of Meier's work, which already was circulating among some of our colleagues. Both manuscripts were submitted to the *Journal of the American Statistical Association*, which recommended a joint paper. Much correspondence over four years was required to reconcile our differing approaches, and we were concerned that meanwhile someone else might publish the idea.

"The nonparametric estimate specifies a discrete distribution, with all the probability concentrated at a finite number of points, or else (for a large sample) an actuarial approximation thereto, giving the probability in each of a number of successive intervals. This paper considers how such estimates are affected when some of the lifetimes are unavailable (*censored*) because the corresponding items have been lost to observation, or their lifetimes are still in progress when the data are analyzed. Such items cannot simply be ignored because they may tend to be longer-lived than the average.

"The result is that every item r has an age $t_r'$ associated with it, but some of the $t_r'$ correspond to deaths (or failures), and some to *losses* (from observation). Now let the $t_r'$ be listed and labeled in order of increasing magnitude, so that one has $0 \leq t_1' \leq t_2' \leq ... t_N'$. Then the product-limit estimate of the proportion surviving beyond the age t is the product $\hat{P}(t) = \Pi_r[(N-r)/(N-r+1)]$, where r assumes those values (from 1 to N) such that $t_r' \leq t$ and $t_r'$ corresponds to a death.

"This means that to each age t' at which a death is observed, there is assigned the probability $\hat{P}(t'-0) - \hat{P}(t'+0)$, the amount by which the step-function $\hat{P}(t)$ decreases at the point t'. The most intuitive derivation of the formula is as a product of the conditional survival probabilities $(N-r)/(N-r+1)$, which would be obtained by the actuarial method if the intervals were made so short that each contained only one death.

"Meier's 1975 paper[2] obtained asymptotic properties for $\hat{P}(t)$ considered as a stochastic process. Three more recent publications[3-5] are listed below.

"Presumably this paper is frequently cited because it gives a good presentation of a simple solution to a problem often encountered by researchers. (It has also been used in a seminar intended to introduce students to the use of the literature.) Similar objectives have motivated my recent book, *Mathematical Programming and Games*."[6]

---

1. Greenwood M. *The natural duration of cancer*. London, England: His Majesty's Stationery Office, 1926. Reports on Public Health and Medical Subjects, No. 33.
2. Meier P. Estimation of a distribution function from incomplete observations. (Gani J, ed.) *Perspectives in probability and statistics*. New York: Academic Press, 1975. p. 67-87.
3. Kalbfleisch J D & Prentice R L. *The statistical analysis of failure time data*. New York: Wiley, 1980. 321 p.
4. Efron B. Censored data and the bootstrap. *J. Amer. Statist. Assn.* 76:312-19, 1981.
5. Chen Y Y, Hollander M & Langberg N A. Small-sample results for the Kaplan-Meier estimator. *J. Amer. Statist. Assn.* 77:141-4, 1982.
6. Kaplan E L. *Mathematical programming and games*. New York: Wiley, 1982. 588 p.

# This Week's Citation Classic

Bross I D J. How to use ridit analysis. *Biometrics* 14:18-38, 1958.
[Dept. Public Health and Preventive Med., Cornell Univ. Medical College, New York, NY]

In biological and behavioral sciences, variables often have more information than dichotomous classifications but less than refined physical measurements. Distributional peculiarities can affect the validity or power of standard statistical methods. After simple empirical probability transformations of the variables, ridit analysis ensures robustness of standard methods. [The *Science Citation Index®* (*SCI®*) and the *Social Sciences Citation Index®* (*SSCI®*) indicate that this paper has been cited over 190 times since 1961.]

Irwin D.J. Bross
Department of Biostatistics
Roswell Park Memorial Institute
Buffalo, NY 14263

May 12, 1981

"In 1950s studies of crash-injuries in highway accidents, the response variable used a graded scale (e.g., none, minor, moderate, severe, fatal). The common practice in analysis of contingency table data then (and sometimes now) was to avoid empty cells by collapsing to a dichotomous scale (e.g., nonfatal, fatal). In an effort to avoid losing information in this way, I invented ridit analysis, which involves a simple empirical cumulative probability transformation of the entire scale. There may be two main reasons for the citations. First, the procedure proved practical in a variety of applications and was cited as methodology. Second, mathematical statisticians have been interested in proving unique optimum properties for various models.

"An interesting point about citation practices in statistics and other areas is this: widely-used techniques tend to undergo name changes because minor (often insignificant) variants become fashionable. Variants are often named for the authors and only the latest variant may be cited. This tends to obscure the historical evolution of a discipline and to confuse students with a multiplicity of names for the same method.

"Because the rationale for ridit analysis was an acronym ('*R*elative to an *Id*entified *D*istribution') plus the productive suffix '-it' which denotes a transformation, this may have avoided this confusion. A short and simple name seems to have survival value and to be preferred to personal names. Actually, however, ridit analysis was named for my wife, Rida.

"As the title 'How to use ridit analysis' suggests, my original purpose was to explain the use of the method rather than to derive it mathematically. The designation as *Citation Classic* suggests that the method has proved practical in a variety of applications and has been useful to researchers for a quarter of a century. However, I suspect it would be just about impossible to publish a 'how to' article for a useful new method in the current statistical literature. The editorial processes today in biostatistics and in my other disciplines focus on highly technical issues that are of interest (if to anyone) only to coteries in the discipline. Material in a discipline that would be of interest or of use to scientists in other disciplines is almost unpublishable. Some recent references have appeared in the *American Journal of Epidemiology and Annals of Statistics*."[1-4]

1. Brockett P L & Levine A. On a characterization of ridits. *Ann. Statist.* 5:1245-8, 1977.
2. Selvin S. A further note on the interpretation of ridit analysis. *Amer. J. Epidemiol.* 105:16-20, 1977.
3. Bross I D J. Comment on ridit analysis by Dr. Bross. (Letter to the editor). *Amer. J. Epidemiol.* 107:263-4, 1978.
4. ——————. Reply by Dr. Bross: on biases in judging statistical methods. *Amer. J. Epidemiol.* 109:30-2, 1979.

# This Week's Citation Classic

CC/NUMBER 39
SEPTEMBER 24, 1984

Pelz D C & Andrews F M. *Scientists in organizations: productive climates for research and development.* New York: Wiley, 1966. 318 p.
[Survey Research Center, Inst. for Social Research, Univ. Michigan, Ann Arbor, MI]

In 11 R&D settings in industry, government, and academia, the technical performance of scientists and engineers was related to the nature of their interactions with colleagues and superiors, type of work, autonomy, influence, and motivations. [The *Science Citation Index®* (*SCI®*) and the *Social Sciences Citation Index®* (*SSCI®*) indicate that this book has been cited in over 435 publications since 1966.]

---

Donald C. Pelz
Center for Research on Utilization
of Scientific Knowledge
Institute for Social Research
University of Michigan
Ann Arbor, MI 48109

July 6, 1984

"This book grew out of a project commissioned by the National Institutes of Health (NIH) in the early 1950s. NIH was expanding rapidly and had added a clinical center, and its directors wanted to assess staff morale and productivity. Of particular interest were results on technical performance of intramural scientists as judged by panels of peers. Would similar factors relate to performance in university and industrial laboratories? F.M. Andrews joined me in the late 1950s for a study of 11 organizations—5 industrial and 5 government labs, and several departments of a university—resulting in the 1966 book.

"Many parallels appeared for basic and developmental laboratories, and for PhDs and non-PhDs. When I sought to extract some basic principles for an article in *Science*[1] (which became the introduction for a revised version of the book 10 years later[2]), technical performance appeared to flourish in the presence of conditions that seemed antithetical—hence the concept of 'creative tensions' or 'creative contradictions.' Effective scientists and engineers needed both some source of 'security' or *protection* from external disruption and some source of 'challenge' or *exposure* to external demands. Security could be provided by autonomy, influence, or specialization; challenge by frequent communication, multiple R&D functions, or colleague diversity. I was reminded of a passage in Emerson's essay on 'Self-Reliance' a century earlier: 'It is easy in the world to live after the world's opinion; it is easy in solitude to live after our own; but the great man [read: effective scientist] is he who in the midst of the crowd keeps with perfect sweetness the independence of solitude.'

"Why has the book been frequently cited? In part, we suspect, because of its timing. It pioneered in the postwar and post-Sputnik wave of concern for the management of research and development. We like to think, too, that the format helped. The book presented a complex matrix of factors in a clear and readable manner, with meaningful charts and simple tables, and concluded each chapter with a dialogue between the authors and a hypothetical reader on practical implications.

"The book has been translated into Japanese and Russian, and, we are told, it has been widely read in both areas. In the early 1970s, it stimulated a cross-national study of research teams under UNESCO auspices. Andrews served as technical adviser for the first round in six European countries—Austria, Belgium, Finland, Hungary, Poland, and Sweden—and edited a book.[3] In second and third rounds, the methodology was extended to South America, Africa, Asia, and Russia—a total of 16 countries, including some repeats. The series will be the subject of a conference in January 1985 in Rio de Janeiro. Andrews found keen interest in the research results among an audience of several hundred in a 1983 visit to Beijing.

"The UNESCO studies have sustained many of the findings from the American data, even though the UNESCO studies focus on performance of R&D *teams* rather than on individuals as in the initial study. The UNESCO data found performance to be higher under conditions of strong personal dedication, diversity in several factors (number of roles, projects, skill areas, funding sources, and disciplines), and frequent communication with many colleagues. Groups were more effective when their members had high influence but moderate autonomy, and were comprised of four to six people who had worked together about 7 to 10 years. In short: conditions governing technical contribution appear to transcend cultural boundaries and political systems."

1. Pelz D C. Creative tensions in the research and development climate. *Science* 157:160-5, 1967. (Cited 30 times.)
2. Pelz D C & Andrews F M. *Scientists in organizations: productive climates for research and development.* Ann Arbor, MI: Institute for Social Research, 1976. 401 p. (Cited 55 times.)
3. Andrews F M, ed. *Scientific productivity, the effectiveness of research groups in six countries.* Cambridge, England: Cambridge University Press, 1979. 469 p.

# Index of Authors

All authors of Citation Classics are listed. An asterisk after a page number indicates that a commentary by the author appears on that page.

Abell LL   24*
Adams CWM   133*
Albright CD   150
Alkjaerisig N   142*
Allen RJL   236*
Allen WM   28*
Alvarez J   198
Aminoff D   9*
Anderson PJ   176
Andres V   179
Andrews FM   267
Anfinsen CB Jr   102
Anton AH   230*
Appelmans F   175
Appleman MM   156
Armitage P   264*
Astrup T   141*
Avron M   210
Axelrod B   32
Axelrod J   166, 234
Axén R   242

Babb AL   224*
Baldwin RL   201
Bandurski RS   32*
Bárány M   152*
Barclay M   34
Bardawill CJ   89
Barka T   176*
Barker SB   44*
Barner HD   81
Barrell BG   67
Battaile J   190
Bauer W   78
Begg G   203*
Bencze WL   131*
Benesch R   115*
Benesch RE   115
Bentley R   12

Bertler Å   228
Björndal H   207
Bligh EG   16*
Bloembergen N   208*
Blundell T   124*
Bodanszky M   244*
Bollum FJ   149*
Bonner J   79
Bonner WM   100*
Bonnichsen R   163*
Born GVR   80*
Böttiger LE   40
Boucher R   106*
Boyer PD   129*
Boyle AJ   25
Bradbury EM   110*
Bradley RM   187
Brady RO   187*
Bragdon JH   125
Bray GA   199*
Brewbaker JL   191*
Brodie BB   24
Bross IDJ   266*
Brown AH   7*
Brown GW Jr   165
Brown JB   29*
Brownlee GG   67
Bruchovsky N   183*
Brumm AF   65
Brunauer S   211
Bunting S   186
Burgoyne LA   75
Burkhalter A   229
Burnett H   26
Burny A   160
Burtner HJ   165
Burton K   68*
Bush IE   27*
Butcher RW   155

Carlson LA   22*, 40*
Carlsson A   228*
Castelli WP   39
Chalkley R   98
Chambers DC   132
Chen PS   238*
Chen RF   246*
Cheng CS   164
Cherney PJ   26
Chrambach A   91*
Christopher TG   224
Cleaver JE   82*
Clegg JB   114*
Cleland WW   195*, 243*
Cleveland DW   101*
Cochran WG   251*
Cohen SS   81*
Cohen W   143
Cohn VH Jr   229
Conney AH   172*
Connolly TN   205
Courtney KD   179
Cox RH Jr   231
Coyle JT   235*
Cram DJ   248*
Cronbach LJ   263*
Cuatrecasas P   105*
Curzon G   233*

Dahlqvist A   168*
Dalgarno A   214*
Dalziel K   194*
Darnell JE   64*
David MM   89
Davidson ER   216*
Davidson N   71
Dawson RMC   33*
de Champlain J   106
de Duve C   175*

# Index of Authors

De Souza E 164
de la Haba GL 137
Denhardt DT 73*
Deutsch DG 112*
Dexter DL 212*
Dickenman RC 26
Dodson G 124
Dounce AL 62
Dozy AM 113
Dreyer WJ 102
du Vigneaud V 244
Duncan DB 257*
Duncombe WG 48*
Dunker AK 97*
Dyer WJ 16

Ebel RL 262*
Eder HA 125
Edman P 203
Edmonds M 74*
Elhafez FAA 248
Ellman GL 179*
Emmett PH 211*
Engelund A 43
Erlanger BF 143*
Ernback S 242
Estabrook RW 171
Estes EH Jr 17

Fahrney DE 144*
Fano U 219*
Farquhar MG 177
Farr AL 87
Fasman GD 107
Featherstone RM 179
Ferreira SH 50*
Finney DJ 253*
Fischer SG 101
Flaks JG 81
Fleischer S 36
Fleischman JB 128*
Fletcher AP 142
Flodin P 92, 93*
Flower RJ 52*
Fraser DR 167*
Fredrickson DS 127
Fridovich I 189
Friedberg SJ 17
Friedenwald JS 178

Gal AE 187
Gehan EA 255*

Geiger R 136
Gelboin HV 182
Genest J 106
Gianetto R 175
Gill DM 90
Gillespie D 72*
Gillette JR 170
Glenner GG 165*
Gold AM 144
Good NE 205*
Goodwin TW 130*
Gordon HT 11*
Gordon T 39
Gornall AG 89*
Granick S 49
Green AR 233
Green DE 36*
Greenfield N 107*
Greengard P 154, 162
Gregory RA 109*
Grunwald T 210
Gryglewski R 186

Haenszel W 260
Häggendal J 232*
Hakomori S 57*
Hamberg M 55*, 185*
Hanson HP 217*
Harrison JS 10
Harter HL 259*
Hatch FT 126*
Havel RJ 125*
Hedrick JL 96*
Henry D 235
Herman F 217
Hewish DR 75*
Higgs EA 54
Hirt B 63*
Hodgkin D 124
Huang C 38*
Huang RC 79*
Hughes EW 223*
Huisman THJ 113*
Hurlbert RB 65*

Inman WHW 43*
Itzhaki RF 90*
Izawa S 205

Jaffé HH 218*
James AT 45*
Janzen EG 221

Jeffay H 198*
Johansson EDB 31*
Johns EW 110
Johnson D 192

Kacian DL 160*
Kalckar HM 148*
Kannel WB 39*
Kaplan EL 265*
Kaplow LS 174*
Karle IL 222*
Karle J 222*
Kates M 56*
Kato R 170*
Katz AM 102*
Kay ERM 62*
Kebabian JW 162*
Kendall FE 24
Killander J 93
Kinsolving CR 150
Kirschner MW 101
Knox WE 145*
Kodicek E 167
Koelle GB 178*
Kokowsky N 143
König W 136*
Korn ED 180*
Kramer CY 258*
Krebs EG 153
Krebs HA 47
Kuo JF 154*

Laemmli UK 101
Lardy HA 192*
Laskey RA 100, 206*
Laster L 187
Lawlor DP 169
Le Pecq JB 77*
Lea JD 217
Lees RS 126, 127
Leloir LF 5
Lerman LS 76*
Levitt M 146
Levy BB 24
Levy RI 127*
Lewis UJ 95
Lichtenstein J 81
Lieber CS 173
Lieberman J 184*
Lindberg B 207*
Lindell TJ 158*
Lipmann F 42*

# Index of Authors

Litchfield JT  254*
Loeb MR  81
Londos C  157
Long FA  209*
Loomis WD  190*
Lopez JA  237
Lowry OH  87*, 237*

Macdonald T  191
Maickel RP  231*
Mäkinen Y  191
Makita M  12
Malins DC  18*, 19
Mangold HK  18, 19*, 20*
Mantel N  260*, 261*
Margoliash E  121*
Marinetti GV  35*
Martensson E  187
Martin AJP  45
Mauzerall D  49*
Mazur RH  247
McCord JM  189*
McKenzie HA  118*
McMurray WC  192
McNamara PM  39
Meier P  265
Mellanby J  47
Menzie C  111
Mercola D  124
Merritt CR  150
Mertz ET  112
Metcalfe LD  46*
Miller FP  231
Miller GJ  41*
Miller GL  88*
Miller NE  41
Mills AD  206
Mitchell P  193*
Moncada S  54*, 186*
Morgan WTJ  4
Morris DL  6*
Morris L  135
Morris PW  158
Morton RA  130
Müllertz S  141
Murphy BEP  30*, 240*

Nachlas MM  164
Nagatsu T  146*
Nakazato H  74
Nash T  239*
Naughton MA  114

Nebert DW  182*
Nelson N  3*
Nemethy G  108*
Ness NF  215*
Nuss GW  53

O'Brien JS  188*
O'Farrell PH  99*
Ogur M  61*
Okada S  188
Osserman EF  169*

Panyim S  98*
Paoletti C  77
Pattee CJ  240
Paul MA  209*
Pelz DC  267*
Perkins JP  153
Peterson RF  34
Petzold GL  162
Piez KA  120, 135*
Popovich RP  224
Porath J  92*, 242*
Porter RR  128
Post RL  150*
Potter VR  65, 149
Poulik MD  202*
Pound RV  208
Press EM  128
Pressman BC  175
Procter DP  10
Purcell FM  208

Radloff R  78*
Ramachandran GN  119*
Randall RJ  87
Randerath E  66
Randerath K  66*
Ratnoff OD  111*
Reisfeld RA  91, 95*
Reissig JL  5*
Remmer H  171
Risley EA  53
Roberts JD  247*
Robison GA  155*
Rodbell M  157
Roeder RG  158
Rondle CJM  4*
Rosalki SB  181*
Rosebrough NJ  87
Rosen G  61
Rosengren E  228

Rottenberg H  210*
Rubin E  173*
Rueckert RR  97
Russell DH  147*
Russell GA  221*
Rutter WJ  158

Saillant J  231
Salomon Y  157*
Sambrook J  161
Samuelsson B  55, 185*
Sanger F  67*
Sasisekharan V  119
Sayre DF  230
Scearce CS  215
Schejter A  121
Schenkman JB  171*
Scheraga HA  108
Scherrer K  64
Schmid K  131
Schmitz AA  46
Schmitz H  65
Schnabel E  245*
Schneider WC  60*
Scholander Mrs. PF  204*
Scholander PF  204
Scribner BH  224
Seek JB  215
Seligman AM  164
Sharp PA  161*
Shemin D  241
Sherry S  142
Shore PA  229*
Simmons NS  62
Simon EJ  241*
Simpson WT  216
Singh RMM  205
Skeggs LT Jr  200*
Skillman S  217
Skipski VP  34*
Skou JC  151*
Smith AJ  96
Smith DW  116*
Smith RE  177*
Snedecor GW  251
Snyder SH  147, 234*
Sperry WM  23*
Spiegelman S  72, 160
Spies JR  132*
Spiro RG  122*
Steel RGD  252*
Stewart RF  216

# Index of Authors

Strom ET 221
Strominger JL 5
Sugden B 161
Summerson WH 44
Sundaralingam M 70*
Sutherland EW 155
Svensson H 104
Svensson J 55
Svensson S 207
Sweeley CC 12*

Tanford C 123*
Teller E 211
Temin HM 159*
Theorell H 163*
Thompson WJ 156*
Thornburg W 11
Toribara TY 238
Torrie JH 252
Tracy HJ 109
Traub W 120*
Trevelyan W 10*
Trout DL 17*
Tsou KC 164*
Tuttle LC 42

Udenfriend S 146
Upadhya MD 191

Van Handel E 21*
van Holde KE 201*
Van Hove L 220*
Vandenheuvel FA 37*
Vane JR 50, 51*, 54, 186
Vaughan MH Jr 74
Vessey MP 43
Vesterberg O 104*
Veyrat R 106
Vinograd J 78
Vogt M 227*

Waddell WJ 86*
Walsh DA 153*
Wang KT 103
Warner H 238
Warren L 8*
Warshaw AL 187
Watson KF 160
Wattiaux R 175
Weatherall DJ 114
Webb M 23
Weil CS 256*
Weinberg F 158
Wells WW 12
Werum LN 11
Westerholm B 43

Wetlaufer DB 117*
Wetmur JG 71*
Whitaker JR 94*
White EG 26
Wilcoxon FA 254
Williams DE 95
Williams RJP 116
Williamson DH 47*
Wilson AJC 213*
Wilson JD 183
Winget GD 205
Winter CA 53*
Winter W 205
Woessner JF Jr 134*
Woods KR 103*
Wurtman RJ 166*
Wyatt GR 69*
Wyckoff M 91

Yarmolinsky MB 137*

Zaccari J 91
Zak B 25*, 26*
Zilversmit DB 21
Zlatkis A 25
Zweig M 234

# Index of Subjects

Adrenaline, determination of 227, 228, 232
Amino acids (*see* Proteins and amino acids)
Atomic polarizabilities and shielding factors 214
Atomic scattering factors calculated from Hartree-Fock-Slater wave functions 217
Avian myeloblastosis virus, DNA polymerase of 160

$t$-Butyloxycarbonyl amino acids 245

Carbohydrates
   colorimetric method for estimation of N acetylamino sugars 5
   detection of sugars on paper chromatograms 10
   determination of free sialic acid in the presence of the bound sugar 9
   determination of glucosamine and galactosamine 4
   determination of pentose in the presence of large quantities of glucose 7
   mass spectrometry of partially methylated alditol acetates 207
   photometric adaptation of the Somogyi method for glucose determination 3
   quantitative determination with Dreywood's anthrone reagent 6
   rapid paper chromatography of carbohydrates and related compounds 11
   separation and estimation by gas-liquid chromatography of trimethylsilyl derivatives 12
   thiobarbituric acid assay of sialic acids 8
Catecholamine
   aluminum oxide-trihydroxyindole procedure for 230
   assay in brain extracts 235
Chemical analysis and preparative methods
   aluminum oxide-trihydroxyindole procedure for catecholamine analysis 230
   assay for catecholamines in brain extracts 235

   chemical coupling of proteins to agarose 242
   colorimetric estimation of formaldehyde 239
   determination of 5-hydroxytryptamine and 5-hydroxyindoleacetic acid 233
   determination of adrenaline and noradrenaline 227, 228, 232
   determination of inorganic phosphate in presence of labile phosphate esters 237
   determination of serotonin and norepinephrine in rat brain 231
   determination of thyroxine 240
   dithiothreitol as protective reagent for SH groups 243
   estimation of phosphorus 236
   fluorescence assay for tissue serotonin 234
   fluorimetric determination of adrenaline and noradrenaline 232
   fluorometric assay of histamine in tissues 229
   interconversion reactions of small-ring compounds 247
   microdetermination of phosphorus 238
   preparation of S-succinyl coenzyme A 241
   removal of fatty acids from serum albumin by charcoal treatment 246
   steric control of asymmetric induction in synthesis of acyclic systems 248
   synthesis of long peptide chains 244
   synthesis of tert-butyloxycarbonyl amino acids 245
Chloroplasts, determination of $\Delta$pH in 210
Cholesterol determination, methods for 23, 24, 25, 26
Clot dissolution by plasmin 142
Colorimetric analysis, automatic method for 144
Crystal structures, refining by use of the least-squares procedure 223
Crystals, phase determination for 222

Dithiothreitol as protective reagent for SH groups 243

Dose-effect experiments, evaluation of 254
Duncan multiple range rule 257, 259

Electron-transfer processes 221
Energy transfer between atoms, probability of 212
Environment productive of research and development 267
Enzymes
 RNA-dependent DNA polymerase in virions of Rous sarcoma virus 159
 adenosine triphosphatase in sodium and potassium transport 150, 151
 adenosine triphosphatase of myosin correlated with speed of muscle shortening 152
 adenylate cyclase assay 157
 alpha$_1$-antitrypsin deficiency in patients with pulmonary emphysema 184
 assay for monoamine oxidase 166
 assay of intestinal disaccharidases 168
 biosynthesis by kidney of a vitamin D metabolite 167
 ceramidetrihexosidase deficiency in Fabry's disease 187
 characterization of heparin-activated lipoprotein lipase 180
 characterization of superoxide dismutase 189
 chemiosmotic coupling in oxidative and photosynthetic phosphorylation 193
 colorimetric determination of acetylcholinesterase activity 179
 conversion of testosterone by rat prostate 183
 cyclic AMP-dependent protein kinase from animal tissues 153, 154
 cytochemical demonstration of succinic dehydrogenase 164
 determination of serum creatine phosphokinase 181
 dopamine-sensitive adenylate cyclase in caudate nucleus of rat brain 162
 drug interaction with hepatic microsomal cytochrome 171
 effect of starvation on NADPH-dependent enzymes in liver microsomes 170
 enzymes of purine metabolism 148
 evaluation of enzyme-coenzyme-substrate reaction mechanisms 194
 fibrin plate method for estimating fibrinolytic activity 141
 hexosaminidase A deficiency in Tay-Sachs disease 188
 histochemical demonstration of acid phosphatase activity 176
 histochemical demonstration of monoamine oxidase 165
 histochemical localization of cholinesterase activity 178
 inducible microsomal aryl hydroxylase in mammalian cell culture 182
 induction and inhibition of hepatic microsomal enzymes 173
 inhibition of nuclear polymerase II by alpha-amanitin 158
 intracellular distribution patterns of enzymes in rat liver tissue 175
 isoenzyme polymorphism in flowering plants 191
 kinetics of enzyme-catalyzed reactions with two or more substrates or products 195
 leukocyte alkaline phosphatase activity in smears of blood and marrow 174
 liver alcohol dehydrogenase 163
 lysosome function in regulation of secretory process 177
 mechanism of clot dissolution by plasmin 142
 mechanisms which increase liver tryptophan peroxidase activity 145
 mediation of hormone action by cyclic AMP 155
 multiple cyclic nucleotide phosphodiesterases from rat brain 156
 ornithine decarboxylase activity in regenerating tissues 147
 pharmacological implications of microsomal enzyme induction 172
 plant phenolic compounds and isolation of plant enzymes 190
 properties of new chromogenic substrates of trypsin 143
 prostaglandin endoperoxides 185, 186
 purification of DNA polymerase of avian myeloblastosis virus 160
 purification of restriction endonucleases from *H. parainfluenzae* 161
 serum and urinary lysozyme in monocytic and monomyelocytic leukemia 169
 soluble enzymes which convert thymidine to thymidine phosphates 149
 sulfonyl fluorides as inhibitors of esterases 144

Index of Subjects 275

toxic antibiotics as tools for metabolic studies 192
tyrosine hydroxylase as initial step in norepinephrine biosynthesis 146

Fatty acids, removal from serum albumin 246
Fibrinolytic activity, estimation of 141
Formaldehyde, colorimetric estimation of 239

Gases, absorption in multimolecular layers 211

Hammett equation, reexamination of 218
Hartree-Fock-Slater wave functions, atomic scattering factors calculated from 217
Heart disease 39, 40, 41, 43
Histamine, fluorometric assay in tissues 229
Hydrogen atom in the hydrogen molecule, X-ray scattering for 216
Hydrogen ion buffers for biological research 205
5-Hydroxytryptamine and 5-hydroxyindoleacetic acid, determination of 233

Indicator acidity functions 219
Inorganic phosphate, determination in presence of labile phosphate esters 237
Isotopes, quantitative film detection in polyacrylamide gels by fluorography 206

Lipids and related compounds
  analysis of complex lipid mixtures by thin-layer chromatography 18, 20
  bacterial lipids 56
  carrageenin-induced edema as assay for antiinflammatory drugs 53
  colorimetric determination of lactic acid 44
  colorimetric determination of urinary steroids 28
  colorimetric micro-determination of long-chain fatty acids 48
  determination of acyl phosphates 42
  determination of delta-aminolevulinic acid and porphobilinogen in urine 49
  determination of estriol, estrone, and estradiol in urine 29
  determination of serum triglycerides 21, 22
  disappearance from and release into the circulation of prostaglandins 50
  drugs which inhibit prostaglandin biosynthesis 52

  enzymatic determination of acetoacetate and 3-hydroxybutyrate in blood 47
  formation and physical characteristics of phosphatidylcholine vesicles 38
  formation of thromboxanes 55
  fractionation on thin layers of silicic acid 19
  Framingham study of risk of coronary heart disease 39
  gas-liquid partition chromatography for separation of volatile fatty acids 45
  identification and estimation of individual phospholipids 33
  identification of biologically important phosphate esters 32
  inhibition of prostaglandin synthesis by aspirin-like drugs 51
  ischaemic heart disease related to plasma high-density lipoprotein concentration 41
  lipids in the myelin sheath of nerve 37
  measurement of progesterone in peripheral plasma 31
  measurement of steroids in body fluids 30
  methods for cholesterol determination 23, 24, 25, 26
  paper chromatography of steroids 27
  preparation of fatty acid esters for gas chromatography 46
  prostacyclin generation by arterial and venous tissues 54
  quantitative analysis of phospholipids 34
  rapid method of total lipid extraction and purification 16
  rapid permethylation of glycolipid and polysaccharide 57
  role of lipids in mitochondrial electron transfer and oxidative phosphorylation 36
  separation, identification, and analysis of phosphatides 35
  Stockholm study of ischaemic heart disease 40
  thromboembolic disease related to oral contraceptives 43
  titration of free fatty acids of plasma 17
Liquid scintillation counting 142, 143
Liver alcohol dehydrogenase 163

Magnetosphere, discovery by the IMP-1 satellite 215
Mass spectrometry of partially methylated alditol acetates 217

Median effective dose, tables for calculation of 256

Noradrenaline, determination of 227, 228, 232
Norepinephrine
  biosynthesis 146
  determination in rat brain 231
Nucleic acids
  aggregation of blood platelets by adenosine diphosphate 80
  chromatographic separation of acid-soluble nucleotides 65
  digestion of chromatin DNA by nuclear deoxyribonuclease 75
  diphenylamine reaction for colorimetric estimation of DNA 68
  extraction of deoxypentose and pentose nucleic acids from animal tissues 60
  extraction of deoxypentose and pentose nucleic acids from plant tissues 61
  extraction of polyoma DNA from infected mouse cell cultures 63
  fluorescent complex between ethidium bromide and nucleic acids 77
  fractionation procedure for radioactive nucleotides 67
  improved preparation of sodium deoxyribonucleate 62
  interaction of DNA and acridines 76
  ion-exchange chromatography of nucleotides 66
  isolation of closed circular DNA in HeLa cells 78
  kinetics of renaturation of DNA 71
  membrane-filter technique for detection of complementary DNA 73
  mode of action of 5-fluorouracil and its derivatives 81
  polyadenylic acid sequences in HeLa cells 74
  purine and pyrimidine composition of deoxypentose nucleic acids 69
  quantitative assay for DNA-RNA hybrids 72
  repair replication of mammalian cell DNA 82
  sedimentation characteristics of rapidly labeled RNA from HeLa cells 64
  stereochemistry of nucleic acids and their constituents 70
  suppression of chromosomal RNA synthesis by histone 79

Peptide chains, long, synthesis of 244
Phosphate esters, identification of 32
Phosphatides, analysis of 35
Phosphatidylcholine vesicles, characteristics of 38
Phospholipids, identification and estimation of 33, 34
Phosphorus, measurement of 236, 238
Physical analysis and instrumentation
  Hartree-Fock-Slater atomic scattering factors 217
  absorption of gases in multimolecular layers 211
  atomic polarizabilities and shielding factors 214
  automatic method for colorimetric analysis 200
  crystal structure of melamine 223
  determination of $\Delta$pH in chloroplasts 210
  determination of absolute from relative X-ray intensity data 213
  determination of phases of X-rays scattered by a crystal 222
  effects of configuration interaction on intensities and phase shifts 219
  electron-transfer processes 221
  generation of the square meter-hour hypothesis 224
  hydrogen ion buffers for biological research 205
  indicator acidity functions 219
  initial results of the IMP-1 magnetic field experiment 215
  liquid scintillation counting 142, 143
  mass spectrometry of partially methylated alditol acetates 207
  protein sequenator 203
  quantitative film detection of isotopes in polyacrylamide gels by fluorography 206
  rapid attainment of sedimentation equilibrium 201
  reexamination of the Hammett equation 218
  relaxation effects in nuclear magnetic resonance absorption 208
  scattering process 220
  starch gel electrophoresis in a discontinuous system of buffers 202
  theory of sensitized luminescence in solids 212
  volumetric gas analyzer 204
  x-ray scattering for the hydrogen atom in the hydrogen molecule 216

Plant enzymes, isolation of 190
Plants, flowering, isoenzyme polymorphism in 191
Probit analysis 253
Productive climates for research and development 267
Progesterone, measurement in peripheral plasma 31
Prostaglandins 50, 51, 52
Protein sequenator 203
Proteins and amino acids
　acrylamide gel electrophoresis of histones 98
　anion-exchange chromatography of hemoglobins 113
　arrangement of peptide chains in gammaglobulin 128
　automatic analysis of amino acid 135
　characteristics of lipoproteins isolated from serum 125
　chemistry and structure of collagen 120
　computed circular dichroism spectra for proteins 107
　conformation of polypeptides and proteins 119
　cytochrome *c* 121
　desalting and group separation by gel filtration 92
　detection of tritium-labeled proteins and nucleic acids in polyacrylamide gels 100
　determination of fibrinogen in small samples of plasma 111
　determination of hydroxyproline in tissues and proteins 134
　determination of molecular weights of proteins 94, 96, 97
　determination of serum proteins by the biuret reaction 89
　determination of tryptophan in proteins 130, 131, 132, 133
　determination of tyrosine in proteins 130, 131
　disc electrophoresis of basic proteins and peptides on polyacrylamide gels 95
　effect of organic phosphates on allosteric properties of hemoglobin 115
　fat transport in lipoproteins 127
　fractionation of serum proteins by gel filtration 93
　glycoproteins 122
　isoelectric focusing of proteins 104
　isolation and properties of gastrins 109
　measurement of plasma angiotensin and renin activity 106
　method for synthesis of peptides 136
　micro-biuret method for estimating proteins 90
　model for thermodynamic properties of liquid water 108
　peptide mapping 101
　peptide separation on filter paper 102
　plasma lipoprotein analysis 126
　preparative methods for histone fractions 110
　procedure for staining protein fractionated by polyacrylamide gel electrophoresis 91
　properties of milk proteins 118
　protein denaturation 123
　protein determination for large numbers of samples 88
　protein measurement with the Folin phenol reagent 87
　protein purification by affinity chromatography 105
　purification of plasminogen from human plasma 112
　puromycin inhibition of amino acid incorporation into protein 137
　reaction of protein sulfhydryl groups with organic mercurials 129
　relation of electrophoretic mobility of proteins to size and charge characteristics 96
　separation and characterization of abnormal hemoglobins 114
　separation of dansyl-amino acids by polyamide layer chromatography 103
　spectra of ferric haems and haemoproteins 116
　structure of insulin 124
　two-dimensional electrophoresis of proteins 99
　ultraviolet spectra of proteins and amino acids 117
　ultraviolet spectrophotometric method for protein determination 86
Proteins, chemical coupling to agarose 242
Purine metabolism, enzymes of 148

Relaxation effects in nuclear magnetic resonance absorption 208
Research and development, productive climates for 267
Restriction endonucleases, purification of 161

## Index of Subjects

Ridit analysis, how to use  266
Rous sarcoma virus, DNA polymerase in  159

Scattering process  220
Serotonin, measurement of  231, 234
Serum triglycerides, determination of  21, 22
Square meter-hour hypothesis, generation of  224
Starch gel electrophoresis in a discontinuous system of buffers  202
Statistical methods
  alpha and the internal structure of tests  263
  applied to experiments in agriculture and biology  251
  critical values for Duncan's multiple range test  259
  estimation of the reliability of ratings  262
  generalized Wilcoxon test for comparing arbitrarily singly-censored samples  255
  how to use ridit analysis  266
  in analysis of data from retrospective studies of disease  260
  in evaluating dose-effect experiments  254
  in evaluation of survival data  261
  multiple range and multiple F tests  257
  multiple range tests for group means with unequal numbers of replications  258
  nonparametric estimation for incomplete observations  265
  principles and procedures of  252
  probit analysis  253
  restricted sequential procedures  264
  tables for calculation of median effective dose  256
Steroids, determination of  27, 28, 30
Succinyl coenzyme A, preparation of  241
Survival data, evaluation of  261

Thyroxine, determination of  240
Trypsin, new chromogenic substrates of  143

Volumetric gas analyzer  204

Wilcoxon test for comparing arbitrarily singly censored samples  255

X-ray intensity, determination of absolute from relative  213

# Index of Institutions

The institutions listed are those at which the work reported in the Citation Classic was done.

Agricultural Research Council, Institute of Animal Physiology, Cambridge, England 33
American Cyanamid Company, Stamford, Connecticut 254
Armour Industrial Chemical Company, McCook, Illinois 46
Australian National University, Canberra, Australia 118

Bayer AG, Elberfeld, Federal Republic of Germany 245
Birkbeck College, London, England 255
Borden Company, Baltimore, Maryland 164
Brandeis University, Waltham, Massachusetts 107
Brooklyn College, Brooklyn, New York 61
Burroughs Wellcome & Co. (USA), Inc., Tuckahoe, New York 172

California Institute of Technology, Pasadena 32, 71, 78, 79, 223
California Packing Corp., Emeryville, California 11
Canada Department of Agriculture, Ottawa, Canada 37
Carlsberg Foundation, Copenhagen, Denmark 141
Case Western Reserve University, Cleveland, Ohio 70
Cavendish Laboratory, Cambridge, England 213
Chalmers Institute of Technology, Gothenburg, Sweden 104
City of Detroit Receiving Hospital, Detroit, Michigan 25, 26
City of Hope Medical Center, Duarte, California 184
City University of New York, New York, New York 173

Cold Spring Harbor Laboratory, Cold Spring Harbor, New York 161
Columbia University, New York, New York 144
Columbia University College of Physicians and Surgeons, New York, New York 23, 115, 143, 160, 169, 241
Committee on Safety of Drugs, Queen Anne's Gate, London, England 43
Cornell University, Ithaca, New York 108, 209, 244
Cornell University Medical College, New York, New York 34, 44, 266

Danish National Health Service, Brønshøj, Denmark 43
Detroit Memorial Hospital, Detroit, Michigan 26
Distillers Company, Ltd., Epson, Surrey, England 10
Duke University Medical Center, Durham, North Carolina 17, 123, 189

Elliott P. Joslin Research Laboratory, Boston, Massachusetts 122

Fisheries Research Board of Canada, Halifax, Nova Scotia 16
Flinders University of South Australia, Bedford Park, South Australia 75
Francis Delafield Hospital, New York, New York 169

Geophysics Corporation of America, Bedford, Massachusetts 214
Glynn Research Laboratories, Bodmin, Cornwall, England 193
Goddard Space Flight Center, Greenbelt, Maryland 215

## Index of Institutions

Goldwater Memorial Hospital, Roosevelt Island, New York 24

Harpur College, Endicott, New York 209
Harvard Medical School, Boston, Massachusetts 66, 122, 131
Harvard University, Cambridge, Massachusetts 73, 208
Heart Disease Epidemiology Study, Framingham, Massachusetts 39
Hoechst AG, Frankfurt am Main, Federal Republic of Germany 136
Hôtel-Dieu Hospital, Montreal, Canada 106

Indiana University, Bloomington 231
Indiana University School of Medicine, Indianapolis 117
Inorganic Chemistry Laboratory, Oxford, England 116
Institut Gustave-Roussy, Villejuif, France 77
Institute for Advanced Study, Princeton, New Jersey 220
Institute for Muscle Disease, New York, New York 152
Institute of Neurology, London, England 233
Instituto de Investigaciones Bioquimicas, Buenos Aires, Argentina 5
Iowa State University, Ames 221, 262

Johns Hopkins University, Baltimore, Maryland 137, 178
Johns Hopkins University School of Medicine, Baltimore, Maryland 111, 114, 147, 164

Karolinska Hospital, Stockholm, Sweden 22, 40
Karolinska Institutet, Stockholm, Sweden 55, 185

Laboratory of Molecular Biophysics, Oxford, England 124
Lister Institute of Preventive Medicine, London, England 4
Llandough Hospital, Penarth, South Wales 41
Lockheed Missiles and Space Company, Sunnyvale, California 217
London School of Hygiene and Tropical Medicine, London, England 264
London University, London, England 133

Massachusetts General Hospital, Boston 42, 131
Massachusetts Institute of Technology, Cambridge 64, 247
May Institute for Medical Research of the Jewish Hospital, Cincinnati, Ohio 3
McGill University, Montreal, Canada 30, 240
Mead Johnson & Company, Evansville, Indiana 6
Medical College of Georgia, Augusta 113
Medical Research Council, Cambridge, England 167
Medical Research Council Laboratory of Molecular Biology, Cambridge, England 67
Mellon Institute, Pittsburgh, Pennsylvania 256
Merck Institute for Therapeutic Research, West Point, Pennsylvania 53
Merck Sharp and Dohme Research Laboratories, Rahway, New Jersey 95
Michigan State University, East Lansing 205
Mount Sinai Hospital of New York, New York, New York 176

National Bureau of Standards, Washington, D.C. 219
National Cancer Institute, Bethesda, Maryland 91, 182, 260, 261
National Heart and Lung Institute, Bethesda, Maryland 39
National Heart Institute, Bethesda, Maryland 102, 125, 127, 146, 170, 180, 199, 229, 246
National Institute for Medical Research, Mill Hill, London, England 27, 45
National Institute of Allergy and Infectious Diseases, Bethesda, Maryland 91
National Institute of Arthritis and Metabolic Diseases, Bethesda, Maryland 5, 8, 105, 187
National Institute of Arthritis, Metabolism, and Digestive Diseases, Bethesda, Maryland 157, 165
National Institute of Dental Research, Bethesda, Maryland 120, 135
National Institute of Mental Health, Bethesda, Maryland 166, 234, 235
National Institute of Neurological Diseases and Blindness, Bethesda, Maryland 187
National Institutes of Health, Bethesda, Maryland 165, 255
National Research Council, Ottawa, Canada 56